Atomically Precise Nanoclusters

Atomically Precise Nanoclusters

edited by

Yan Zhu

Rongchao Jin

JENNY STANFORD
PUBLISHING

Published by

Jenny Stanford Publishing Pte. Ltd.
Level 34, Centennial Tower
3 Temasek Avenue
Singapore 039190

Email: editorial@jennystanford.com
Web: www.jennystanford.com

British Library Cataloguing-in-Publication Data
A catalogue record for this book is available from the British Library.

Atomically Precise Nanoclusters

ISBN 978-981-4877-08-4 (Hardcover)
ISBN 978-1-003-11990-6 (eBook)

Contents

Preface

Atomically precise metal nanoclusters are composed of an exact number of metal atoms and, when in solution phase, they are protected by a specific number of ligands. Such nanoparticles are unique in that their atomic-level packing structures can be determined by X-ray crystallography; for example, a range of atomically precise gold nanoclusters from tens to hundreds of atoms have been structurally characterized recently. These metal nanoclusters possess precise total structures (i.e., metal core plus surface ligands) and constitute a new class of nanomaterials that have found applications in optics, electronics, catalysis, sensing, etc. They offer us a unique platform to assess the quantum nature of the physicochemical properties and provide unprecedented opportunities to establish precise relationships between the structures and properties at the atomic level and even at the single-electron level (i.e., the charged state of the metal core). In the past years, researchers have made significant advances in the synthesis toward one pot for one size and in structural determination at the atomic level as well as in correlating the physicochemical properties with atomic-level structures. This book illustrates the research progress in the field of ligand-protected, atomically precise metal nanoclusters.

The book contains 3 parts with 10 chapters. Part I covers nanocluster synthesis and structural characterization. Specifically, Chapters 1–3 summarize the chemical synthesis and stable structures of the atomically precise gold nanoclusters. Chapter 4 describes the structural prediction of gold nanoclusters based on the density functional theory (DFT). Part II focuses on the electronic and optical properties of nanoclusters. Chapter 5 introduces the basics of the electronic structures of gold nanoclusters at the DFT level. Chapter 6 analyzes the size-dependent evolution from plasmon to exciton state in ligand-protected gold nanoclusters by performing ultrafast spectroscopic studies. Chapter 7 reports the theoretical study on the optical properties of gold nanoclusters. Part III is devoted to the catalytic application of gold nanoclusters, illustrated in three chapters. Chapters 8 and 9 discuss the important chemical reactions, such as oxidation, hydrogenation, and coupling reactions. In addition, electrocatalytic and photocatalytic applications of gold-based nanoclusters are also mentioned. Chapter 10 gives a brief introduction to the application of the DFT methods in exploring the structure–activity relationship in catalysis of atomically precise metal nanoclusters.

This book could not have been completed without the contributing authors' great effort, and we are grateful for their dedication and enthusiasm. It will be of interest to readers from various fields of study, including (nano)cluster chemistry, noble metal chemistry, nanoscience and nanotechnology, materials science and engineering, and catalysis.

Yan Zhu

Rongchao Jin

Autumn 2020

About Editors

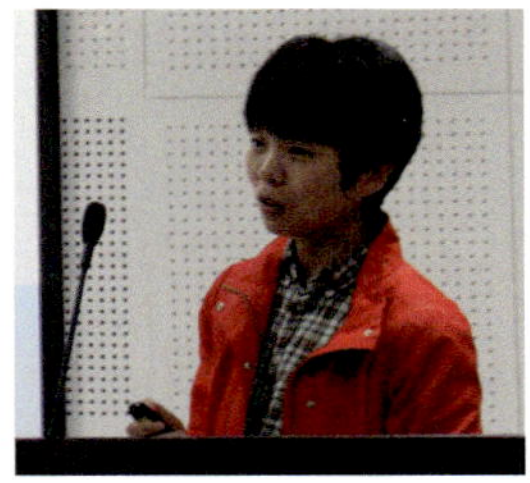

Yan Zhu is a full professor of physical chemistry at Nanjing University, China. She earned her PhD from Nanjing University and performed her postdoctoral research at Keio University, Japan, and Carnegie Mellon University, USA. She joined the chemistry faculty of the Chinese Academy of Sciences in 2011 and moved to Nanjing University in 2017. Her current research focuses on the design of atomically precise nanoclusters for catalytic applications.

Rongchao Jin is a professor of chemistry at Carnegie Mellon University, USA. He received his BS in chemical physics from the University of Science and Technology of China (USTC, Hefei) in 1995; his MS in physical chemistry/catalysis in 1998 from Dalian Institute of Chemical Physics, Chinese Academy of Sciences; and his PhD in chemistry from Northwestern University, Evanston, Illinois, USA, in 2003. He then performed his postdoctoral research at the University of Chicago. He joined the chemistry faculty of Carnegie Mellon University in 2006 and was promoted to associate professor in 2012 and full professor in 2015. His research interests include atomically precise nanoparticles, optics of nanoparticles, and catalysis.

Part I
Nanocluster Synthesis and Structural Characterization

Chapter 1

Chemical Synthesis and Physical Isolation of Metal Nanoclusters

Nan Yan and Zhikun Wu

Key Laboratory of Materials Physics, Anhui Key Laboratory of Nanomaterials and Nanostructures, CAS Center for Excellence in Nanoscience, Institute of Solid State Physics, Chinese Academy of Sciences, 350 Shushanhu Road, Hefei 230031, China

Institute of Physical Science and Information Technology, Anhui University, Hefei 230601, China

zkwu@issp.ac.cn

Synthesis of atomically precise metal nanoparticles has long been a dream goal of chemists, which is the precondition to determine the total structures and understand the properties of nanomaterials at the atomic level. This goal has been realized through the research and development of atomically precise metal nanoclusters in the range from 1 to 3 nm. Various synthesis methods have been developed to prepare atomically precise metal nanoclusters with molecular purity in recent years. In this chapter, we introduce the basic principles in the synthesis of atomically precise metal nanoclusters, especially thiolated gold nanoclusters, which can be described as "kinetic control and thermodynamic selection." By kinetic control, a proper

Atomically Precise Nanoclusters
Edited by Yan Zhu and Rongchao Jin

ISBN 978-981-4877-08-4 (Hardcover), 978-1-003-11990-6 (eBook)
www.jennystanford.com

size-distributed gold NCs (nanoclusters) mixture is generated and then size focused in the thermodynamic selection process. Guided by this principle, even the "one-pot one-cluster" synthesis is achieved. Furthermore, subtle isolation is still necessary in some other cases. Thus, several commonly used isolation (separation) methods are reviewed in this chapter.

1.1 Introduction

It is a hot topic in the field of nanoscience to elucidate a clear correlation between the precise structure/composition and the physicochemical properties of metal nanoparticles (NPs) [1]. Therefore, significant effort has been devoted to the development of new techniques for synthesizing or processing exceptionally monodisperse NPs [2–4]. However, the dispersive quality of NPs is always reflected in their size distribution characterized by transmission electron microscopy [5]. High-quality metal NPs with "monodisperse" size are not atomically precise since there can be a difference from hundreds to thousands of metal atoms in each NP, let alone the ambiguous surface ligands that cannot be observed by electron microscope. Hence, synthesis of atomically precise metal NPs with molecular purity has long been a major dream of nanochemists, which is the precondition to determine the total structures and understand the properties of nanomaterials at the atomic level. This dream has been realized with the development of atomically precise metal NCs with a size of less than 3 nm [6–8]. The atomically precise metal NC represents an intermediate state between metal atom (or complex) and metal NP (larger than 3 nm). Owing to their ultrasmall size, metal NCs exhibit quantum confinement effects and possess some intriguing molecular-like properties differing from those of the bulk metal, such as HOMO–LUMO (highest occupied and lowest unoccupied molecular orbitals) electronic transitions [9, 10], quantized charging [11, 12], and enhanced photoluminescence [13, 14]. This cluster state is also an important stage in which many transitions from the molecular-like properties to the bulk properties occur [15]. The chemical formula of the metal NCs can be written as M_nL_m, where M and L refer to a certain type of metal and protecting ligand, respectively, and *n* (few to a few hundreds) and *m* are the number of metal atoms and protecting ligands, respectively. Such a formula can be determined by mass spectrometry or single-crystal X-ray crystallography [16]. Even the latter can be employed to unravel the total structures of atomically precise metal NCs [17].

To date, various atomically precise metal NCs [e.g., M = gold (Au), silver (Ag), and copper (Cu)] have been successfully synthesized and characterized [18–20]. Among these metal NCs, Au NCs have emerged as a paradigm system due to their relative robustness. In early studies (beginning in the 1960s), phosphine was employed as a protective ligand, including $[Au_{11}(PPh_3)_8Cl_2]^+$, $[Au_{13}(PMe_2Ph)_8Cl_2]^{3+}$, $[Au_{39}(PPh_3)_{14}Cl_6]^{2+}$, and $Au_{55}(PPh_3)_{12}Cl_6$ [21]. However, these NCs were found to be unstable in solution, which restricted their practical applications. In 1994, Brust et al. added thiols to protect gold NPs and obtained relatively stable gold NPs not only in solid but also in solution owing to the formation of strong bonds between thiolate ligands and gold atoms [22]. Their pioneer work prompted the research on thiolated metal NCs. After that, the synthesis of gold clusters protected by other chalcogenates (selenolate (SeR) or tellurolate (TeR)), by alkynes, or by mixture ligands were also reported [23–26]. Unfortunately, in the initial stage, the Au NCs synthesized by the Brust method had a wide distribution in the number of constituent atoms and were not atomically precise at all. Thus, obtaining Au NCs with atomic monodispersity needed high-resolution isolation techniques, such as polyacrylamide gel electrophoresis (PAGE) [27]. With the improvement of the synthesis method, scientists tried to tackle the issue of the "one-pot one-cluster" synthesis of Au NCs by doing away with complicated post-synthetic size-separation steps [28, 29]. However, "one-pot one-cluster" is not achieved for every cluster synthesis and the subtle isolation is still necessary in some events.

In this chapter, we will focus on the synthesis principles rather than the synthesis details and some isolation (separation) methods or techniques commonly used in the research of gold NCs. Since the surface ligands of gold NCs include thiolate, selenolate, tellurolate, phosphine, alkynes, mixed ligands, and so on, we focus mainly on those protected by thiolate, which is most widely studied.

1.2 Synthesis Principles of Gold Nanoclusters

At the early stage of thiolated gold NCs research, the yield of gold NCs was very low, which inhibited the subsequent research on structures, properties, and applications of gold NCs. Therefore, the most emergent thing at that time was the synthesis improvement, which however needed some guidance from the synthesis principles. At the infant stage of NCs study, many principles, including the synthesis principles, were unknown. Fortunately, with the advances of wet chemistry, Wu and Jin proposed the synthesis principles after years of efforts, which is discussed next [30].

As mentioned previously, the Brust method, which was reported in 1994, was the earliest chemical method for synthesizing thiolated gold NCs [22]. Specifically, this method included two steps. In the first step, Au(III) salt was dissolved in an aqueous solution that served as gold precursor and further transferred to an organic solvent using a phase transfer reagent such as tetraoctylammonium bromide (TOABr). Then protecting thiolate ligand (SR) was added to the organic solvent to form Au(I)–SR complex. In the second step, the reducing agent sodium borohydride ($NaBH_4$) was added to the mixture, which led to the reduction of Au ions and formation of small Au NPs in organic phase. Murray and co-workers modified the synthetic procedure and finally obtained $Au_{38}(SC_2H_4Ph)_{24}$ NCs (finally adjusted as $Au_{25}(SC_2H_4Ph)_{18}$ with the advances of mass spectrometric techniques) [31]. Subsequently, the conventional Brust method was also modified to synthesize thiolate-protected NCs in water. For example, Tsukuda et al. synthesized a series of glutathione (GSH) protected Au NCs, including $Au_{10}(SG)_{10}$, $Au_{15}(SG)_{13}$, $Au_{18}(SG)_{14}$, $Au_{22}(SG)_{16}$, $Au_{22}(SG)_{17}$, $Au_{25}(SG)_{18}$, $Au_{29}(SG)_{20}$, $Au_{33}(SG)_{22}$, and $Au_{39}(SG)_{24}$, but the as-obtained NCs were a mixture prior to gel isolation and the yields of individual clusters were low [32]. Kornberg's group synthesized $Au_{102}(SR)_{44}$ NC in water/methanol mixture solvent and determined the crystal structure of such a thiolate-protected Au NC for the first time in 2007 [17].

Basing on the early works in the synthesis of thiolate-protected Au NCs by Whetten et al. [31–33], Zhu et al. continuously pursued the two-phase synthesis of Au NCs and successfully obtained monodisperse $Au_{25}(SC_2H_4Ph)_{18}$ with high yield (40% Au atom basis) by controlling the reaction intermediates for the first time [34]. Similar to the classical Brust method, this method involved two steps: (i) reduction of Au(III) (e.g., $HAuCl_4$) to Au(I) by thiols, forming an intermediate of Au(I):SR complexes, and (ii) further reduction of Au(I) to Au(0) by a strong reducing agent ($NaBH_4$) (Fig. 1.1). They found that the kinetics for the formation of Au(I):SR intermediate in the first step was very critical for the high-yield synthesis of $Au_{25}(SC_2H_4Ph)_{18}$ NCs. Specifically, by controlling the reaction temperature (0 °C) and stirring speed (~30 rpm), a particular aggregation state of Au(I):SR intermediates that leads to exclusive formation of $Au_{25}(SC_2H_4Ph)_{18}$ NCs could be generated. In contrast, the Au(I):SR aggregates formed at the room temperature led to a low yield of $Au_{25}(SC_2H_4Ph)_{18}$ NCs, which revealed the importance of controlling the reaction intermediates. Although the yield of $Au_{25}(SC_2H_4Ph)_{18}$ had been greatly improved, the "one-pot one-cluster" dream was not fulfilled, which inspired the coming enormous efforts.

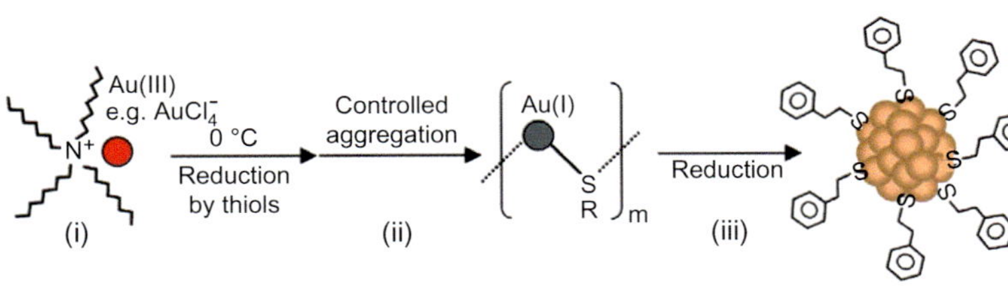

Figure 1.1 Basic steps in the synthesis of $Au_{25}(SC_2H_4Ph)_{18}$ NCs. Reprinted with permission from Ref. [34], Copyright 2008, American Chemical Society.

By replacing the solvent with tetrahydrofuran (THF), which can dissolve both $HAuCl_4$ and HSC_2H_4Ph, Wu et al. reported the one-phase synthesis of $Au_{25}(SR)_{18}$ with multiple types of functionalized thiols. Very importantly, they observed an interesting phenomenon, which was termed as size focusing [35]. They found that the initial polydisperse product eventually converted to the monodisperse Au_{25} NC when the reaction time was extended, as evidenced by the evolution of the UV–Vis spectrum (Fig. 1.2). Dass and co-workers subsequently provided matrix-assisted laser desorption/ionization (MALDI) mass spectrometric evidence for the size-focusing process of Au_{25} (Fig. 1.3) [36]. The special phenomenon follows the basic principle of "survival of the robustest": The as-formed, differently sized Au NCs are vastly different in stability, which leads to a spontaneous size focusing during the aging process, and eventually only the robustest size survives in the size-focusing process. Therefore, the "one-pot one-cluster" dream came true for the first time after employing such a law. The "size-focusing" process is not only tolerant to various ligands, but also applicable to the synthesis of other NCs (note that "size focusing" is universal, but it does not mean "one-pot one-cluster" in every case). For example, Qian et al. subsequently improved the $Au_{38}(SR)_{24}$ yield by the "size-focusing" laws [37]. In this method, a crude mixture of glutathionate-capped $Au_n(SG)_m$ NPs was first made; the mixed NPs were then subjected to a thermal thiol etching (size-focusing) process in a two-phase (water/organic) system (Fig. 1.4). They found that different size distributions of the $Au_n(SG)_m$ mixture synthesized in methanol and acetone influenced the final product. The acetone-mediated synthesis of $Au_n(SG)_m$ NPs produced a dominant size range from 8 to 18 kDa (with n from ~38 to ~102), while the methanol system produced a dominant size range below 8 kDa. A high yield of Au_{38} NC was obtained with the higher mass $Au_n(SG)_m$ NPs prepared in acetone. After that, by tuning the starting $HAuCl_4$ to thiol ratio into 1:3, followed by reduction in toluene, the size range of initial NCs was controlled to 26–36 kDa, which was further focused into the pure $Au_{144}(SR)_{60}$ [38]. The "size-focusing" law was also employed to synthesize non-gold metal NCs. For example, Wu et al. employed this strategy to synthesize Ag_7 cluster, which is the first thiolated silver cluster precisely characterized by mass spectrometry [39]. It is worth noting that the revealing of "size focusing" not only boosts the synthesis of metal NCs, but also promotes the structure, property, and application research of metal NCs due to the obtaining of high-quality cluster product in high yield. For instances, Wu et al. introduced nuclear paramagnetic resonance (NMR) to elucidate the structure of glutathione-protected Au_{25}, which was otherwise difficult to obtain by single crystal X-ray crystallography, conducted the cluster composition (structure)–fluorescence correlation research, applied the well-defined metal NCs for chemical sensing, and so on [13, 40, 41].

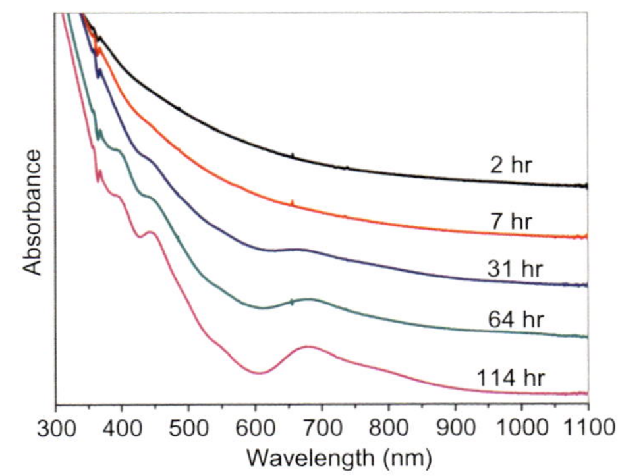

Figure 1.2 Evolution of the UV–Vis spectra of the crude product with aging time. The spectra are vertically shifted for the ease of comparison. Reprinted with permission from Ref. [35], Copyright 2009, The Royal Society of Chemistry.

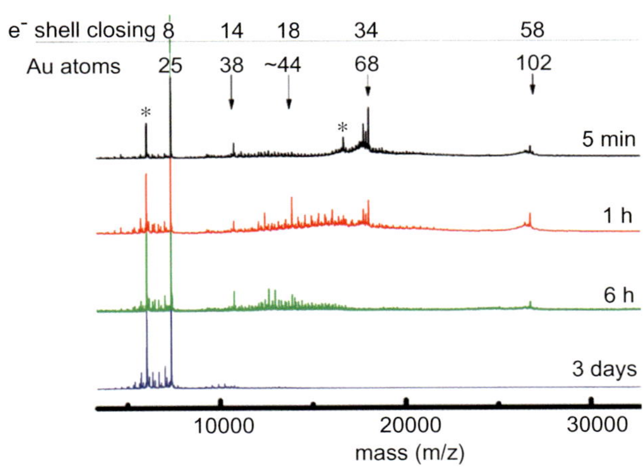

Figure 1.3 Mass spectrometric evidence of size focusing in the one-phase synthesis of $Au_{25}(SC_2H_4Ph)_{18}$ NCs. Peaks marked by asterisks are fragments. Reprinted with permission from Ref. [36], Copyright 2009, American Chemical Society.

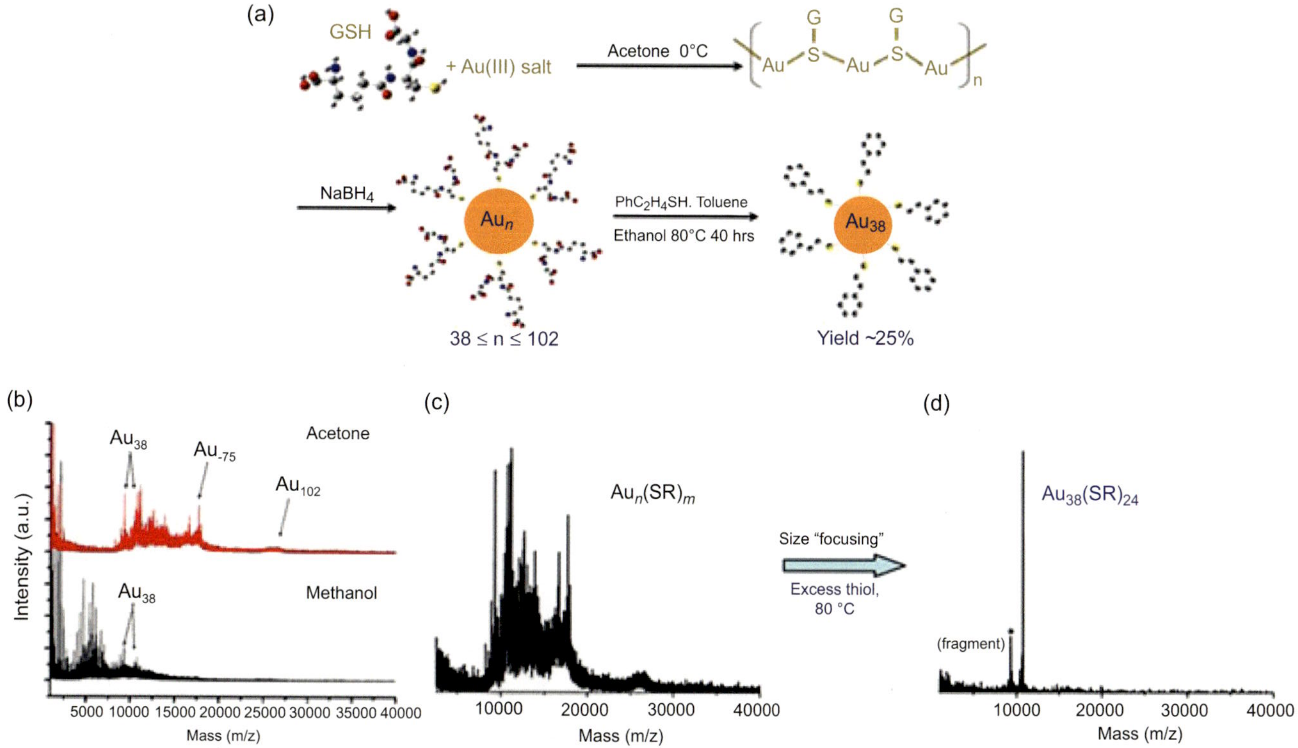

Figure 1.4 (a) Scheme of the size-focusing synthesis of $Au_{38}(SR)_{24}$ NCs; (b) the effect of solvent on the size range of the initial crude gold NCs; (c, d) illustration of size focusing of polydisperse NCs into pure $Au_{38}(SR)_{24}$. Reprinted with permission from Ref. [37], Copyright 2009, American Chemical Society.

With the increasing understanding of the fundamentals of synthesis, Wu et al. synthesized an atomically monodisperse, novel gold NC-$Au_{19}(SC_2H_4Ph)_{13}$ through tuning the reduction kinetics by simply replacing the strong reductant ($NaBH_4$) with a relatively weaker one (borane-*tert*-butylamine complex) while keeping all other conditions the same in the synthesis of $Au_{25}(SC_2H_4Ph)_{18}$ NC. Basing on this fact and some of the previous works, they proposed the synthesis principles, that is, the combined application of kinetic control and thermodynamic selection (Fig. 1.5) [30]. Specifically, the "kinetic control" means the control of reaction conditions (solvent, Au/S ratio, temperature, etc.), which aims to obtaining a proper size-distributed gold NCs mixture that greatly influences the "size-focusing" result. The "thermodynamic selection" means fulfilling "size focusing" by employing the thermodynamic differences of different NCs in the proper size-distributed NCs mixture, that is, by making some conditions to let the most stable ones survive, while the others are decomposed or transferred. Note that kinetic and thermodynamic influences are everywhere in all chemical reactions, but herein the terms "kinetic control" and "thermodynamic selection" have special implications as indicated above. These principles can be applied to most of the wet chemical syntheses.

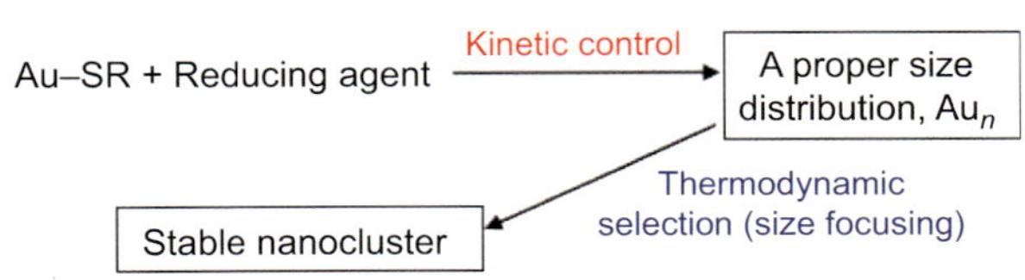

Figure 1.5 Kinetic control and thermodynamic selection in the synthesis of atomically monodisperse NCs. Reprinted with permission from Ref. [30], Copyright 2011, American Chemical Society.

For example, Xie and co-workers devised a synthesis route by utilizing carbon monoxide (CO) as a mild reducing agent and adjusting the pH value for the size-tunable syntheses of $Au_{15}(SR)_{13}$, $Au_{18}(SR)_{14}$, $Au_{25}(SR)_{18}$, and so on (Fig. 1.6) [42]. Thanks to the slow reduction kinetics, they revealed the growth process of $Au_{25}(SR)_{18}$ by monitoring the evolution of Au(I) precursor and Au NC intermediate species in later work [43]. They also reported a NaOH-mediated $NaBH_4$ reduction method for $Au_{25}(SR)_{18}$ synthesis in which NaOH was used to tune the formation kinetics by decreasing the reducibility of $NaBH_4$ and enhancing the etching ability of thiol [44]. By replacing the THF solvent with CH_2Cl_2 in the synthesis of $Au_{25}(SC_2H_4Ph)_{18}$, Tian et al. synthesized a low-temperature-stable structure of $Au_{38}(SC_2H_4Ph)_{24}$, which constitutes the first pair of structural isomers of metal NCs with high temperature–stable $Au_{38}(SR)_{24}$ [45].

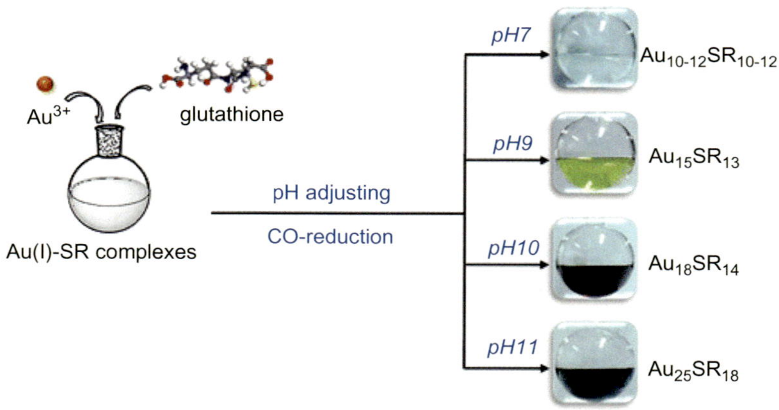

Figure 1.6 Size-controlled synthesis of aqueous NCs through CO reduction. Reprinted with permission from Ref. [42], Copyright 2013, American Chemical Society.

Zeng et al. reported the synthesis of $Au_{44}(SR)_{28}$ and $Au_{52}(SR)_{32}$ under similar conditions except for the heating temperature difference: $Au_{44}(SR)_{28}$ was size focused at 60 °C, while $Au_{52}(SR)_{32}$ was generated at a higher temperature of 80 °C [46, 47]. Not only the heating temperature can influence the final product, but also the heating time under the same temperature can affect the final product, as shown in the synthesis of $Au_{44}(SR)_{26}$ and $Au_{49}(SR)_{27}$ reported by Liao et al. [48, 49]. The ligand type can also affect the final products. For example, when the same synthetic protocols are employed, $Au_{144}(SC_2H_4Ph)_{60}$ and $Au_{64}(S\text{-}c\text{-}C_6H_{11})_{32}$ can be obtained when 2-phenylethanethiol and cyclohexanethiol are used as the protecting ligands, respectively [50]. In another case, the protecting cyclohexanethiolate leads to the forming of $[Au_{23}(S\text{-}c\text{-}C_6H_{11})_{16}]^{-1}$, while the protecting adamantanethiolate (S-Adm) leads to the producing of $Au_{30}(S\text{-}Adm)_{18}$, with the other synthesis protocols remaining the same [51, 52]. Chen et al. reported the surprising result of employing isomeric methylbenzenethiols (MBTs) to control the size of $Au_n(SR)_m$ under similar reaction conditions. By simply adjusting the position of methyl group from *para*- to *meta*- to *ortho*-, the most stable size changes from $Au_{130}(p\text{-MBT})_{50}$ to $Au_{104}(m\text{-MBT})_{41}$ and to $Au_{40}(o\text{-MBT})_{24}$, respectively (Fig. 1.7) [53]. The decreasing size trend is interestingly in line with the increasing hindrance of the methyl group with respect to the interfacial Au–S bond. The addition of chemical reagents is also an effective way to tune the kinetics and thermodynamics. For example, Zhuang et al. developed an acid induction method in which acid (nitric acid, acetic acid, etc.) was added to the reaction solution after the addition of the reducing reagent $NaBH_4$. The acid can accelerate the hydrolysis of $NaBH_4$ and strengthen the reactivity of $NaBH_4$, hence accelerating the reaction kinetics. However, it can weaken the interaction between Au and thiolate, reducing the reactivity of thiolate and changing the thermodynamic selection, thus influencing both the kinetics and the thermodynamics of the synthesis. Several novel Au NCs have been obtained by this method, including $Au_{52}(SC_2H_4Ph)_{32}$, $Au_{42}(SPh\text{-}t\text{-}Bu)_{26}$, $Au_{44}(SPh\text{-}t\text{-}Bu)_{26}$, and $Au_{48}(SPh\text{-}t\text{-}Bu)_{28}$ [54–56]. Further, Zhuang et al. introduced Cd^{2+} in replacement of acid and synthesized a non-fcc-structured $Au_{42}(SPh\text{-}t\text{-}Bu)_{26}$ NC, which is the structural isomer of the existing fcc-structured $Au_{42}(SPh\text{-}t\text{-}Bu)_{26}$ (where fcc is face-centered cubic) [57]. The two Au_{42} NCs are the second pair of genuine structural isomers by far. The addition of Cd

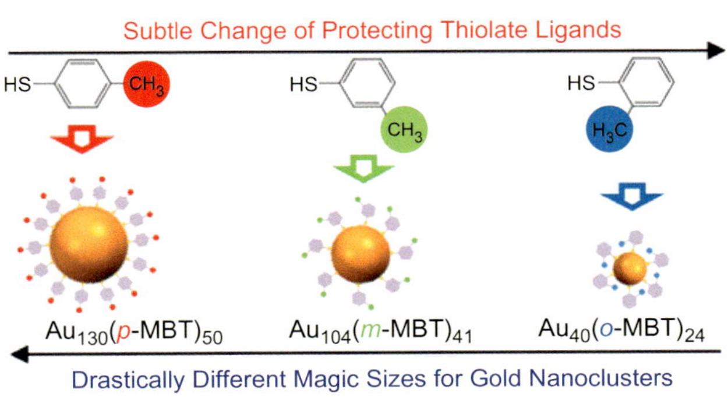

Figure 1.7 Isomeric methylbenzenethiols (MBTs) for size-controlled synthesis of $Au_n(SR)_m$ NCs. Reprinted with permission from Ref. [53], Copyright 2015, American Chemical Society.

might influence the kinetics and thermodynamics by forming some unstable Au/Cd intermediates, and the novel synthesis method can be dubbed the ion induction method. It is to be noted that the synthesis principles can be also applied to many other reactions that are not discussed herein, including the highly selective ligand exchange–induced size/structure transformations (LEIST) [29, 58, 59] and anti-galvanic reactions [60–62].

1.3 Isolation of Gold Nanoclusters

When a product is generated, the isolation (separation) of the product from the reaction mixture may be problematical or even challenging, especially for those cases in which multiple products with similar chemical and physical properties coexist. In this section, we will mainly review the physical techniques and methods commonly used in the gold NCs isolation (separation).

1.3.1 Fractionated Precipitation

Fractionated precipitation is an effective, facile separation method for those NCs with obvious solubility difference [63–66]. The larger or relatively solubility-poor NCs will be precipitated out first after the addition of poor solvent into the NCs mixture solution. By such a principle, Whetten and co-workers isolated a series of Au NPs in 1.5–3.5 nm range, with the core masses to be 93–92 k, ~57 k, 46–45 k, and 29–27 k (k = 1000 amu or kDa) determined by laser desorption/ionization mass spectrometry (LDIMS), containing from ~1300 to ~100 Au atoms [33]. The separations are performed starting from nearly saturated toluene solutions of NPs mixtures (typically 10 mL of 20 mg/mL concentration) and slowly adding a miscible poor solvent (usually acetone) by passive vapor transfer until the solution volume is increased by a predetermined amount. The solution is then isolated, allowed an extended equilibration period while stirring, and finally centrifuged and decanted to remove the soluble fraction from the precipitated fraction, each of which is analyzed by MS. The procedure is then repeated separately on both fractions, and so on, to generate many further fractions, until a handful of highly purified fractions are obtained. In the subsequent work, even smaller Au NCs were obtained through repeated fractional crystallization, such as species with core masses of 15–14 k (1.3 nm, ~75 atoms) and 8 k (1.1 nm, ~38 atoms) [67].

In addition to the separation of oil-soluble NCs reported by the Whetten group, the isolation of water-soluble NCs and NPs can be also conducted by this method. For example, Wu et al. isolated three glutathione (-SG) capped gold nanoparticles, including $Au_{25}(SG)_{18}$ NC, 2- and 4-nm NPs in one pot by precipitation with different amounts of methanol [68]. Larger NPs are readily precipitated out with the addition of only a small amount of methanol, leaving smaller particles in the supernatant. By controlling the amount of methanol added, the three predominant species (**1–3**, in order of the decreasing particle size) were separated (Fig. 1.8). Yang and co-workers separated water-soluble gold NCs stabilized with penicillamine ligands in aqueous medium by sequential size-selective precipitation. Au NCs were precipitated out and separated successively from larger to smaller ones by progressively increasing the concentration of acetone in the aqueous Au NCs solution [69]. The Dass group even used the fractionated precipitation method to isolate the Au–Ag alloy NCs, such as $Au_{38\text{-}n}Ag_n(SR)_{24}$ [70] and $Au_{25\text{-}n}Ag_n(SR)_{24}$ [71].

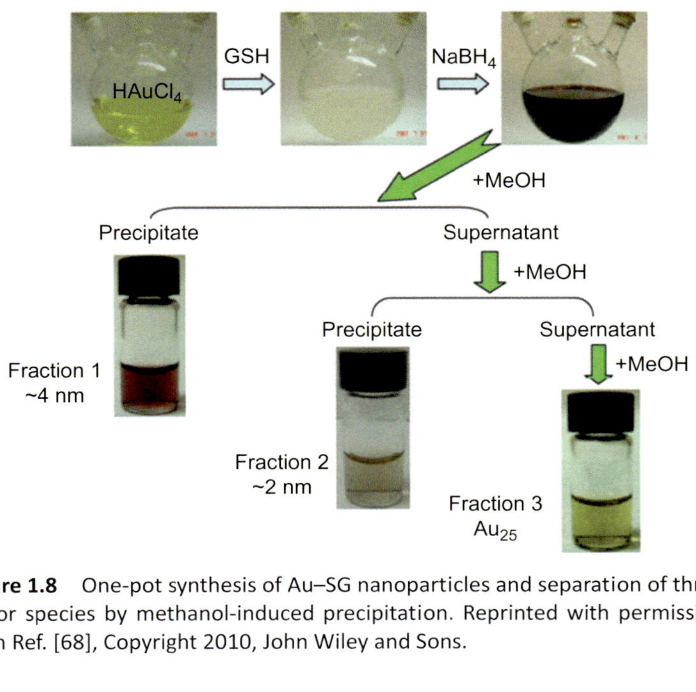

Figure 1.8 One-pot synthesis of Au–SG nanoparticles and separation of three major species by methanol-induced precipitation. Reprinted with permission from Ref. [68], Copyright 2010, John Wiley and Sons.

1.3.2 Recrystallization

Recrystallization, which means repeated crystallization, is widely used in industry and laboratory, and it is also used in the purification of NCs. For instance, Schaaff et al. carried out the recrystallization of 29 kDa Au:SR compounds (confirmed as $Au_{144}(SR)_{60}$ lately) and obtained the microcrystals from slow precipitation out of a dilute solvent–nonsolvent interaction [72]. Specifically, the Au:SR compounds were precipitated out of concentrated toluene solutions by the addition of excess ethanol and allowed to stand at room temperature for 12–15 h. The precipitate was then filtered, redissolved in toluene, and reprecipitated twice to ensure that the excess disulfide (RSSR) and unreacted thiol (RSH) were removed and the finely crystalline black powder was obtained (Fig. 1.9). The Murray group prepared Au–Pd alloy NCs and purified them by recrystallization from the acetonitrile/methanol mixture for several times [73]; Wu and co-workers obtained pure $Au_{25}(SG)_{18}$ by recrystallization in water–methanol solution [35, 40, 41, 74]; Quinn's group utilized precipitation/dispersion cycles to purify Au_{38} NCs [75].

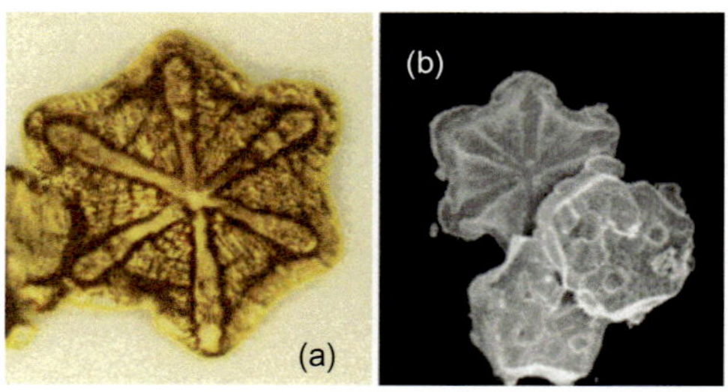

Figure 1.9 (a) Optical micrograph and (b) SEM micrograph show the typical crystal of 29 kDa Au:SR compounds that is grown from slow precipitation out of a dilute solvent–nonsolvent interaction. Reprinted with permission from Ref. [72], Copyright 2001, American Chemical Society.

1.3.3 Solvent Extraction

Solvent extraction is an inverse process of crystallization that takes advantage of the good solubility of target NCs to isolate them from the crude product and transfer them to the solution, usually in conjunction with recrystallization. The negative $Au_{25}(SR)_{18}$ shows exceptional solubility in acetonitrile, and thus the Murray and Jin groups extracted pure $Au_{25}(SR)_{18}$ from the crude product and successfully grew the single crystals [76, 77]. Generally, the solvents used for extraction were acetonitrile, acetone, dichloromethane, toluene, and so on. By the same method, Jin's group obtained a series of pure Au NCs, including Au_{24} [78], Au_{38} [37], Au_{44} [46], Au_{130} [79], Au_{144} [38], and Au_{246} [80]. Negishi et al. successfully extracted $Au_{25-n}Ag_n(SC_{12}H_{25})_{18}$ with acetone from the crude product [81]. A special example is the isolation of Au_{30}(S-Adm)$_{18}$, which is dissolvable only in benzene and even insoluble in dichloromethane, a widely used solvent for oleophilic clusters. This peculiar solubility of Au_{30}(SAdm)$_{18}$ is indeed important as it allows one to purify NC by removing some by-products with dichloromethane first and then subject the extract from the resulting residue [52]. Besides, hydrophilic ligand-protected Au NCs can also be extracted by a suitable solvent such as methanol [82].

1.3.4 Polyacrylamide Gel Electrophoresis

For subtle separation of the NCs mixture with close polarity, the above-mentioned methods generally do not work well. An earliest solution for that is PAGE, which requires two kinds of gels: a stacking gel and a separating gel. Normally, sample solutions are loaded onto the stacking gel and then eluted for a long time at a constant voltage to get sufficient separation. The NCs are separated based on their size and aqueous solubility, and PAGE often works well for those NCs that are protected by hydrophilic ligands, such as glutathione (GSH), mercaptobenzoic acid (MBA), and captopril (Cap). Separation can easily be observed by looking at the gels with different colors, which are then cut and extracted in appropriate solvents to get the purified NC.

In 1998, Whetten and co-workers fractionated Au NCs protected by monolayers of GSH by using PAGE and identified the most abundant species as $Au_{28}(SG)_{16}$ by mass spectrometry, which was lately corrected as $Au_{25}(SG)_{18}$ [83]. After that, a series of giant Au NCs composed of a gold core and a glutathione (GSH) adsorbate layer have been prepared from Au(I)SG polymers and separated by gel electrophoresis [84]. Tsukuda's group isolated a series of magic-numbered Au NCs [27]. In 2005, they significantly improved the PAGE separation of the aqueous Au–SG NCs and obtained high-purity $Au_n(SG)_m$ NCs. Compared to the earlier electrospray ionization mass spectrometry (ESI-MS) work, very clean ESI-MS spectra were obtained for the first time by suppressing the fragmentation of $Au_n(SG)_m$ NCs in ESI-MS (Fig. 1.10) [32]. Through the precise determination of masses of the isolated species, distinct NCs were identified unequivocally, including $Au_{15}(SG)_{13}$, $Au_{18}(SG)_{14}$, $Au_{22}(SG)_{16}$, $Au_{22}(SG)_{17}$, $Au_{25}(SG)_{18}$, $Au_{29}(SG)_{20}$, $Au_{33}(SG)_{22}$, and $Au_{39}(SG)_{24}$. Xie and co-workers isolated a red-emitting Au NC with the precise molecular formula of $Au_{22}(SG)_{18}$ (Fig. 1.11) [85]. Other research groups have also used this method to obtain isolated gold NCs [86–91].

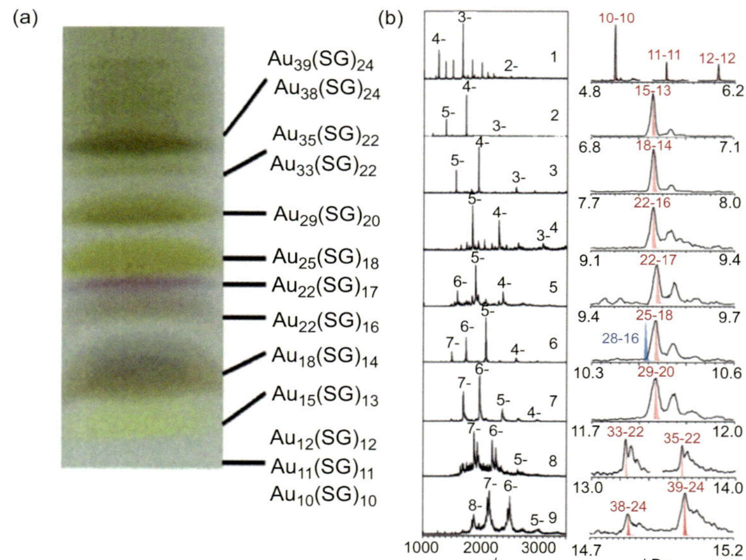

Figure 1.10 (a) Separation of $Au_n(SG)_m$ NCs by PAGE. (b) Low-resolution ESI mass spectra of the fractionated Au:SG clusters (left). The mass spectra reproduced from the most intense peaks in the high-resolution spectra (right). The calculated spectra for $Au_n(SG)_m$ are shown by the colored peaks with the corresponding *n-m* values. Reprinted with permission from Ref. [32], Copyright 2005, American Chemical Society.

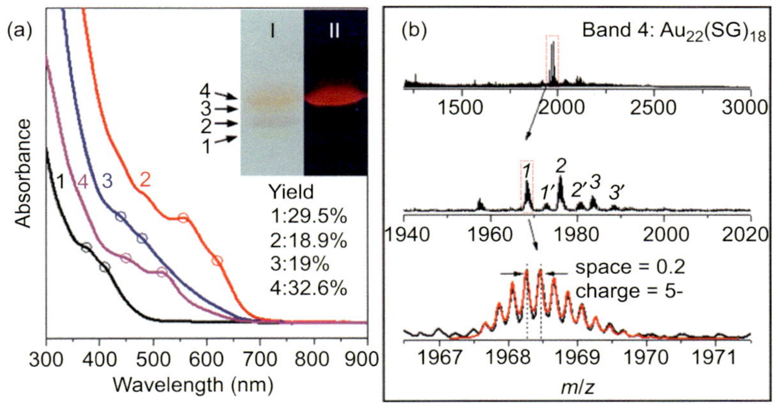

Figure 1.11 (a) UV–Vis absorption spectra of Au NCs separated from bands 1–4 in the native PAGE gel (insets). (b) ESI mass spectra of species 4. The red lines are the simulated isotope pattern of $[Au_{22}(SG)_{18}{}^{0}\text{-}5H]^{5-}$. Reprinted with permission from Ref. [85], Copyright 2014, American Chemical Society.

In addition to Au NCs, Ag NCs were also isolated by PAGE. For example, Kumar et al. synthesized glutathione-stabilized silver NCs by a modified Brust method and isolated them by PAGE using homemade polyacrylamide gels. There were at least 16 naked eye discernible PAGE bands with different colors, strongly indicating the multiple species involved in the as-synthesized Ag:SG NCs. Importantly, they found that the positions and colors of the PAGE bands were independent of the reaction conditions, which suggested that only the magic-numbered silver NCs were formed under different experimental conditions [92]. It should be noted that the preparation of polyacrylamide is very important for the successful separation of NCs. By changing acrylamide concentration and other gel conditions, different core sizes can be separated.

1.3.5 Size Exclusion Chromatography

Size exclusion chromatography (SEC) is an important chromatographic technique in which porous hydrophobic microgels are employed to separate differently sized molecules or particles. The basic principle of SEC is simple: large molecules (particles) cannot penetrate the pores of the column media as effectively as small ones; as a result, they spend less time exploring the pore structure of the column and elute through the column faster than small ones [93]. In some of the earlier literature, it was also called gel permeation chromatography (GPC) [94, 95]. For example, Tsunoyama et al. reported the size separation of Au:SC_x NCs using GPC with high reproducibility, resolution, and throughput. Figure 1.12 shows the typical recycling chromatogram of Au:SC_{18} NCs. It can be seen that two separated peaks **I** and **II** can be obtained after repeated passages through the column. After further GPC separation of the two crude fractions, four fractions can be obtained at different retention times as shown in Fig. 1.12b. From the core size characterization by LDIMS (Fig. 1.12c), the NC size decreases from fraction **1** to **4**. Fraction **2** is composed of only 11 kDa clusters, while fractions **1** and **3** are mixed with ~16 and ~8 kDa NCs, respectively, and fraction **4** is dominated by the smallest 8 kDa clusters. Figure 1.12d clearly shows that the fractions **2** and **4** are mixed by different-sized Au NCs (53 ± 10 and 35 ± 6 Au atoms). In the subsequent work, they also used SEC to isolate Au NCs and further characterized them by MALDI mass spectrometry and thermogravimetric analysis [96, 97]. Qian et al. isolated $Au_{40}(SC_2H_4Ph)_{24}$ from a mixture with $Au_{38}(SC_2H_4Ph)_{24}$. At first, they attempted solvent extraction approach but were unsuccessful due to the close size and solubility of the component NCs. The hydrophobicity of these NCs also precluded their isolation via PAGE. They finally resorted to SEC and found that $Au_{38}(SC_2H_4Ph)_{24}$ and $Au_{40}(SC_2H_4Ph)_{24}$ were indeed separable by SEC [98]. In later work by Knoppe et al., they separated $Au_{40}(SC_2H_4Ph)_{24}$ NC using SEC on a semipreparative scale [99]. Moreover, Wu et al. used SEC to check the purity of as-prepared $Au_{19}(SC_2H_4Ph)_{13}$ [30]. Except for these small size NCs, several large NCs have been purified using this method in recent years, such as Au_{102} [100], Au_{130} [101], and Au_{144} [102]. It is worth noting that SEC can also be applied to isolate alloy NCs. For example, Qian et al. successfully separated $Au_{25}(SC_2H_4Ph)_{18}$ and $Pt_1Au_{24}(SC_2H_4Ph)_{18}$ although the two NCs have very similar sizes [103]. The different charge states in $Au_{25}(SC_2H_4Ph)_{18}$ (negative) and $Pt_1Au_{24}(SC_2H_4Ph)_{18}$ (neutral) may be responsible for their separation by SEC. Negishi and co-workers synthesized $Au_{25-n}Cu_n(SC_2H_4Ph)_{18}$ NCs by reacting a gold salt ($HAuCl_4$) and a copper salt ($CuCl_2$) with HSC_2H_4Ph in methanol and reducing the resulting complexes with $NaBH_4$. After that, $Au_{25-n}Cu_n(SC_2H_4Ph)_{18}$ NCs were extracted by acetonitrile from the as-prepared product and then separated by GPC [104].

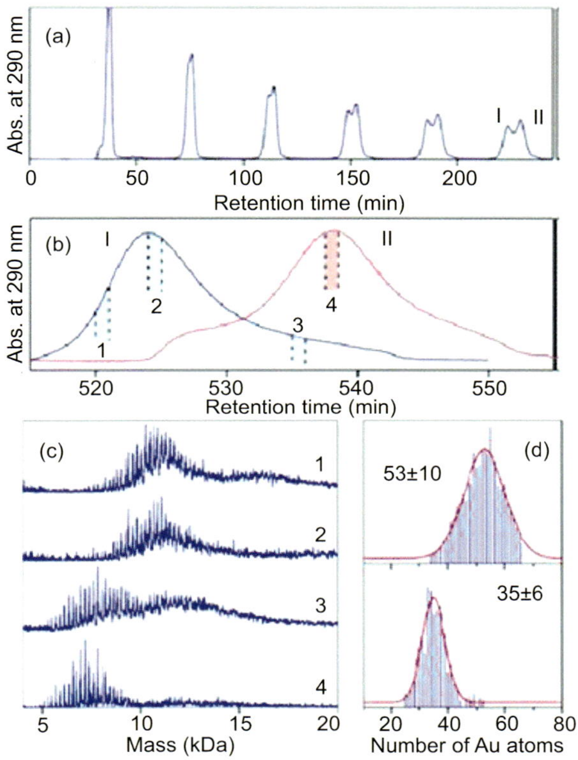

Figure 1.12 (a) Recycling gel permeation chromatography (GPC) chromatogram of Au:SC_{18} NCs. Toluene was used as the eluent at a flow rate of 3.5 mL min^{-1}. (b) Recycling chromatograms of the fractions **I** and **II**. (c) LDI mass spectra of fractions **1–4** in the positive mode. (d) Histograms of the core numbers for fractions **2** and **4**. Reprinted with permission from Ref. [95], Copyright 2006, American Chemical Society.

1.3.6 High-Performance Liquid Chromatography

High-performance liquid chromatography (HPLC) is known to allow the high-resolution separation of a wide variety of chemical compounds. The driving principle behind HPLC is chromatography, which separates substances basing on the differences in solute affinity for stationary and mobile phases. As a liquid chromatography technique, HPLC is based on the use of a liquid compressed at high pressures as the mobile phase. There is an extensive body of the literature on HPLC separations of metal NCs [105, 106], which can be classified into several categories: (1) separation of NCs depending on core sizes (the number of metal atoms in the NC), (2) separation of NCs depending on the charge (keeping the NC size constant), (3) separation of doped NCs, (4) separation of NCs depending on the ligand composition, (5) separation of the coordination isomer, (6) separation of enantiomers of intrinsically chiral NCs. Below is the detailed review.

1.3.6.1 Separation of NCs depending on core sizes

Negishi and co-workers prepared a mixture of $Au_n(SC_{12}H_{25})_m$ NCs using the Brust method and subsequently separated them by reversed-phase high-performance liquid chromatography (RP-HPLC), using a C8 column in series with a phenyl column, and a mobile phase containing tetrabutylammonium perchlorate ($TBAClO_4$) at a concentration of 10 mM [107]. Multiple peaks were clearly presented in the resulting chromatograms (Fig. 1.13a), indicating the high-resolution separation of these NCs, which was further confirmed by mass spectrometry. By such a method, they successfully isolated $Au_{38}(SC_{12}H_{25})_{24}$, $Au_{104}(SC_{12}H_{25})_{45}$, $Au_{130}(SC_{12}H_{25})_{50}$, $Au_{144}(SC_{12}H_{25})_{60}$, $Au_{187}(SC_{12}H_{25})_{68}$, $Au_{\sim226}(SC_{12}H_{25})_{\sim76}$, $Au_{\sim253}(SC_{12}H_{25})_{\sim90}$, $Au_{329}(SC_{12}H_{25})_{84}$, $Au_{\sim356}(SC_{12}H_{25})_{\sim112}$, and $Au_{\sim520}(SC_{12}H_{25})_{\sim130}$.

1.3.6.2 Separation of NCs depending on the charge

The solubility of $Au_n(SR)_m$ NCs varies depending on the charge state. For example, neutral $[Au_{25}(SC_6H_{13})_{18}]^0$ exhibits a lower solubility in polar solvents than anionic $[Au_{25}(SC_6H_{13})_{18}]^-$. Accordingly, these two $Au_{25}(SR)_{18}$ NCs, with different charge states, showed different retention times in the RP-HPLC chromatogram [108]. Negishi and co-workers synthesized $[Au_{25}(SC_{12}H_{25})_{18}]^-$ and $[Au_{25}(SC_{12}H_{25})_{18}]^0$ and then measured their chromatograms [109]. Figure 1.14 shows the chromatograms of $[Au_{25}(SC_{12}H_{25})_{18}]^-$ and $[Au_{25}(SC_{12}H_{25})_{18}]^0$. $[Au_{25}(SC_{12}H_{25})_{18}]^-$ displays a peak at a retention time of 10.6 min, whereas $[Au_{25}(SC_{12}H_{25})_{18}]^0$ displays a peak at a longer retention time of 11.9 min. It can be concluded that $[Au_{25}(SC_{12}H_{25})_{18}]^0$ features stronger interactions with the stationary phase when compared with $[Au_{25}(SC_{12}H_{25})_{18}]^-$ because of the lower solubility of the former cluster in polar solvents, resulting in a longer retention time. These results imply that $Au_n(SR)_m$ NCs with different charge states can be separated with high resolution by RP-HPLC.

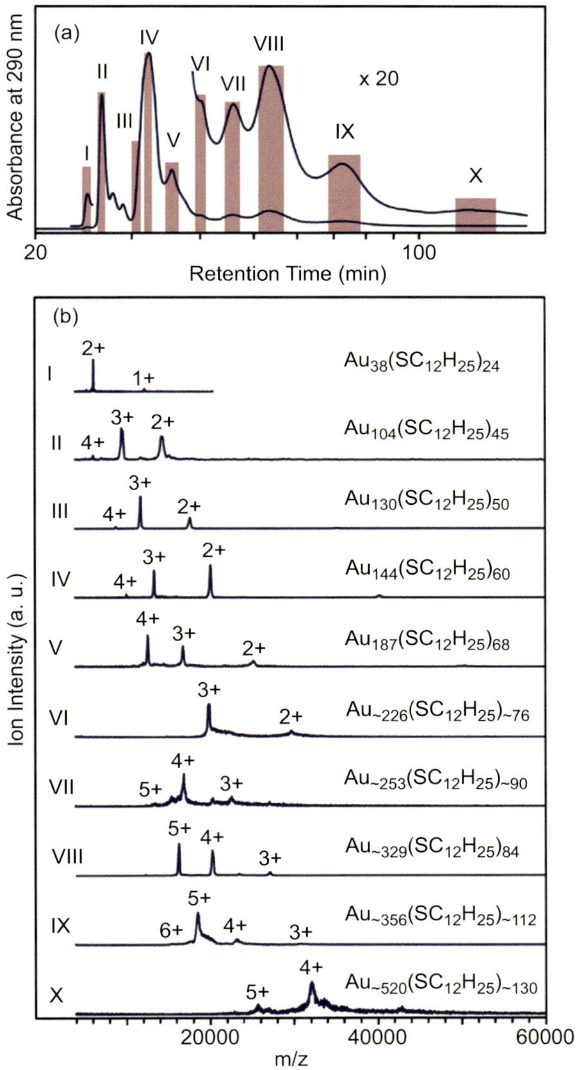

Figure 1.13 Separation of a series of $Au_n(SC_{12}H_{25})_m$ clusters; (a) chromatogram of crude sample of $Au_n(SC_{12}H_{25})_m$ clusters and (b) positive ion ESI mass spectra of fractions **I–X**. Reprinted with permission from Ref. [107], Copyright 2015, American Chemical Society.

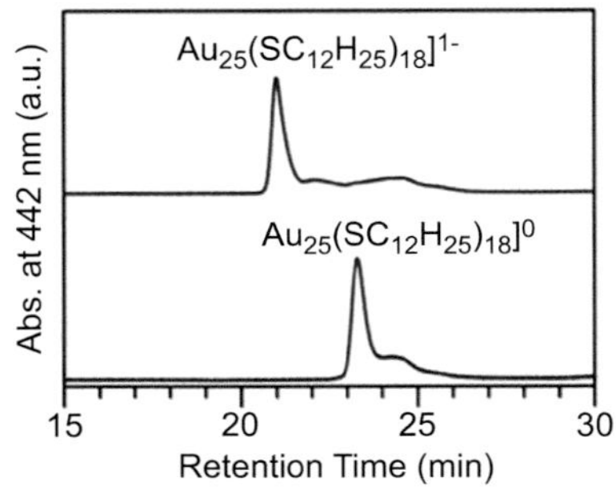

Figure 1.14 Reversed-phase HPLC chromatograms of $[Au_{25}(SC_{12}H_{25})_{18}]^-$ and $[Au_{25}(SC_{12}H_{25})_{18}]^0$. Reproduced with permission from Ref. [109], Copyright 2010, the Royal Society of Chemistry.

1.3.6.3 Separation of doped NCs

Because Au, Ag, Pt, and Pd have different electronegativities, partial charge transfer occurs from other metal atom(s) to Au atom(s). This charge transfer also influences the charge density of SR ligands, which are directly connected to metal atoms. Therefore, doped Au NCs with different numbers of substitutions should have different polarities. On the basis of this prediction, Niihori et al. enabled separation of $[Au_{25-x}Ag_x(SC_4H_9)_{18}]^-$ (x = 0–4) with atomic precision (Fig. 1.15) [110]. This separation method can also be applied to separate other alloy NCs, including $Au_{38-x}Ag_x(SR)_{24}$ (R = C_4H_9 or C_2H_4Ph) and $Au_{24}M(SC_4H_9)_{18}$ (M = Pt or Pd).

1.3.6.4 Separation of NCs depending on the ligand composition

The polarity of $Au_n(SR)_m$ NCs depends on the ligands surrounding the metal core as well. Thus, NCs with different ligand compositions can also be separated by HPLC. The Negishi and Pradeep groups collaborated to study the separation of $Au_{24}Pd$ NCs protected with mixed ligands [106]. In their study, $Au_{24}Pd(SR_1)_{18-x}(SR_2)_x$ NCs protected by two different thiolate ligands (SR_1 and SR_2) were prepared by incomplete ligand exchange. The resulting mixture was successfully separated by RP-HPLC with high purity using a C18 column. To investigate the versatility of this separation method, similar experiments were conducted for separating $Au_{24}Pd$ NCs with a varying ligand composition. Figure 1.16 shows the chromatograms of $Au_{24}Pd(SR_1)_{18-x}(SR_2)_x$ (x = 0–18). All chromatograms showed well-resolved and relatively intense peaks, which were well consistent with the MALDI mass spectrometry, indicating the universality of this method [111].

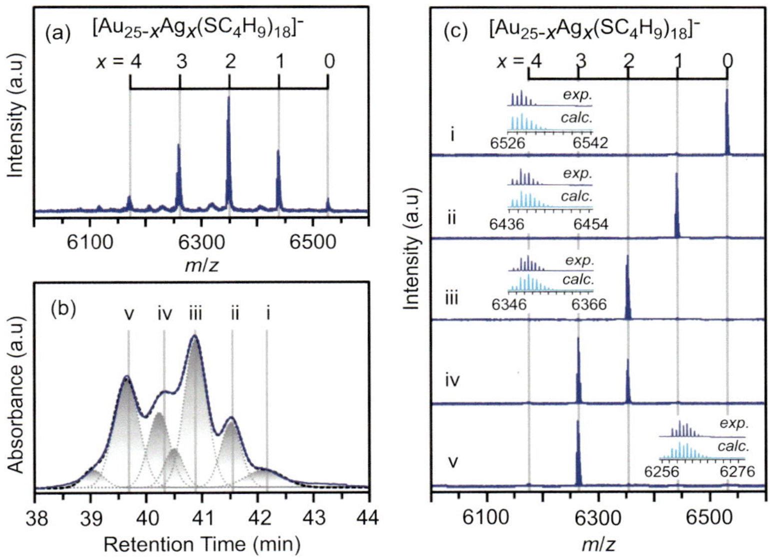

Figure 1.15 Results obtained for $[Au_{25-x}Ag_x(SC_4H_9)_{18}]^-$ (x = 0–4) synthesized by anti-galvanic reaction. (a) Negative-ion ESI mass spectrum and (b) UV chromatogram and the fitting result of a NC mixture. (c) Negative-ion ESI mass spectra and isotope patterns for peaks in (b). Reprinted with permission from Ref. [110], Copyright 2018, American Chemical Society.

1.3.6.5 Separation of coordination isomer

The $Au_{24}Pd(SR)_{18}$ discussed in the previous section has a geometric structure in which six [–S(R)–Au–S(R)–Au–S(R)–] staples surround the $Au_{12}Pd$ metal core. Two types of SR units exist in such a geometric structure, that is, SR that is bound directly to the metal core (core site) and SR that is positioned at the center of the staple (apex site). Thus, two coordination isomers can be considered for $Au_{24}Pd(SR_1)_{17}(SR_2)$ that was prepared by the ligand exchange reaction (Fig. 1.17a) [112]. The peak with a shorter retention time was derived from isomers exchanged at the core site. The peak with a longer retention time was attributed to isomers exchanged at the apex site. As shown in

Fig. 1.17a, major products of $[Au_{24}Pd(SC_2H_4Ph)_{17}(SC_{12}H_{25})]^0$ were the NCs in which the ligand was exchanged at the core site. This result demonstrated that the first ligand exchange reaction occurred preferentially at the core site. A similar reaction preference was also observed in the second ligand exchange reaction (Fig. 1.17b).

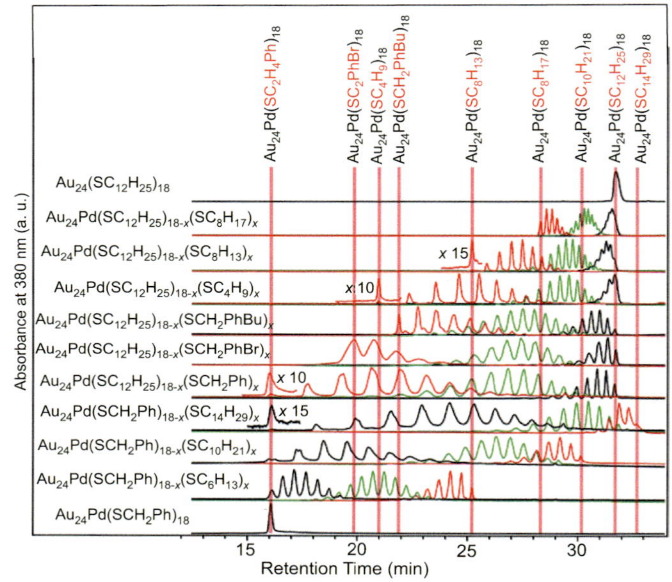

Figure 1.16 Chromatograms obtained for $Au_{24}Pd(SR_1)_{18-x}(SR_2)_x$ (x = 0–18) with various ligand combinations (SR_1, SR_2), $Au_{24}Pd(SC_{12}H_{25})_{18}$ and $Au_{24}Pd(SC_2H_4Ph)_{18}$ for comparison purposes. Reprinted with permission from Ref. [111], Copyright 2014, The Royal Society of Chemistry.

1.3.6.6 Separation of enantiomers of intrinsically chiral NCs

The chiral enantiomers of NCs can be separated by HPLC when using a chiral analytical HPLC column. Dolamic et al. reported the first separation of the enantiomers of a gold NC protected by achiral thiolates, $Au_{38}(SC_2H_4Ph)_{24}$, achieved by chiral HPLC [113]. Two peaks well separated were observed at 8.45 and 17.45 min (enantiomers **1** and **2**, respectively, according to increasing elution times). The second peak is broadened and less intense compared to the first one, but integration gives identical peak areas within the accuracy of the measurement. The UV spectra of both peaks clearly show the distinct signature of $Au_{38}(SR)_{24}$ NC. Zeng et al. also used chiral HPLC to separate the enantiomers of $Au_{28}(SR)_{20}$ [114].

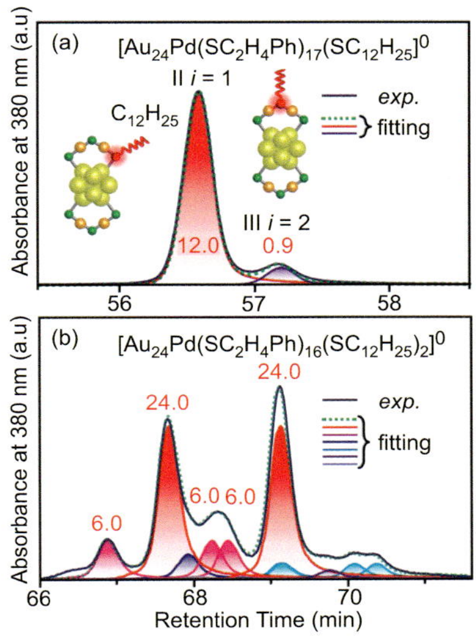

Figure 1.17 Chromatograms representing the expanded regions for (a) $[Au_{24}Pd(SC_2H_4Ph)_{17}(SC_{12}H_{25})]^0$ and (b) $[Au_{24}Pd(SC_2H_4Ph)_{16}(SC_{12}H_{25})_2]^0$, together with curve fitting results. Reprinted with permission from Ref. [112], Copyright 2015, American Chemical Society.

1.3.7 Thin-Layer Chromatography

As the simplest chromatographic technique, thin-layer chromatography (TLC) has been attempted for the separation of metal NCs. The separation mechanism is similar to that of PAGE and SEC; that is, the NCs of different polarities and sizes will reach different positions of chromatoplate during the elution process, making the separation efficient and visual. When the separation is complete, different bands of NCs are cut and extracted with appropriate solvents to get the purified NCs. In early 2014, Wu's group reported the recycling of $Au_{25}(SC_2H_4Ph)_{18}$ by column chromatography packed by silica gel [115]. Then, Pradeep's group and Wu's group independently illustrated the subtle separation of thiolated noble metal NCs by TLC [116, 117]. Furthermore, Wu's group employed preparative thin-layer chromatography (PTLC) to isolate an isomer of $Au_{38}(SC_2H_4Ph)_{24}$, namely, Au_{38T}, a full family of tiara-like thiolated Pd NCs, and a series of other metal NCs [45, 48–49, 54, 118–122]. The separation method can also be used to monitor the reaction process and analyze the reaction mechanism of NCs chemistry. For example, Wu's group utilized TLC to show the diversity of products in the ion-precursor and ion-dose dependent anti-galvanic reduction (Fig. 1.18) [123]. Different silver ion precursors (including $AgNO_3$, Ag-EDTA, Ag-PET, and Ag-DTZ; EDTA (ethylenediamine tetraacetic acid disodium salt), and DTZ (dithizone)) in the reaction with $Au_{25}(SC_2H_4Ph)_{18}$ under similar conditions lead to various products, as illustrated by TLC in combination with mass spectrometry (Fig. 1.18a,c). Besides, in the reaction of $Au_{25}(PET)_{18}$ and Ag-DTZ, as the

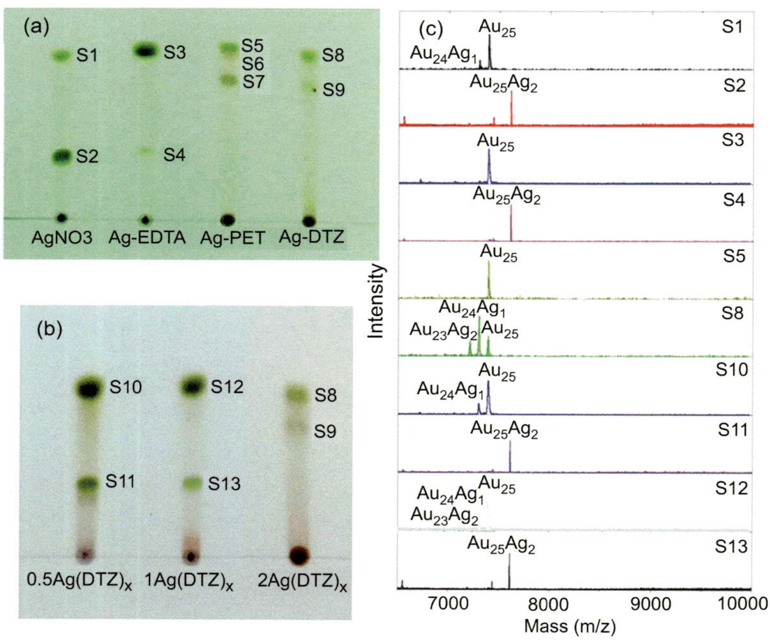

Figure 1.18 Monitoring the product by TLC: (a) $AgNO_3$, Ag-EDTA, Ag-PET, and Ag-DTZ were employed as the precursors with the Au:Ag ratio of 1:2 and (b) different Au:Ag ratios, 2:1, 1:1, and 1:2, in the case of Ag-DTZ. (c) Mass spectra of the products S1–S5, S8, and S10–S13. The spectra of S1, S3, S5, S8, S10, and S12 were acquired in negative ionization mode, while the others were collected in positive ionization mode. Reproduced with permission from Ref. [123], Copyright 2015, The Royal Society of Chemistry.

Ag-DTZ dose increases, the content of $Au_{24}Ag(PET)_{18}$ in the product mixture increases, while the content of $Au_{25}Ag_2(PET)_{18}$ decreases and is finally negligible when the Ag:Au atomic ratio reaches 2:1 (Fig. 1.18b,c). Recently, the same group introduced PTLC to quantitatively monitor the NCs "size-focusing" process to reveal that mainly the ~3 nm NPs formed in the reaction process promote the transformation from Au_{44} to Au_{36} and to optimize the reaction parameters, thus improving the yields of Au_{44} and Au_{36} [124].

1.3.8 Other Separation Methods

In addition to the above-mentioned, widely used separation methods, there are some other methods as well. For example, Choi et al. demonstrated the efficacy of the ion-pair chromatography for separations of samples of charged, polydisperse, water-soluble Au NCs protected by monolayers of *N*-acetyl-L-cysteine and of tiopronin ligands [125]. Lo et al. developed an effective capillary electrophoretic technique for separating samples of negatively charged, polydisperse, water-soluble Au NCs [126]. Bakr and co-workers demonstrated the efficacy of ultracentrifugation for quantitative determination of the ligand-to-gold ratio in $Au_n(SR)_m$ clusters [127].

1.4 Summary

As a kind of rising star materials, gold NCs have attracted extensive interest in recent years. However, the nanocluster research is still in its infant stage and lots of work needs to be done. In particular, the synthesis needs to be greatly improved since it is the cornerstone and lays foundations for the subsequent research on the structure, property, and application. Before that, the fundamentals of synthesis should be understood. For this purpose, we present the synthesis principles proposed recently, which may have important implications for future efforts in improving the NCs synthesis. In the event that "one-pot one-cluster" syntheses were not achieved, the subtle isolation is critical for obtaining atomically monodisperse NCs; therefore, the common isolation methods and techniques are reviewed in this chapter, too, which may provide help for researchers to choose or even develop appropriate methods (techniques) for NCs isolation. It is expected that the synthesis and isolation research is still active in this emerging field and novel synthesis (isolation) methods or techniques are developed in the near future.

Acknowledgments

This work was supported by Natural Science Foundation of China (No. 21501181, 21925303, 21771186, 21829501, 21222301, 21171170, and 21528303), the CASHIPS Director's Fund (BJPY2019A02), Key Program of 13th Five-Year Plan, CASHIPS (KP-2017-16), Innovative Program of Development Foundation of Hefei Center for Physical Science and Technology (2017FXCX002), and the CAS/SAFEA International Partnership Program for Creative Research Teams.

References

1. Jin, R., Zeng, C., Zhou, M. and Chen, Y. (2016). Atomically precise colloidal metal nanoclusters and nanoparticles: Fundamentals and opportunities, *Chem. Rev.*, **116**, pp. 10346–10413.
2. Yang, X., Yang, M., Pang, B., Vara, M. and Xia, Y. (2015). Gold nanomaterials at work in biomedicine, *Chem. Rev.*, **115**, pp. 10410–10488.
3. Park, J., An, K. J., Hwang, Y. S., Park, J. G., Noh, H. J., Kim, J. Y., Park, J. H., Hwang, N. M. and Hyeon, T. (2004). Ultra-large-scale syntheses of monodisperse nanocrystals, *Nat. Mater.*, **3**, pp. 891–895.
4. Sun, S. H., Murray, C. B., Weller, D., Folks, L. and Moser, A. (2000). Monodisperse FePt nanoparticles and ferromagnetic FePt nanocrystal superlattices, *Science*, **287**, pp. 1989–1992.
5. Jin, R. C., Cao, Y. W., Mirkin, C. A., Kelly, K. L., Schatz, G. C. and Zheng, J. G. (2001). Photoinduced conversion of silver nanospheres to nanoprisms, *Science*, **294**, pp. 1901–1903.
6. Qian, H. F., Zhu, M. Z., Wu, Z. K. and Jin, R. C. (2012). Quantum sized gold nanoclusters with atomic precision, *Acc. Chem. Res.*, **45**, pp. 1470–1479.
7. Jin, R. (2010). Quantum sized, thiolate-protected gold nanoclusters, *Nanoscale*, **2**, pp. 343–362.
8. Jin, R. (2015). Atomically precise metal nanoclusters: Stable sizes and optical properties, *Nanoscale*, **7**, pp. 1549–1565.

9. Gao, Y., Shao, N. and Zeng, X. C. (2008). Ab initio study of thiolate-protected Au_{102} nanocluster, *ACS Nano*, **2**, pp. 1497–1503.
10. Li, G., Abroshan, H., Liu, C., Zhuo, S., Li, Z., Xie, Y., Kim, H. J., Rosi, N. L. and Jin, R. (2016). Tailoring the electronic and catalytic properties of Au_{25} nanoclusters via ligand engineering, *ACS Nano*, **10**, pp. 7998–8005.
11. Murray, R. W. (2008). Nanoelectrochemistry: Metal nanoparticles, nanoelectrodes, and nanopores, *Chem. Rev.*, **108**, pp. 2688–2720.
12. Laaksonen, T., Ruiz, V., Liljeroth, P. and Quinn, B. M. (2008). Quantised charging of monolayer-protected nanoparticles, *Chem. Soc. Rev.*, **37**, pp. 1836–1846.
13. Wu, Z. and Jin, R. (2010). On the ligand's role in the fluorescence of gold nanoclusters, *Nano Lett.*, **10**, pp. 2568–2573.
14. Luo, Z., Yuan, X., Yu, Y., Zhang, Q., Leong, D. T., Lee, J. Y. and Xie, J. (2012). From aggregation-induced emission of Au(I)-thiolate complexes to ultrabright Au(0)@Au(I)-thiolate core-shell nanoclusters, *J. Am. Chem. Soc.*, **134**, pp. 16662–16670.
15. Zhou, M., Zeng, C., Chen, Y., Zhao, S., Sfeir, M. Y., Zhu, M. and Jin, R. (2016). Evolution from the plasmon to exciton state in ligand-protected atomically precise gold nanoparticles, *Nat. Commun.*, **7**, p. 13240.
16. Lu, Y. and Chen, W. (2015). Application of mass spectrometry in the synthesis and characterization of metal nanoclusters, *Anal. Chem.*, **87**, pp. 10659–10667.
17. Jadzinsky, P. D., Calero, G., Ackerson, C. J., Bushnell, D. A. and Kornberg, R. D. (2007). Structure of a thiol monolayer-protected gold nanoparticle at 1.1 Å resolution, *Science*, **318**, pp. 430–433.
18. Diez, I. and Ras, R. H. A. (2011). Fluorescent silver nanoclusters, *Nanoscale*, **3**, pp. 1963–1970.
19. Lu, Y. and Chen, W. (2012). Sub-nanometre sized metal clusters: From synthetic challenges to the unique property discoveries, *Chem. Soc. Rev.*, **41**, pp. 3594–3623.
20. Chakraborty, I. and Pradeep, T. (2017). Atomically precise clusters of noble metals: Emerging link between atoms and nanoparticles, *Chem. Rev.*, **117**, pp. 8208–8271.
21. Schmid, G. (2008). The relevance of shape and size of Au_{55} clusters, *Chem. Soc. Rev.*, **37**, pp. 1909–1930.
22. Brust, M., Walker, M., Bethell, D., Schiffrin, D. J. and Whyman, R. (1994). Synthesis of thiol-derivatised gold nanoparticles in a two-phase liquid–liquid system, *J. Chem. Soc. Chem. Commun.*, pp. 801–802.

23. Shichibu, Y., Negishi, Y., Watanabe, T., Chaki, N. K., Kawaguchi, H. and Tsukuda, T. (2007). Biicosahedral gold clusters $[Au_{25}(PPh_3)_{10}(SC_nH_{2n+1})_5Cl_2]^{2+}$ (n = 2–18): A stepping stone to cluster-assembled materials, *J. Phys. Chem. C,* **111**, pp. 7845–7847.

24. Kurashige, W., Yamaguchi, M., Nobusada, K. and Negishi, Y. (2012). Ligand-induced stability of gold nanoclusters: Thiolate versus selenolate, *J. Phys. Chem. Lett.*, **3**, pp. 2649–2652.

25. Maity, P., Takano, S., Yamazoe, S., Wakabayashi, T. and Tsukuda, T. (2013). Binding motif of terminal alkynes on gold clusters, *J. Am. Chem. Soc.*, **135**, pp. 9450–9457.

26. Kurashige, W., Yamazoe, S., Yamaguchi, M., Nishido, K., Nobusada, K., Tsukuda, T. and Negishi, Y. (2014). Au_{25} clusters containing unoxidized tellurolates in the ligand shell, *J. Phys. Chem. Lett.*, **5**, pp. 2072–2076.

27. Negishi, Y., Takasugi, Y., Sato, S., Yao, H., Kimura, K. and Tsukuda, T. (2004). Magic-numbered Au_n clusters protected by glutathione monolayers (n = 18, 21, 25, 28, 32, 39): Isolation and spectroscopic characterization, *J. Am. Chem. Soc.*, **126**, pp. 6518–6519.

28. Jin, R., Qian, H., Wu, Z., Zhu, Y., Zhu, M., Mohanty, A. and Garg, N. (2010). Size focusing: A methodology for synthesizing atomically precise gold nanoclusters, *J. Phys. Chem. Lett.*, **1**, pp. 2903–2910.

29. Zeng, C., Chen, Y., Das, A. and Jin, R. (2015). Transformation chemistry of gold nanoclusters: From one stable size to another, *J. Phys. Chem. Lett.*, **6**, pp. 2976–2986.

30. Wu, Z., MacDonald, M. A., Chen, J., Zhang, P. and Jin, R. (2011). Kinetic control and thermodynamic selection in the synthesis of atomically precise gold nanoclusters, *J. Am. Chem. Soc.*, **133**, pp. 9670–9673.

31. Donkers, R. L., Lee, D. and Murray, R. W. (2004). Synthesis and isolation of the molecule-like cluster $Au_{38}(PhCH_2CH_2S)_{24}$, *Langmuir*, **20**, pp. 1945–1952.

32. Negishi, Y., Nobusada, K. and Tsukuda, T. (2005). Glutathione-protected gold clusters revisited: Bridging the gap between gold(I)–thiolate complexes and thiolate-protected gold nanocrystals, *J. Am. Chem. Soc.*, **127**, pp. 5261–5270.

33. Whetten, R. L., Khoury, J. T., Alvarez, M. M., Murthy, S., Vezmar, I., Wang, Z. L., Stephens, P. W., Cleveland, C. L., Luedtke, W. D. and Landman, U. (1996). Nanocrystal gold molecules, *Adv. Mater.*, **8**, pp. 428–433.

34. Zhu, M., Lanni, E., Garg, N., Bier, M. E. and Jin, R. (2008). Kinetically controlled, high-yield synthesis of Au_{25} clusters, *J. Am. Chem. Soc.*, **130**, pp. 1138–1139.

35. Wu, Z., Suhan, J. and Jin, R. (2009). One-pot synthesis of atomically monodisperse, thiol-functionalized Au-25 nanoclusters, *J. Mater. Chem.*, **19**, pp. 622–626.

36. Dharmaratne, A. C., Krick, T. and Dass, A. (2009). Nanocluster size evolution studied by mass spectrometry in room temperature $Au_{25}(SR)_{18}$ synthesis, *J. Am. Chem. Soc.*, **131**, pp. 13604–13605.

37. Qian, H., Zhu, Y. and Jin, R. (2009). Size-focusing synthesis, optical and electrochemical properties of monodisperse $Au_{38}(SC_2H_4Ph)_{24}$ nanoclusters, *ACS Nano*, **3**, pp. 3795–3803.

38. Qian, H. and Jin, R. (2009). Controlling nanoparticles with atomic precision: The case of $Au_{144}(SCH_2CH_2Ph)_{60}$, *Nano Lett.*, **9**, pp. 4083–4087.

39. Wu, Z., Lanni, E., Chen, W., Bier, M. E., Ly, D. and Jin, R. (2009). High yield, large scale synthesis of thiolate-protected Ag_7 clusters, *J. Am. Chem. Soc.*, **131**, pp. 16672–16673.

40. Wu, Z., Gayathri, C., Gil, R. R. and Jin, R. (2009). Probing the structure and charge state of glutathione-capped $Au_{25}(SG)_{18}$ clusters by NMR and mass spectrometry, *J. Am. Chem. Soc.*, **131**, pp. 6535–6542.

41. Wu, Z., Wang, M., Yang, J., Zheng, X., Cai, W., Meng, G., Qian, H., Wang, H. and Jin, R. (2012). Well-defined nanoclusters as fluorescent nanosensors: A case study on $Au_{25}(SG)_{18}$, *Small*, **8**, pp. 2028–2035.

42. Yu, Y., Chen, X., Yao, Q., Yu, Y., Yan, N. and Xie, J. (2013). Scalable and precise synthesis of thiolated Au_{10-12}, Au_{15}, Au_{18}, and Au_{25} nanoclusters via pH controlled CO reduction, *Chem. Mater.*, **25**, pp. 946–952.

43. Luo, Z., Nachammai, V., Zhang, B., Yan, N., Leong, D. T., Jiang, D.-e. and Xie, J. (2014). Toward understanding the growth mechanism: Tracing all stable intermediate species from reduction of Au(I)–thiolate complexes to evolution of Au_{25} nanoclusters, *J. Am. Chem. Soc.*, **136**, pp. 10577–10580.

44. Yuan, X., Zhang, B., Luo, Z., Yao, Q., Leong, D. T., Yan, N. and Xie, J. (2014). Balancing the rate of cluster growth and etching for gram-scale synthesis of thiolate-protected Au_{25} nanoclusters with atomic precision, *Angew. Chem. Int. Ed.*, **53**, pp. 4623–4627.

45. Tian, S., Li, Y.-Z., Li, M.-B., Yuan, J., Yang, J., Wu, Z. and Jin, R. (2015). Structural isomerism in gold nanoparticles revealed by X-ray crystallography, *Nat. Commun.*, **6**, p. 9667.

46. Zeng, C., Chen, Y., Iida, K., Nobusada, K., Kirschbaum, K., Lambright, K. J. and Jin, R. (2016). Gold quantum boxes: On the periodicities and the

quantum confinement in the Au_{28}, Au_{36}, Au_{44}, and Au_{52} magic series, *J. Am. Chem. Soc.*, **138**, pp. 3950–3953.

47. Zeng, C., Chen, Y., Liu, C., Nobusada, K., Rosi, N. L. and Jin, R. (2015). Gold tetrahedra coil up: Kekulé-like and double helical superstructures, *Sci. Adv.*, **1**, p. e1500425.
48. Liao, L., Zhuang, S., Wang, P., Xu, Y., Yan, N., Dong, H., Wang, C., Zhao, Y., Xia, N., Li, J., Deng, H., Pei, Y., Tian, S.-K. and Wu, Z. (2017). Quasi-dual-packed-kerneled $Au_{49}(2,4\text{-DMBT})_{27}$ nanoclusters and the influence of kernel packing on the electrochemical gap, *Angew. Chem. Int. Ed.*, **56**, pp. 12644–12648.
49. Liao, L., Zhuang, S., Yao, C., Yan, N., Chen, J., Wang, C., Xia, N., Liu, X., Li, M.-B., Lo, L., Bao, X. and Wu, Z. (2016). Structure of chiral $Au_{44}(2,4\text{-DMBT})_{26}$ nanocluster with an 18-electron shell closure, *J. Am. Chem. Soc.*, **138**, pp. 10425–10428.
50. Zeng, C., Chen, Y., Li, G. and Jin, R. (2014). Magic size $Au_{64}(S\text{-}c\text{-}C_6H_{11})_{32}$ nanocluster protected by cyclohexanethiolate, *Chem. Mater.*, **26**, pp. 2635–2641.
51. Das, A., Li, T., Nobusada, K., Zeng, C., Rosi, N. L. and Jin, R. (2013). Nonsuperatomic $[Au_{23}(SC_6H_{11})_{16}]^-$ nanocluster featuring bipyramidal Au_{15} kernel and trimeric $Au_3(SR)_4$ motif, *J. Am. Chem. Soc.*, **135**, pp. 18264–18267.
52. Higaki, T., Liu, C., Zeng, C., Jin, R., Chen, Y., Rosi, N. L. and Jin, R. (2016). Controlling the atomic structure of Au_{30} nanoclusters by a ligand-based strategy, *Angew. Chem. Int. Ed.*, **55**, pp. 6694–6697.
53. Chen, Y., Zeng, C., Kauffman, D. R. and Jin, R. (2015). Tuning the magic size of atomically precise gold nanoclusters via isomeric methylbenzenethiols, *Nano Lett.*, **15**, pp. 3603–3609.
54. Zhuang, S., Liao, L., Li, M.-B., Yao, C., Zhao, Y., Dong, H., Li, J., Deng, H., Li, L. and Wu, Z. (2017). The fcc structure isomerization in gold nanoclusters, *Nanoscale*, **9**, pp. 14809–14813.
55. Zhuang, S., Liao, L., Yuan, J., Wang, C., Zhao, Y., Xia, N., Gan, Z., Gu, W., Li, J., Deng, H., Yang, J. and Wu, Z. (2018). Kernel homology in gold nanoclusters, *Angew. Chem. Int. Ed.*, **57**, pp. 15450–15454.
56. Zhuang, S., Liao, L., Zhao, Y., Yuan, J., Yao, C., Liu, X., Li, J., Deng, H., Yang, J. and Wu, Z. (2018). Is the kernel-staples match a key-lock match? *Chem. Sci.*, **9**, pp. 2437–2442.
57. Zhuang, S., Liao, L., Yuan, J., Xia, N., Zhao, Y., Wang, C., Gan, Z., Yan, N., He, L., Li, J., Deng, H., Guan, Z., Yang, J. and Wu, Z. (2019). Fcc versus non-fcc structural isomerism of gold nanoparticles with kernel atom

packing dependent photoluminescence, *Angew. Chem. Int. Ed.*, **58**, pp. 4510–4514.

58. Zeng, C., Liu, C., Pei, Y. and Jin, R. (2013). Thiol ligand-induced transformation of $Au_{38}(SC_2H_4Ph)_{24}$ to Au_{36}(SPh-*t*-Bu)$_{24}$, *ACS Nano*, **7**, pp. 6138–6145.

59. Zeng, C., Liu, C., Chen, Y., Rosi, N. L. and Jin, R. (2014). Gold–thiolate ring as a protecting motif in the $Au_{20}(SR)_{16}$ nanocluster and implications, *J. Am. Chem. Soc.*, **136**, pp. 11922–11925.

60. Wu, Z. (2012). Anti-galvanic reduction of thiolate-protected gold and silver nanoparticles, *Angew. Chem. Int. Ed.*, **51**, pp. 2934–2938.

61. Li, M.-B., Tian, S.-K., Wu, Z. and Jin, R. (2015). Cu^{2+} induced formation of $Au_{44}(SC_2H_4Ph)_{32}$ and its high catalytic activity for the reduction of 4-nitrophenol at low temperature, *Chem. Commun.*, **51**, pp. 4433–4436.

62. Yao, C., Tian, S., Liao, L., Liu, X., Xia, N., Yan, N., Gan, Z. and Wu, Z. (2015). Synthesis of fluorescent phenylethanethiolated gold nanoclusters via pseudo-AGR method, *Nanoscale*, **7**, pp. 16200–16203.

63. Hicks, J. F., Templeton, A. C., Chen, S., Sheran, K. M., Jasti, R., Murray, R. W., Debord, J., Schaaff, T. G. and Whetten, R. L. (1999). The monolayer thickness dependence of quantized double-layer capacitances of monolayer-protected gold clusters, *Anal. Chem.*, **71**, pp. 3703–3711.

64. Ingram, R. S., Hostetler, M. J., Murray, R. W., Schaaff, T. G., Khoury, J. T., Whetten, R. L., Bigioni, T. P., Guthrie, D. K. and First, P. N. (1997). 28 kDa alkanethiolate-protected Au clusters give analogous solution electrochemistry and STM coulomb staircases, *J. Am. Chem. Soc.*, **119**, pp. 9279–9280.

65. Price, R. C. and Whetten, R. L. (2005). All-aromatic, nanometer-scale, gold-cluster thiolate complexes, *J. Am. Chem. Soc.*, **127**, pp. 13750–13751.

66. Chen, S., Ingram, R. S., Hostetler, M. J., Pietron, J. J., Murray, R. W., Schaaff, T. G., Khoury, J. T., Alvarez, M. M. and Whetten, R. L. (1998). Gold nanoelectrodes of varied size: Transition to molecule-like charging, *Science*, **280**, pp. 2098–2101.

67. Schaaff, T. G., Shafigullin, M. N., Khoury, J. T., Vezmar, I., Whetten, R. L., Cullen, W. G., First, P. N., Gutiérrez-Wing, C., Ascensio, J. and Jose-Yacamán, M. J. (1997). Isolation of smaller nanocrystal Au molecules: Robust quantum effects in optical spectra, *J. Phys. Chem. B*, **101**, pp. 7885–7891.

68. Wu, Z., Chen, J. and Jin, R. (2010). One-pot synthesis of $Au_{25}(SG)_{18}$ 2-and 4-nm gold nanoparticles and comparison of their size-dependent properties, *Adv. Funct. Mater.*, **21**, pp. 177–183.

69. Yang, X., Su, Y., Paau, M. C. and Choi, M. M. F. (2012). Mass spectrometric identification of water-soluble gold nanocluster fractions from sequential size-selective precipitation, *Anal. Chem.*, **84**, pp. 1765–1771.

70. Kumara, C. and Dass, A. (2012). AuAg alloy nanomolecules with 38 metal atoms, *Nanoscale*, **4**, pp. 4084–4086.

71. Kumara, C., Aikens, C. M. and Dass, A. (2014). X-ray crystal structure and theoretical analysis of $Au_{25-x}Ag_x(SCH_2CH_2Ph)_{18}^-$ alloy, *J. Phys. Chem. Lett.*, **5**, pp. 461–466.

72. Schaaff, T. G., Shafigullin, M. N., Khoury, J. T., Vezmar, I. and Whetten, R. L. (2001). Properties of a ubiquitous 29 kDa Au:SR cluster compound, *J. Phys. Chem. B*, **105**, pp. 8785–8796.

73. Fields-Zinna, C. A., Crowe, M. C., Dass, A., Weaver, J. E. F. and Murray, R. W. (2009). Mass spectrometry of small bimetal monolayer-protected clusters, *Langmuir*, **25**, pp. 7704–7710.

74. Wang, M., Wu, Z., Yang, J., Wang, G., Wang, H. and Cai, W. (2012). $Au_{25}(SG)_{18}$ as a fluorescent iodide sensor, *Nanoscale*, **4**, pp. 4087–4090.

75. Toikkanen, O., Ruiz, V., Rönnholm, G., Kalkkinen, N., Liljeroth, P. and Quinn, B. M. (2008). Synthesis and stability of monolayer-protected Au_{38} clusters, *J. Am. Chem. Soc.*, **130**, pp. 11049–11055.

76. Heaven, M. W., Dass, A., White, P. S., Holt, K. M. and Murray, R. W. (2008). Crystal structure of the gold nanoparticle $[N(C_8H_{17})_4]$ $[Au_{25}(SCH_2CH_2Ph)_{18}]$, *J. Am. Chem. Soc.*, **130**, pp. 3754–3755.

77. Zhu, M., Aikens, C. M., Hollander, F. J., Schatz, G. C. and Jin, R. (2008). Correlating the crystal structure of a thiol-protected Au_{25} cluster and optical properties, *J. Am. Chem. Soc.*, **130**, pp. 5883–5885.

78. Das, A., Li, T., Li, G., Nobusada, K., Zeng, C., Rosi, N. L. and Jin, R. (2014). Crystal structure and electronic properties of a thiolate-protected Au_{24} nanocluster, *Nanoscale*, **6**, pp. 6458–6462.

79. Higaki, T., Liu, C., Zhou, M., Luo, T.-Y., Rosi, N. L. and Jin, R. (2017). Tailoring the structure of 58-electron gold nanoclusters: $Au_{103}S_2(S\text{-}Nap)_{41}$ and its implications, *J. Am. Chem. Soc.*, **139**, pp. 9994–10001.

80. Zeng, C., Chen, Y., Kirschbaum, K., Lambright, K. J. and Jin, R. (2016). Emergence of hierarchical structural complexities in nanoparticles and their assembly, *Science*, **354**, pp. 1580–1584.

81. Negishi, Y., Iwai, T. and Ide, M. (2010). Continuous modulation of electronic structure of stable thiolate-protected Au_{25} cluster by Ag doping, *Chem. Commun.*, **46**, pp. 4713–4715.

82. Kumar, S. and Jin, R. (2012). Water-soluble $Au_{25}(Capt)_{18}$ nanoclusters: synthesis, thermal stability, and optical properties, *Nanoscale*, **4**, pp. 4222–4227.

83. Schaaff, T. G., Knight, G., Shafigullin, M. N., Borkman, R. F. and Whetten, R. L. (1998). Isolation and selected properties of a 10.4 kDa gold:glutathione cluster compound, *J. Phys. Chem. B*, **102**, pp. 10643–10646.

84. Schaaff, T. G. and Whetten, R. L. (2000). Giant gold–glutathione cluster compounds: intense optical activity in metal-based transitions, *J. Phys. Chem. B*, **104**, pp. 2630–2641.

85. Yu, Y., Luo, Z., Chevrier, D. M., Leong, D. T., Zhang, P., Jiang, D.-e. and Xie, J. (2014). Identification of a highly luminescent $Au_{22}(SG)_{18}$ nanocluster, *J. Am. Chem. Soc.*, **136**, pp. 1246–1249.

86. Pyo, K., Thanthirige, V. D., Kwak, K., Pandurangan, P., Ramakrishna, G. and Lee, D. (2015). Ultrabright luminescence from gold nanoclusters: rigidifying the Au(I)–thiolate shell, *J. Am. Chem. Soc.*, **137**, pp. 8244–8250.

87. Li, Q., Pan, Y., Chen, T., Du, Y., Ge, H., Zhang, B., Xie, J., Yu, H. and Zhu, M. (2018). Design and mechanistic study of a novel gold nanocluster-based drug delivery system, *Nanoscale*, **10**, pp. 10166–10172.

88. Marjomäki, V., Lahtinen, T., Martikainen, M., Koivisto, J., Malola, S., Salorinne, K., Pettersson, M. and Häkkinen, H. (2014). Site-specific targeting of enterovirus capsid by functionalized monodisperse gold nanoclusters, *Proc. Natl. Acad. Sci. U.S.A.*, **111**, pp. 1277–1281.

89. Yao, H. and Yaomura, S. (2013). Emergence of large chiroptical responses by ligand exchange cross-linking of monolayer-protected gold clusters with chiral dithiol, *Langmuir*, **29**, pp. 6444–6451.

90. Ackerson, C. J., Jadzinsky, P. D., Sexton, J. Z., Bushnell, D. A. and Kornberg, R. D. (2010). Synthesis and bioconjugation of 2 and 3 nm-diameter gold nanoparticles, *Bioconjugate Chem.*, **21**, pp. 214–218.

91. Gautier, C. and Bürgi, T. (2006). Chiral *N*-isobutyryl-cysteine protected gold nanoparticles: preparation, size selection, and optical activity in the UV–vis and infrared, *J. Am. Chem. Soc.*, **128**, pp. 11079–11087.

92. Kumar, S., Bolan, M. D. and Bigioni, T. P. (2010). Glutathione-stabilized magic-number silver cluster compounds, *J. Am. Chem. Soc.*, **132**, pp. 13141–13143.

93. Wilcoxon, J. P., Martin, J. E. and Provencio, P. (2000). Size distributions of gold nanoclusters studied by liquid chromatography, *Langmuir*, **16**, pp. 9912–9920.

94. Wellsted, H., Sitsen, E., Caragheorgheopol, A. and Chechik, V. (2004). Polydisperse composition of mixed monolayer-protected, spin-labeled Au nanoparticles, *Anal. Chem.*, **76**, pp. 2010–2016.

95. Tsunoyama, H., Negishi, Y. and Tsukuda, T. (2006). Chromatographic isolation of "missing" Au_{55} clusters protected by alkanethiolates, *J. Am. Chem. Soc.*, **128**, pp. 6036–6037.

96. Tsunoyama, H., Nickut, P., Negishi, Y., Al-Shamery, K., Matsumoto, Y. and Tsukuda, T. (2007). Formation of alkanethiolate-protected gold clusters with unprecedented core sizes in the thiolation of polymer-stabilized gold clusters, *J. Phys. Chem. C*, **111**, pp. 4153–4158.

97. Tsunoyama, R., Tsunoyama, H., Pannopard, P., Limtrakul, J. and Tsukuda, T. (2010). MALDI mass analysis of 11 kDa gold clusters protected by octadecanethiolate ligands, *J. Phys. Chem. C*, **114**, pp. 16004–16009.

98. Qian, H., Zhu, Y. and Jin, R. (2010). Isolation of ubiquitous $Au_{40}(SR)_{24}$ clusters from the 8 kDa gold clusters, *J. Am. Chem. Soc.*, **132**, pp. 4583–4585.

99. Knoppe, S., Boudon, J., Dolamic, I., Dass, A. and Bürgi, T. (2011). Size exclusion chromatography for semipreparative scale separation of $Au_{38}(SR)_{24}$ and $Au_{40}(SR)_{24}$ and larger clusters, *Anal. Chem.*, **83**, pp. 5056–5061.

100. Chen, Y. D., Wang, J., Liu, C., Li, Z. M. and Li, G. (2016). Kinetically controlled synthesis of $Au_{102}(SPh)_{44}$ nanoclusters and catalytic application, *Nanoscale*, **8**, pp. 10059–10065.

101. Ren, X. Q., Fu, X. M., Lin, X. Z., Liu, C., Huang, J. H. and Yan, J. H. (2018). Highly efficient synthesis of $Au_{130}(SPh\text{-}Br)_{50}$ nanocluster, *Chem. Res. Chin. Univ.*, **34**, pp. 719–722.

102. Liu, C., Yan, C. Y., Lin, J. Z., Yu, C. L., Huang, J. H. and Li, G. (2015). One-pot synthesis of $Au_{144}(SCH_2Ph)_{60}$ nanoclusters and their catalytic application, *J. Mater. Chem. A*, **3**, pp. 20167–20173.

103. Qian, H., Jiang, D.-e., Li, G., Gayathri, C., Das, A., Gil, R. R. and Jin, R. (2012). Monoplatinum doping of gold nanoclusters and catalytic application, *J. Am. Chem. Soc.*, **134**, pp. 16159–16162.

104. Negishi, Y., Munakata, K., Ohgake, W. and Nobusada, K. (2012). Effect of copper doping on electronic structure, geometric structure, and stability of thiolate-protected Au_{25} nanoclusters, *J. Phys. Chem. Lett.*, **3**, pp. 2209–2214.

105. Niihori, Y., Yoshida, K., Hossain, S., Kurashige, W. and Negishi, Y. (2019). Deepening the understanding of thiolate-protected metal clusters using high-performance liquid chromatography, *Bull. Chem. Soc. Jpn.*, **92**, pp. 664–695.

106. Niihori, Y., Uchida, C., Kurashige, W. and Negishi, Y. (2016). High-resolution separation of thiolate-protected gold clusters by reversed-phase high-performance liquid chromatography, *Phys. Chem. Chem. Phys.*, **18**, pp. 4251–4265.

107. Negishi, Y., Nakazaki, T., Malola, S., Takano, S., Niihori, Y., Kurashige, W., Yamazoe, S., Tsukuda, T. and Häkkinen, H. (2015). A critical size for emergence of nonbulk electronic and geometric structures in dodecanethiolate-protected Au clusters, *J. Am. Chem. Soc.*, **137**, pp. 1206–1212.

108. Negishi, Y., Chaki, N. K., Shichibu, Y., Whetten, R. L. and Tsukuda, T. (2007). Origin of magic stability of thiolated gold clusters: a case study on $Au_{25}(SC_6H_{13})_{18}$, *J. Am. Chem. Soc.*, **129**, pp. 11322–11323.

109. Negishi, Y., Kurashige, W., Niihori, Y., Iwasa, T. and Nobusada, K. (2010). Isolation, structure, and stability of a dodecanethiolate-protected Pd_1Au_{24} cluster, *Phys. Chem. Chem. Phys.*, **12**, pp. 6219–6225.

110. Niihori, Y., Koyama, Y., Watanabe, S., Hashimoto, S., Hossain, S., Nair, L. V., Kumar, B., Kurashige, W. and Negishi, Y. (2018). Atomic and isomeric separation of thiolate-protected alloy clusters, *J. Phys. Chem. Lett.*, **9**, pp. 4930–4934.

111. Niihori, Y., Matsuzaki, M., Uchida, C. and Negishi, Y. (2014). Advanced use of high-performance liquid chromatography for synthesis of controlled metal clusters, *Nanoscale*, **6**, pp. 7889–7896.

112. Niihori, Y., Kikuchi, Y., Kato, A., Matsuzaki, M. and Negishi, Y. (2015). Understanding ligand-exchange reactions on thiolate-protected gold clusters by probing isomer distributions using reversed-phase high-performance liquid chromatography, *ACS Nano*, **9**, pp. 9347–9356.

113. Dolamic, I., Knoppe, S., Dass, A. and Bürgi, T. (2012). First enantioseparation and circular dichroism spectra of Au_{38} clusters protected by achiral ligands, *Nat. Commun.*, **3**, p. 798.

114. Zeng, C., Li, T., Das, A., Rosi, N. L. and Jin, R. (2013). Chiral structure of thiolate-protected 28-gold-atom nanocluster determined by X-ray crystallography, *J. Am. Chem. Soc.*, **135**, pp. 10011–10013.

115. Li, M.-B., Tian, S.-K. and Wu, Z. (2014). Catalyzed formation of α,β-unsaturated ketones or aldehydes from propargylic acetates by a recoverable and recyclable nanocluster catalyst, *Nanoscale*, **6**, pp. 5714–5717.

116. Ghosh, A., Hassinen, J., Pulkkinen, P., Tenhu, H., Ras, R. H. A. and Pradeep, T. (2014). Simple and efficient separation of atomically precise noble metal clusters, *Anal. Chem.*, **86**, pp. 12185–12190.

117. Yao, C., Chen, J., Li, M.-B., Liu, L., Yang, J. and Wu, Z. (2015). Adding two active silver atoms on Au_{25} nanoparticle, *Nano Lett.*, **15**, pp. 1281–1287.

118. Chen, J., Liu, L., Liu, X., Liao, L., Zhuang, S., Zhou, S., Yang, J. and Wu, Z. (2017). Gold-doping of double-crown Pd nanoclusters, *Chem. Eur. J.*, **23**, pp. 18187–18192.

119. Gan, Z., Chen, J., Liao, L., Zhang, H. and Wu, Z. (2018). Surface single-atom tailoring of a gold nanoparticle, *J. Phys. Chem. Lett.*, **9**, pp. 204–208.

120. Dong, H., Liao, L., Zhuang, S., Yao, C., Chen, J., Tian, S., Zhu, M., Liu, X., Li, L. and Wu, Z. (2017). A novel double-helical-kernel evolution pattern of gold nanoclusters: alternate single-stranded growth at both ends, *Nanoscale*, **9**, pp. 3742–3746.

121. Yan, N., Xia, N., Liao, L., Zhu, M., Jin, F., Jin, R. and Wu, Z. (2018). Unraveling the long-pursued Au_{144} structure by x-ray crystallography, *Sci. Adv.*, **4**, p. eaat7259.

122. Chen, J., Liu, L., Weng, L., Lin, Y., Liao, L., Wang, C., Yang, J. and Wu, Z. (2015). Synthesis and properties evolution of a family of tiara-like phenylethanethiolated palladium nanoclusters, *Sci. Rep.*, **5**, p. 16628.

123. Tian, S., Yao, C., Liao, L., Xia, N. and Wu, Z. (2015). Ion-precursor and ion-dose dependent anti-galvanic reduction, *Chem. Commun.*, **51**, pp. 11773–11776.

124. Liao, L., Yao, C., Wang, C., Tian, S., Chen, J., Li, M.-B., Xia, N., Yan, N. and Wu, Z. (2016). Quantitatively monitoring the size-focusing of Au nanoclusters and revealing what promotes the size transformation from $Au_{44}(TBBT)_{28}$ to $Au_{36}(TBBT)_{24}$, *Anal. Chem.*, **88**, pp. 11297–11301.

125. Choi, M. M. F., Douglas, A. D. and Murray, R. W. (2006). Ion-pair chromatographic separation of water-soluble gold monolayer-protected clusters, *Anal. Chem.*, **78**, pp. 2779–2785.

126. Lo, C. K., Paau, M. C., Xiao, D. and Choi, M. M. F. (2008). Application of capillary zone electrophoresis for separation of water-soluble gold monolayer-protected clusters, *Electrophoresis*, **29**, pp. 2330–2339.

127. Carney, R. P., Kim, J. Y., Qian, H., Jin, R., Mehenni, H., Stellacci, F. and Bakr, O. M. (2011). Determination of nanoparticle size distribution together with density or molecular weight by 2D analytical ultracentrifugation, *Nat. Commun.*, **2**, p. 335.

Chapter 2

Nanoparticles with Atomic Resolution: Synthesis and Stable Structures of Atomically Precise Gold Nanoclusters

Tatsuya Higaki and Rongchao Jin
Department of Chemistry, Carnegie Mellon University, 4400 Fifth Avenue, Pittsburgh, PA 15213, USA
rongchao@andrew.cmu.edu

Precision at the nanoscale has long been a challenging yet critical issue in nanoscience. Recently, atomically precise metal nanoclusters have paved new avenues for the approach by structure determination by X-ray crystallography. The crystal structures of nanoclusters have provided atomic-level insights into the chemistry of nanoparticles, and such progress has also impacted nanoparticle assembly, self-assembled monolayers, catalysis, and so on. Especially, the structural complexity in larger sized nanoclusters hampers the prediction by experiment and theory other than crystallography. In this chapter, we focus on the synthesis and structure determination of atomically precise metal nanoclusters in the large size regime (i.e., >100 metal atoms). The determined structures have provided

Atomically Precise Nanoclusters
Edited by Yan Zhu and Rongchao Jin

ISBN 978-981-4877-08-4 (Hardcover), 978-1-003-11990-6 (eBook)
www.jennystanford.com

atomic insights into the fundamental questions of nanoscience, for example, structural evolution and transition from non-metallic to metallic states. The observed structures show unique features that are unprecedented in the conventional theory to predict thermodynamically stable structures. The unique structures are protected by gold–thiolate oligomers called "staple" motifs, which stabilize the nanostructures with aesthetic surface patterns. In the last part of the chapter, we will discuss the remaining challenges and problems to be overcome by structure determination as well as the future perspective of nanochemistry at the atomic level of precision.

2.1 Introduction

Atomically precise metal nanoclusters have gained great attention because of their unique properties [1–9]. Unlike conventional nanoparticles, metal nanoclusters can be synthesized with atomic-level precision without any size distribution [1]. Therefore, molecular techniques (e.g., mass spectrometry, X-ray crystallography) can be applied for characterization of nanoclusters in order to determine their molecular formulas and atomic structures [1]. Recent progress in structure determination of metal nanoclusters has unveiled unique geometrical features that are unlike those of bulk or plasmonic nanoparticles [1–9]. For example, $Au_{30}(SR)_{18}$ [10] and $Au_{38}S_2(SR)_{20}$ [11] exhibited hexagonal close-packed (hcp) and body-centered cubic (bcc) structures, respectively. These exotic crystalline phases are unprecedented for gold as the bulk or plasmonic nanoparticles of gold adopt the face-centered cubic (fcc) structure. The new capability of structural control has led to the observations of unique electronic and optical properties in such phases [12]. According to the classical theory, the stable structures of metal nanoparticles are determined by their geometrical stability (e.g., surface energy) [13, 14]. The most stable structures typically change from fcc to decahedron (D_h) to icosahedron (I_h) with decreasing size in order to minimize the surface energy by reducing the surface dangling bonds (Fig. 2.1). However, this model considers only the core cohesive energy and bare surfaces, while the stabilization by surface ligand protection is not taken into consideration.

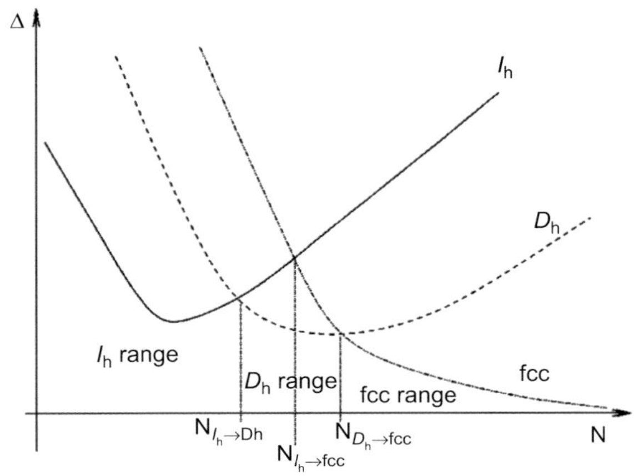

Figure 2.1 Thermodynamic stability of metal nanoparticles with different crystalline phases according to the classical theory. Reprinted with permission from Ref. [14], Copyright 2005, American Physical Society.

This classical theory also assumes band structure with continuous energy levels for the electronic structure. However, recent studies have revealed that a sharp transition from the continuous band structure to discrete energy levels in nanogold occurs between two nanoparticle sizes (246 and 279 gold atoms) in the case of spherical particles [15–17]. Below the size of Au_{246}, the stability of metal nanoclusters largely originates from their geometric packing and electronic structure [i.e., a sizable HOMO–LUMO (highest occupied and lowest unoccupied molecular orbitals) gap]. The exotic structures (e.g., hcp or bcc) can be realized with higher surface energy as long as the total energy of electrons is minimized. Surface protection also plays a key role for the emergence of unexpected geometries in metal nanoclusters [1]. X-ray crystallographic analysis has revealed that gold nanoclusters are protected by eight types of surface protection motifs, called "staple" from the analogy of a stationery tool (Fig. 2.2). The largest staple (a closed form) that has been discovered so far is made up of eight gold atoms, which form a ring to wrap up the core of the Au_{20} nanocluster [18]. However, the surface protection pattern by these staple motifs is still difficult to predict, especially for the nanoclusters with more than 100 metal atoms.

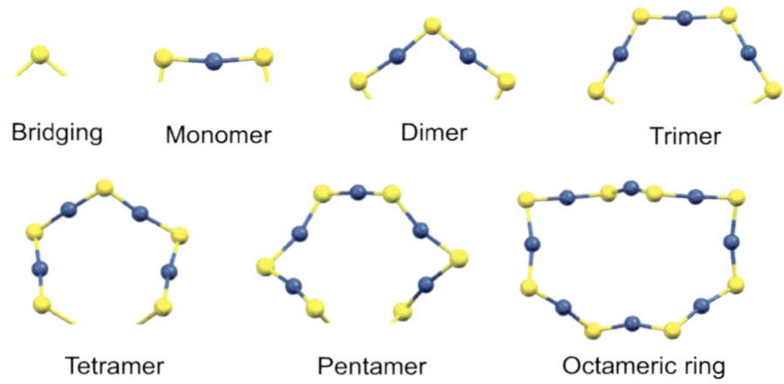

Figure 2.2 Staple motifs of gold–thiolate oligomers for surface protection on gold nanoclusters. Carbon and hydrogen atoms are omitted for clarity [1]. Color code: yellow = S, blue = Au.

In this chapter, we summarize the synthesis and structure determination of atomically precise gold nanoclusters with a focus on the stable sizes of more than 100 Au atoms. Such large-sized nanoclusters are particularly important because they serve as the link to plasmonic nanoparticles and provide the atomic insights into the surface protection pattern and stability of much larger nanoparticles (Fig. 2.3). The atomic insights from these crystal structures could also provide critical information on the quantum state transition from non-metallic to metallic states as well as the structure prediction of much larger gold nanoparticles. Future perspectives include the discussion on the synthesis and structure determination of larger nanoclusters and potential applications in nanoscience.

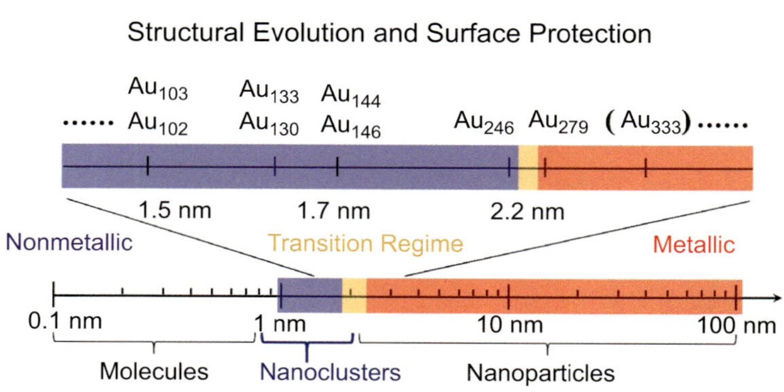

Figure 2.3 Structure evolution and surface protection of thiolate-protected gold nanoclusters near the transition size regime.

2.2 Atomically Precise Synthesis of Metal Nanoclusters

2.2.1 Size-Focusing Methodology

A conventional synthesis of nanoparticles normally yields a product with a size distribution (e.g., ~5% Std Dev even for the highest quality of nanoparticles), which is well known as the problem called "no nanoparticles are the same" [1]. The inherent size distribution of conventionally synthesized nanoparticles has significantly hampered the investigation of size-sensitive properties with high reproducibility. Therefore, preparing nanoparticles at the atomic level of precision has long been the major dream of nanochemists. In this context, recent research has invented the size-focusing methodology to synthesize nanoparticles at the atomic level of precision (Fig. 2.4) [19]. The invented method can provide the highest quality samples with sufficiently high yields for the investigation of their properties, such as optical absorption/emission, catalytic reactivity, magnetism, and others. Recent advances in this technique have demonstrated wide accessibility to different sizes of nanoclusters ranging from 18 to ~900 metal atoms [1], which has resulted in the exploration of the size-dependent properties at the atomic level of precision.

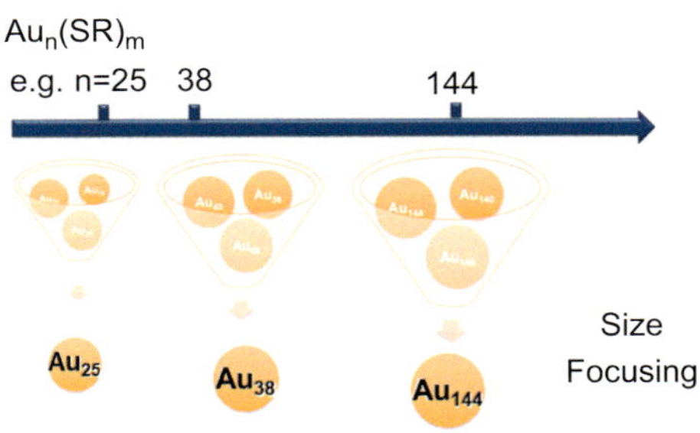

Figure 2.4 Schematic illustration of the size-focusing method to synthesize atomically precise nanoclusters (e.g., with 25, 38, and 144 gold atoms). Reprinted with permission from Ref. [19], Copyright 2010, American Chemical Society.

The size-focusing method is made up of two primary steps: (i) preparation of polydisperse nanoclusters (i.e., $Au_x(SR)_y$) but with a controlled size range and (ii) etching process under harsh conditions (Fig. 2.5A) [20].

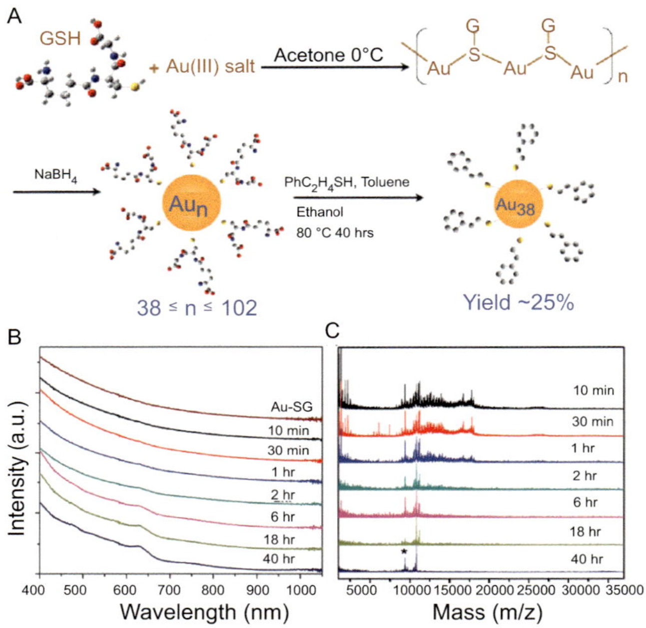

Figure 2.5 The size-focusing method to synthesize atomically precise Au_{38} monitored by UV–Vis absorption spectroscopy and MALDI (matrix-assisted laser desorption/ionization) mass spectrometry. The asterisk indicates a fragment peak due to the destructive MALDI. Reprinted with permission from Ref. [20], Copyright 2009, American Chemical Society.

The first primary step typically involves the dissolution of ionic Au species (e.g., $HAuCl_4 \cdot 3H_2O$) in water, followed by its phase transfer to organic solvent (e.g., toluene) under the aid of organic counterions [e.g., $(C_8H_{17})_4N^+Br^-$]. In the organic phase, Au(III) ions react with thiols (weak reducing agents), forming Au(I)-SR polymers during the reduction of gold from 3+ to 1+. This intermediate is then reduced by $NaBH_4$ to form Au(0) nanoclusters with different sizes [i.e., $Au_x(SR)_y$]. For the synthesis of a highly pure sample in high yield, careful modification of the protocol is needed, including the optimization of the Au/thiol ratio and the reaction temperature, in order to optimize the initial size distribution for the targeted size range.

The second process starts with the addition of excess thiol to the polydisperse product obtained from the first step. The mixture of thiol and $Au_x(SR)_y$ nanoclusters is then treated under high temperatures. During the process, unstable or meta-stable nanoclusters are decomposed or transformed to more stable sizes and structures, and the reaction continues until the mixed sizes converge to the most stable size and structure. This reaction demonstrates "the survival of the most robust," reminiscent of "the survival of the fittest" in the natural law. For example, the size-focusing synthesis of $Au_{38}(PET)_{24}$ (where PET is 2-phenylethanethiolate) starts with the preparation of polydisperse precursors protected by glutathionate (SG), followed by size focusing with excess 2-phenylethanethiol at 80 °C for ~48 h (Fig. 2.5) [20]. The process was monitored by optical absorption spectroscopy and mass spectrometry. The gradual conversion to the single-sized nanocluster from a size-mixed product is confirmed by the emergence of characteristic absorption peaks from the initially featureless absorption profile as well as the observation of a single mass peak (at ~11 kDa) in the mass spectra.

For optimizing reaction conditions, careful selection of solvent was done for the synthesis of $Au_{38}(PET)_{24}$. For example, the boosted reaction yield from 5–10% to 25% was realized by targeting the initial size distribution at around Au_x ($38 < x < 100$) through the modification of reaction solvent from methanol to acetone for the first step [20].

The size-focusing process can proceed spontaneously when a specific combination of thiolate and solvent is used for the reaction. For example, $Au_{23}(SR)_{16}$ [21] and $Au_{30}(SR)_{18}$ [10] can be synthesized using cyclohexanethiol and 1-adamantanethiol, respectively, as the protecting ligands. The spontaneous size focusing proceeds at room temperature in one pot while stirring the methanol solution of as-synthesized $Au_x(SR)_y$.

2.2.2 Ligand-Exchange-Induced Size/Structure Transformation

The size-focusing method has provided feasible syntheses of atomically precise nanoclusters for a number of sizes with sufficient yields. It has also led to the invention of the transformation reaction from one stable size to another, the so-called ligand-exchange-induced size/structure transformation (LEIST). In LEIST, single-sized nanoclusters prepared by the size-focusing method are used for the benchmarks to explore the new stable size(s) near the starting nanoclusters [22]. For example, $Au_{25}(PET)_{18}$ [23, 24], $Au_{38}(PET)_{24}$ [25], and $Au_{144}(PET)_{60}$ [26] are reacted with excess *tert*-butyl benzenethiol (TBBT) in toluene at 80 °C and transformed to $Au_{28}(TBBT)_{20}$ [27], $Au_{36}(TBBT)_{24}$ [28], and $Au_{133}(TBBT)_{52}$ [29], respectively (Fig. 2.6).

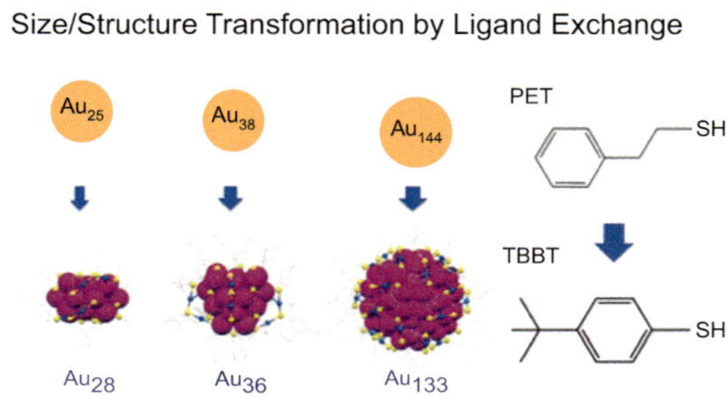

Figure 2.6 LEIST from Au_{25}, Au_{38}, and Au_{144} to Au_{28}, Au_{36}, and Au_{133}, respectively [22].

The crystal structure and the number of gold/thiolate are all changed after the transformation, and thus this approach allows one to access new stable sizes of nanoclusters and also new structures different from the starting sizes. The structural change from Au_{144} to Au_{133} is described in detail as an example in the following section.

2.3 Synthesis and Structure Determination of Large Gold Nanoclusters

2.3.1 Icosahedral Structures

2.3.1.1 Case of $Au_{133}(SR)_{52}$

The synthesis of $Au_{133}(SR)_{52}$ is performed by LEIST from $Au_{144}(SR)_{60}$ [29]. Typically, it starts with the preparation of $Au_{144}(PET)_{60}$ by the size-focusing method. The as-synthesized Au_{144} is then treated with excess TBBT in toluene at 80 °C. After ~4 days, the starting $Au_{144}(PET)_{60}$ is transformed to $Au_{133}(TBBT)_{52}$ with >90% yield [29]. Structure determination by X-ray crystallography has revealed that Au_{133} comprises an icosahedral Au_{107} core protected by 26 monomeric staple motifs (Fig. 2.7) [29]. The core of Au_{133} is made up of a shell-by-shell structure starting from the innermost central Au_1, which is covered by Au_{12} to form the icosahedral Au_{13} (Fig. 2.7A). The icosahedral Au_{13} is further covered by another shell of Au_{42} to form Au_{55} (Fig. 2.7B). The structure of Au_{55} is well known as the Mackay icosahedron (MI) [30], which possesses 20 triangular {111} facets, that is, 20 tetrahedra of fcc structure. Each of the 20 tetrahedra in the Au_{55} is then covered in an *a-b-c-b* manner with additional three atoms for each of the 16 {111} facets (Fig. 2.7E) and in an *a-b-c-a* manner with additional one atom only for each of the remaining four {111} facets (Fig. 2.7F). The MI-Au_{55} is thus covered by 52 atoms in total, forming the Au_{107} core (Fig. 2.7C). The surface protecting monomeric staples either reside on Au_4 facets or bridge two Au atoms on the 52-atom shell (Fig. 2.7D,G).

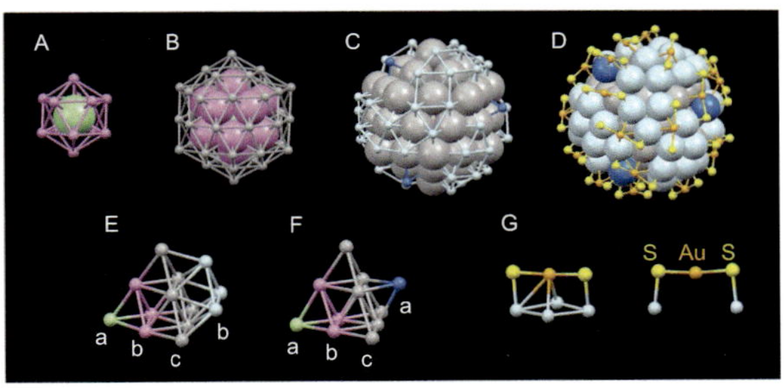

Figure 2.7 Crystal structure of $Au_{133}(SR)_{52}$. (A) Au_{13}, (B) Au_{55}, and (C) Au_{107} core structures of Au_{133} in shell-by-shell illustration. (D) Surface protection of Au_{107} core by 26 monomeric staple motifs. (E,F) Layered structures from inner to outer core in *a-b-c-b* or *a-b-c-a* manners. (G) Monomeric staple motifs on the outermost Au_{107} core. Carbon and hydrogen atoms are omitted for clarity. Adapted with permission from Ref. [29], Copyright 2015, American Association for the Advancement of Science.

The arrangement of staple motifs exhibits helical patterns on the surface of the nanocluster (Fig. 2.8). The helical patterns consist of four strands, each with six monomeric staples (Fig. 2.8A,C). Twenty-four out of 26 staples form the four strands along the C_2 rotation axis. The remaining two motifs are on the equatorial region of the spherical core, reducing the symmetry to C_2 (Fig. 2.8B). The surface helical pattern of monomeric staples induces chirality in the Au_{133} (Fig. 2.8D).

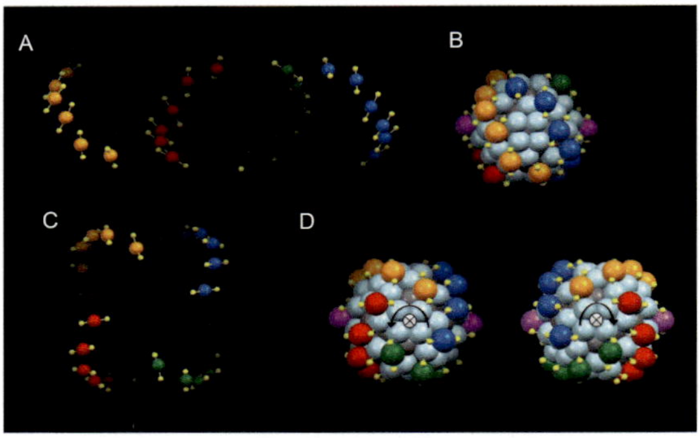

Figure 2.8 Helical arrangement of staples motifs on $Au_{133}(SR)_{52}$. (A) Four strands of six monomeric motifs colored with orange, red, green, and blue for clarity. (B) Four strands and two additional staples (shown in purple) on the Au_{107} core. (C) The four strands from the view along the rotational axis. (D) Pair of enantiomers induced by surface chirality. Carbon and hydrogen atoms are omitted for clarity [29].

2.3.1.2 Case of $Au_{144}(SR)_{60}$

Compared to other large-sized nanoclusters, $Au_{144}(SR)_{60}$ has the longest history from the synthesis to structure determination, with a span of a decade. According to the early mass spectrometric analysis in 2008 [31], the molecular formula of Au_{144} was determined to be $Au_{144}(SR)_{59}$, which was later corrected to be $Au_{144}(SR)_{60}$ in 2009 [26]. The crystal structure of Au_{144} was not available for a long time until the recent report of $Au_{144}(SCH_2Ph)_{60}$ in 2018 [32]. The reported crystallographic formula further supports the corrected chemical formula as $Au_{144}(SR)_{60}$. The crystal structure of $Au_{144}(SCH_2Ph)_{60}$ is made up of a Au_{114} core protected by 30 monomeric staple motifs (Fig. 2.9) [32]. The core of Au_{144} is made up of a shell-by-shell structure, starting from a hollow shell of icosahedral Au_{12}, in contrast to the center-filled icosahedral Au_{13} in Au_{133} (Fig. 2.9A). The icosahedral Au_{12} is covered by another shell of Au_{42} to form Au_{54} (Fig. 2.9B) in contrast to the Au_{55} in Au_{133}. The structure of Au_{54} is a MI with 20 triangular {111} facets except for the missing central atom. The Au_{54} is then covered by additional three atoms each for all the 20 {111} facets (Fig. 2.9C), as opposed to only 16 in the case of Au_{133}. The outermost layer is in *a-b-c* manner to add Au_{60} in total to form the Au_{114} core (Fig. 2.9E). The surface protecting monomeric staples are on Au_4 squares of the exposed 60 Au atoms (Fig. 2.9D,F).

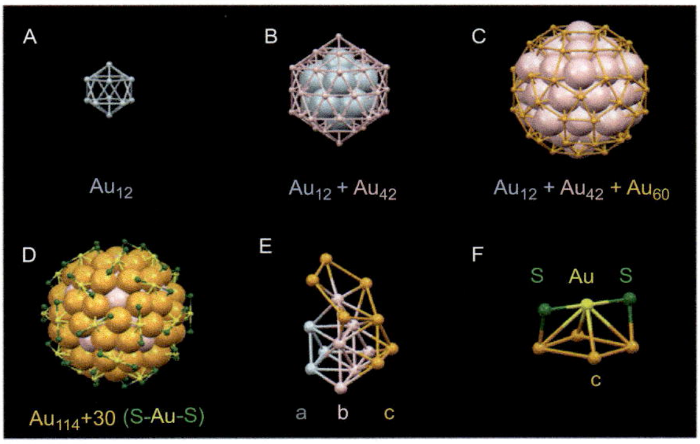

Figure 2.9 Crystal structure of $Au_{144}(SR)_{60}$. (A) Au_{12}, (B) Au_{54}, and (C) Au_{114} core structures of Au_{144} in shell-by-shell illustration. (D) Surface protection of Au_{114} core by 30 monomeric staple motifs. (E) Layered structures from inner to outer core in *a-b-c* manner. (F) Monomeric staple motifs on the outermost Au_{114} core. Carbon and hydrogen atoms are omitted for clarity [32].

The arrangement of staple motifs on the Au_{114} core's surface induces chirality to the Au_{144} cluster (Fig. 2.10) [32]. The monomeric staple motif pattern exhibits symmetry of fivefold rotation (Fig. 2.10B). Except for the equatorial row, five monomeric motifs form a pentagonal circle with different diameters (shown in red and blue in Fig. 2.10C–E). The diameter of the middle circle (shown in yellow) is the largest and made up of 10 monomeric staple motifs.

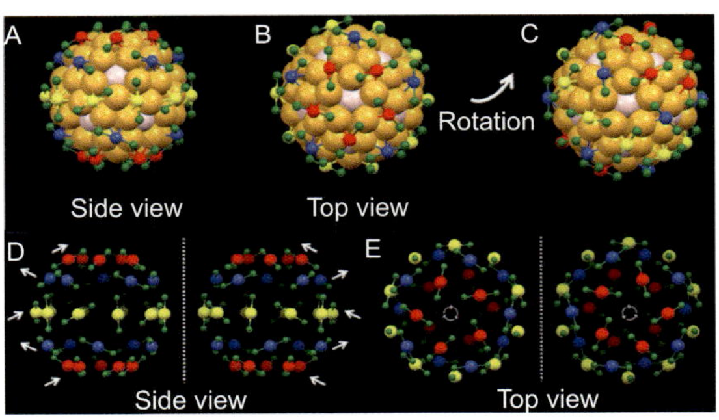

Figure 2.10 Surface protection pattern on $Au_{144}(SR)_{60}$. (A) Side, (B) top, and (C) rotated views of surface protection by monomeric staple motifs. (D) Surface protection of the Au_{114} core by 30 monomeric staple motifs. (E) Layered structures from inner to outer in *a-b-c* manner. (F) Monomeric staple motifs on the outermost Au_{114} core. Carbon and hydrogen atoms are omitted for clarity [32].

2.3.2 Decahedral Structures

2.3.2.1 Case of $Au_{102}(SR)_{44}$

$Au_{102}(SR)_{44}$ is one of the thiolate-protected decahedral gold nanocluster structures determined by X-ray crystallography [33]. This nanocluster possesses 58 nominal valence electrons based on the counting rule of the number of Au (i.e., 102) subtracting the number of ligands/charge (i.e., 44 and 0) [34]. The single crystal was, however, grown from a trace amount of sample, and so detailed studies on the $Au_{102}(SR)_{44}$ properties had to wait for the subsequent development of large-scale synthesis with the atomic level of precision [35].

The synthesis was performed in aqueous phase under the basic condition with *p*-mercaptobenzoic acid (*p*MBA) as the surface ligand. The crystal structure of $Au_{102}(SR)_{44}$ is made up of a Au_{79} core with a shell-by-shell structure starting from the innermost Au_7 decahedron (Fig. 2.11A) [33]. The decahedral Au_7 is covered by an additional layer of Au_{32} to form Au_{39} (Fig. 2.11B). The Au_{39} shows 10 {111} facets on its top and bottom in total and five {100} facets on the waist (Fig. 2.11C). The {100} facets are further each covered by two atoms to form Au_{49}, and the Au_{49} geometry is known as Marks decahedron [13]. The unique decahedral structure exhibits five truncated corners compared to the regular decahedron, and the truncated decahedron shows reduced surface energy according to the classical theory. The 10 {111} facets are each covered by a triangle of three atoms to form Au_{79} (Fig. 2.11C). Then, the top and bottom halves of the Au_{79} core are each protected by five monomeric staples, each bridging two triangular facets (Fig. 2.11D–F). The five monomeric staples either on the top or on the bottom are placed in the symmetry of fivefold rotation. Surface protection on the waist of Au_{102} is made up of nine monomeric and two dimeric staples, which are arranged in the symmetry of twofold rotation along the lateral direction.

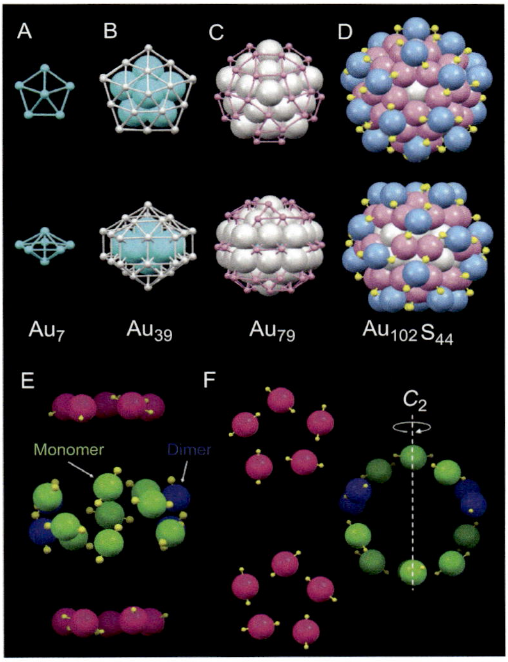

Figure 2.11 Crystal structure of $Au_{102}(SR)_{44}$. (A) Au_7, (B) Au_{39}, and (C) Au_{79} core structures of Au_{102} in shell-by-shell illustration. (D) Surface protection of the Au_{79} core by staple motifs. (E, F) Side and top views of surface staple motifs on Au_{102}. Carbon and hydrogen atoms are omitted for clarity [36].

2.3.2.2 Case of $Au_{103}S_2(SR)_{41}$

$Au_{103}S_2(SR)_{41}$ was synthesized by LEIST from $Au_{99}(SPh)_{42}$ [36] using 2-naphthalenethiolate as exchanged protecting ligands [37]. According to the electron counting rule, $Au_{103}S_2(SR)_{41}$ also possesses 58 valence electrons ($103 - 41 - 2 \times 2$), noting that one sulfido takes two electrons since its valence is two. The transformation from Au_{99} to Au_{103} is completed after 48 h at 80 °C in a mixed solvent of toluene and 4-*tert*-butyltoluene. Interestingly, the crystal structure of $Au_{103}S_2(SR)_{41}$ is made up of exactly the same Marks decahedral Au_{79} core in Au_{102} [33, 37] (see Fig. 2.11B). Both Au_{103} and Au_{102} have 58 valence electrons, and the observation of the same core structure in these two nanoclusters could provide atomic insights into the structural rules of the nanoclusters that share the same number of valence electrons. On the surface of Au_{103}, the top and bottom halves of each Au_{79} core are protected by five monomeric staples, similar to Au_{102} (Fig. 2.12; see Fig. 2.11D–F for details). However, remarkable differences are observed on the waist of Au_{103} [37]. In contrast to nine monomeric and two dimeric staples on Au_{102}, Au_{103} exhibits six monomeric, one dimeric, and two trimeric staples on its waist. The trimeric staple motifs contain one μ_3-sulfido at the position indicated by the small arrow in top middle panel in Fig. 2.12. The overall core/staple geometry shows the symmetry of twofold rotation along the lateral direction.

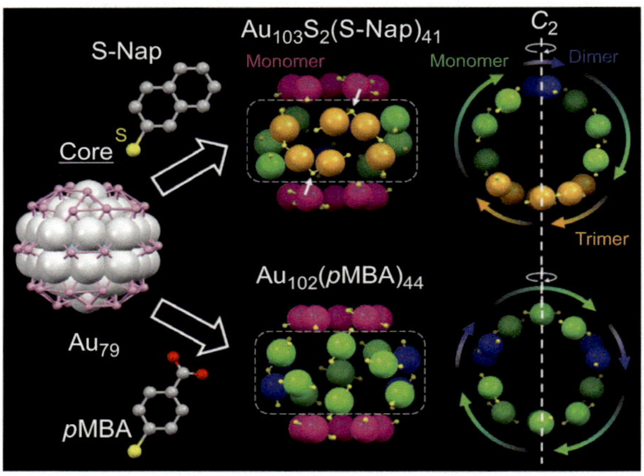

Figure 2.12 Comparison of surface structure between $Au_{103}S_2(SR)_{41}$ and $Au_{102}(SR)_{44}$ [37].

The 2-naphthalene carbon tails on the Au_{103} exhibit surface patterns via C-H···π and π···π interactions (Fig. 2.13) [37]. For example, a dimeric pattern is formed by two naphthalene groups via the edge-to-face C-H···π interaction with an angle of 51° (Fig. 2.13 I,II). A cyclic tetramer pattern is further constructed from a pair of dimeric patterns with a distance of 3.02 or 2.76 Å (Fig. 2.13 I,II). Such tetrameric units of naphthalene ligands resemble the "herringbone pattern" [38] that is often observed in crystal structures of polycyclic aromatic hydrocarbons. In contrast, the bottom of Au_{103} shows π···π stacking between four naphthalene ligands with an angle of 97° with 3.40 ± 0.10 Å spacing (Fig. 2.13 III).

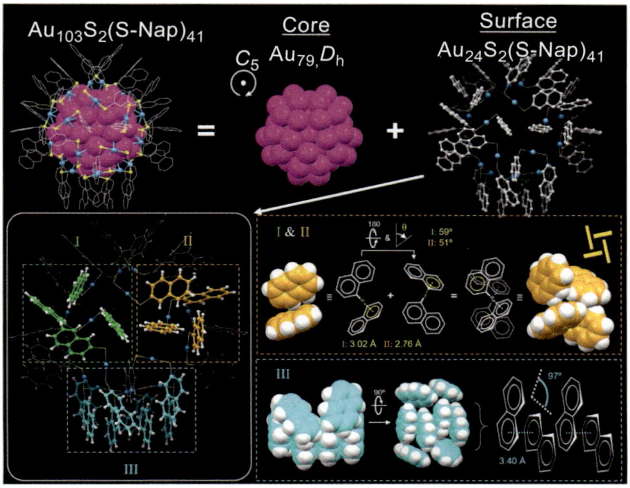

Figure 2.13 Crystal structure of Au_{103}. Total structure of Au_{103} (upper left), Marks decahedral Au_{79} core (upper middle) and surface staple motifs (upper right). Intracluster ligand interactions on the surface (lower left). The anatomy of the tetrameric herringbone structure via C–H···π interactions (I and II) and the parallel tetramer via staggered π···π interactions (III). Color labels: magenta = Au in the kernel, blue = Au in the staple motifs, yellow = S, white = H, and all the other colors are used for C in different positions [37].

2.3.2.3 Case of $Au_{130}(SR)_{50}$

The synthesis of $Au_{130}(SR)_{50}$ is performed by the size-focusing method [39]. The preparation starts with polydisperse nanoclusters protected by *para*-methylbenzenethiolate (*p*MBT). Then the polydisperse precursor is treated with excess thiol at high temperature. In this reaction, the position of the substituent methyl group plays a critical role in controlling the size/structure of final nanoclusters [39]. For example, when *meta*- or *ortho*-methyl benzenethiol (*m*MBT or *o*MBT for short) is used for surface protection, the product is, respectively, Au_{104}(*m*MBT)$_{41}$ or Au_{40}(*o*MBT)$_{24}$ (Fig. 2.14) [39]. After the work-up procedure, the as-synthesized Au_{130}(*p*MBT)$_{50}$ is crystallized by vapor diffusion of acetonitrile into the nanocluster solution in toluene [40].

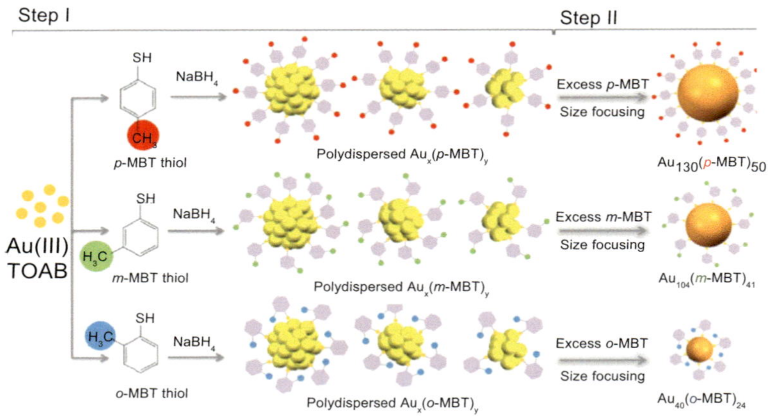

Figure 2.14 Tuning the magic size of thiolate-protected gold nanoclusters through the careful choice of the position of the substituent methyl group [39].

The crystal structure of $Au_{130}(SR)_{50}$ is made up of a Au_{105} core with a shell-by-shell structure starting from the innermost Au_1 center (shown in green in Fig. 2.15) [40]. The inner core is covered by a Au_{12} (in magenta) shell, forming Au_{13}, which is called Ino decahedron (Fig. 2.15A) [41]. The decahedral Au_{13} is covered by a shell of Au_{42} (in gray) to form Au_{55} (Fig. 2.15B). The Au_{55} shows 10 {111} facets on its top and bottom in total and five {100} facets on the waist (Fig. 2.15C). The {100} facets are further covered by four atoms on each facet, and the 10 {111} facets are instead covered by a triangle of three atoms on each facet to form Au_{105} (Fig. 2.15C). The surface on Au_{105} is then protected by 25 monomeric staple motifs.

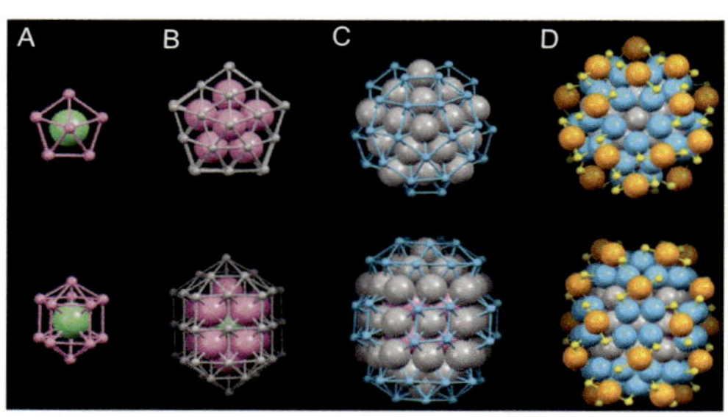

Figure 2.15 Crystal structure of $Au_{130}(pMBT)_{50}$. Top and side views of the core structure made up of (A) Au_{13}, (B) Au_{55}, and (C) Au_{105}. (D) Surface protection of the core by 25 monomeric staples [40].

The monomeric staple motifs are placed with the symmetry of fivefold rotation (Fig. 2.16) [40]. In each row, five monomeric motifs form a pentagonal circle, making up five rows of stripe pattern (Fig. 2.16A). The diameter of the middle circle (shown in blue) is the largest, while the ones on the top and bottom possess the smallest diameter. The overall structure of 25 monomeric motifs has the dimension of 1.9×1.6 nm with a barrel-like shape, where each row of pentagonal circle resembles metal hoops to bind the wooden pieces for a barrel. The core and staple structure of $Au_{130}(SR)_{50}$ shows quasi-D_5 symmetry with chirality due to the difference in orientation of staple motif circles (Fig. 2.16B,C).

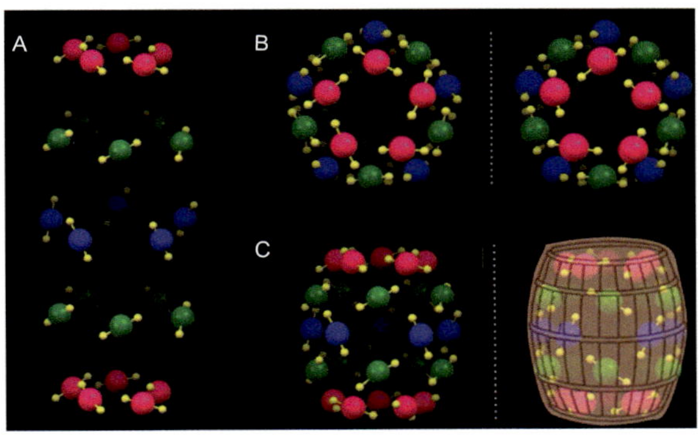

Figure 2.16 Surface protection pattern of $Au_{130}(pMBT)_{50}$. (A) Five pentagon ripples of 25 monomeric staple motifs. (B) Top and (C) side views showing fivefold rotation symmetry [40].

2.3.2.4 Case of $Au_{246}(SR)_{80}$

The synthesis of $Au_{246}(SR)_{80}$ is performed by the size-focusing methodology, followed by solvent fractionation for further purification [15]. During the first size-focusing process, the size distribution of the product is converged to Au_{246} with ~90% purity under thermal treatment with excess thiol. The following solvent fractionation successfully removes the minor product of smaller nanoclusters for molecular purity of Au_{246}. Crystallization of Au_{246} is performed by vapor diffusion of acetonitrile into the nanocluster solution of toluene [15].

The crystal structure of Au_{246} is made up of a decahedral Au_{206} core with a shell-by-shell structure starting from an innermost Au_7

decahedron, which is covered by a layer of Au_{32} to form Au_{39} (Fig. 2.17A) [15]. The Au_{39} is further covered by an additional layer of Au_{77} to form Au_{116} (Fig. 2.17B). The Au_{116} shows 10 {111} facets on its top and bottom in total, as well as five {100} facets on the waist (Fig. 2.17C). The five {100} facets are further covered by six atoms on each facet to form an additional layer on it, and the ten {111} facets are instead covered by a triangle with six atoms on each facet to form Au_{206} (Fig. 2.17C). Then, the top and bottom halves of the Au_{206} core are protected by five surface protecting dimeric staples by bridging across the triangular facets (Fig. 2.17D). The six-atom triangles (i.e., {111} of Au_{206}) are connected by 10 bridging thiolates in total for top and bottom. Additional five monomeric staple motifs are used for the connection between six-atom triangles on top/bottom {111} and six-atom squares on the middle {100}. On the waist of Au_{206}, six-atom squares on {100} are connected by two monomeric staples at each junction.

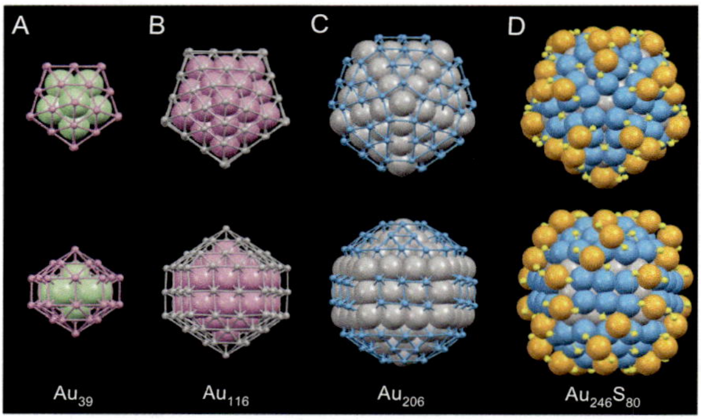

Figure 2.17 Crystal structure of $Au_{246}(p\text{MBT})_{80}$. Top and side views of the core structure made up of (A) Au_{39}, (B) Au_{116}, and (C) Au_{206}. (D) Surface protection of the core by staple motifs [15].

The surface of Au_{246} shows aesthetic patterns of carbon tails (Fig. 2.18) [15]. At the top and bottom sites of Au_{246}, *p*MBT ligands form pentagonal circles, which is defined as α-rotation (Fig. 2.18A,B). The waist of Au_{246} possesses alternating parallel pairs of *p*MBT ligands, which is defined as β-parallel (Fig. 2.18A,C). The overall structure shows chirality induced by these arrangements of *p*MBT ligands on the surface.

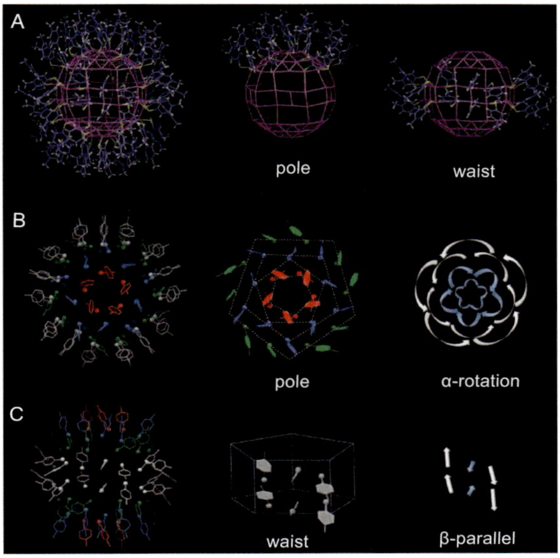

Figure 2.18 Surface patterns of *p*MBT ligands on Au_{246}. (A) Overall structures as well as pole and waist. (B) The α-rotation arrangement of *p*MBT at the top and bottom. (C) The β-parallel packing of *p*MBT at the waist [15].

2.3.3 Face-Centered Cubic Structures

2.3.3.1 Case of $Au_{146}(SR)_{57}$

$Au_{146}(SR)_{57}$ is synthesized via kinetically controlled reduction in aqueous phase [42], similarly to the case of $Au_{102}(SR)_{44}$, where both nanoclusters are protected by *para*-mercaptobenzoic acid. The crystal structure of $Au_{146}(SR)_{57}$ was solved by electron diffraction and synchrotron X-rays [42]. The core of $Au_{146}(SR)_{57}$ is made up of Au_{109} with the twinned fcc structure. The innermost Au_1 (shown in green) and Au_{12} (in magenta) form a Au_{13} anti-cuboctahedron (Fig. 2.19A). The Au_{13} is then covered by a layer of Au_{42} (in gray) to form Au_{55} (Fig. 2.19B). The Au_{55} shows eight {111} facets and six {100} facets (Fig. 2.19C). The {100} facets are then covered by four atoms on each facet. For the two {100} planes in the front side (top, Fig. 2.19C), two pairs of Au_2 are additionally placed on the four-atom layer. The {111} facets are instead covered by incomplete triangles with five atoms on each facet, except for the two {111} facets on the back {111} covered by one atom each at the center of the triangular

{111} facet. The incomplete additional shell makes up a Au_{119} core (Fig. 2.19C). The surface on Au_{119} is then protected by staple motifs as shown in Fig. 2.19D.

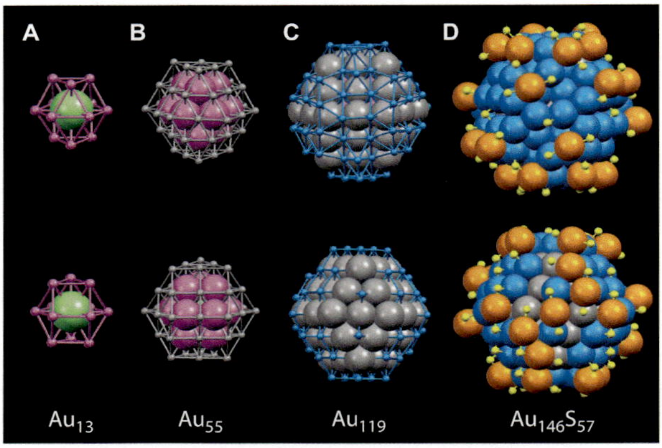

Figure 2.19 Crystal structure of $Au_{146}(p\text{MBA})_{57}$. Front and back views of the core structure made up of (A) Au_{13}, (B) Au_{55}, and (C) Au_{119}. (D) Surface protection of the core by staple motifs. Reprinted with permission from Ref. [42], Copyright 2017, American Chemical Society.

2.3.3.2 Case of $Au_{279}(SR)_{84}$

The synthesis of $Au_{279}(SR)_{84}$ is performed either by size focusing followed by chromatographic isolation [43] or by LEIST from the larger sized $Au_{333}(SR)_{79}$ nanocluster [16]. Crystallization is performed by vapor diffusion of acetonitrile into the nanocluster solution of toluene or vapor diffusion of pentane into CH_2Cl_2 solution of the nanocluster [16, 43]. The core of Au_{279} is made up of Au_{211} with the fcc structure. The innermost Au_1 (shown in dark blue) and Au_{12} (in green) form a Au_{13} cuboctahedron (Fig. 2.20A). The Au_{13} is covered by a layer of Au_{42} (in violet) to form Au_{55} (Fig. 2.20B). The Au_{55} is further covered by another layer of Au_{92} to form Au_{147} (Fig. 2.20C). The Au_{147} shows eight {111} facets and six {100} facets (Fig. 2.20D). Each {100} facet is covered by a square of nine atoms. On the other hand, each {111} facet is covered by a triangle of six atoms, with some distortion on some facets. The incomplete additional shell makes up the Au_{211} core (Fig. 2.20D). The surface on Au_{211} is then protected by staple motifs as shown in Fig. 2.20E.

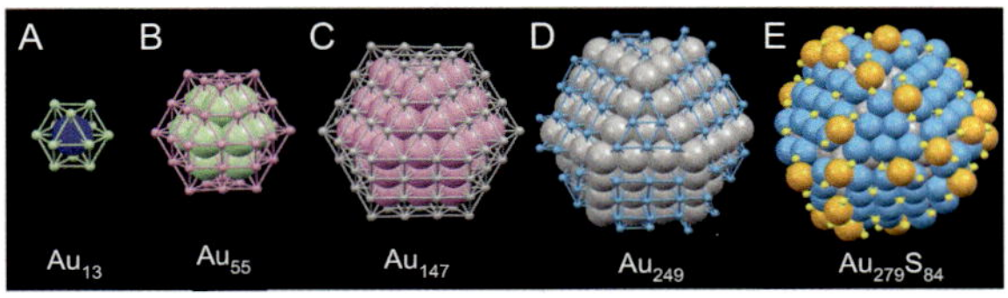

Figure 2.20 Crystal structure of $Au_{279}(TBBT)_{84}$. The fcc core structure made up of (A) Au_{13}, (B) Au_{55}, and (C) Au_{147}, and (D) Au_{249}. (E) Surface protection of the core by staple motifs [43].

2.4 Conclusions and Future Perspectives

In this chapter, we have summarized the synthesis and structure determination of atomically precise gold nanoclusters with more than 100 metal atoms. The discussion includes D_h structures (e.g., $Au_{102}(SR)_{44}$, $Au_{103}S_2(SR)_{41}$, $Au_{130}(SR)_{50}$, and $Au_{246}(SR)_{80}$), I_h structures (e.g., $Au_{133}(SR)_{52}$ and $Au_{144}(SR)_{60}$), and fcc structures (e.g., $Au_{146}(SR)_{57}$ and $Au_{279}(SR)_{84}$). The observed size independence in these large atomic-level structures goes against the theoretical prediction of thermodynamically stable structures based on the core geometry [13, 14]. Surface protection on these nanoclusters contributes to stabilization of their unique structures. The core geometry and the type of surface protection are summarized in Table 2.1.

Table 2.1 Crystal structures of atomically precise gold nanoclusters with >100 atoms

Size	Core	Staple motifs			
		Bridging	Monomer	Dimer	Trimer
Au_{102}	D_h	0	19	2	0
Au_{103}	D_h	0	16	1	2
Au_{130}	D_h	0	25	0	0
Au_{133}	I_h	0	26	0	0
Au_{144}	I_h	0	30	0	0
Au_{146}	fcc	7	19	4	0
Au_{246}	D_h	10	20	10	0
Au_{279}	fcc	30	18	6	0

According to Table 2.1, these larger size gold nanoclusters do not follow the trends predicted by the conventional theory, in which the most stable structures would change from fcc, D_h, to I_h with decrease in size. In contrast, our summary shows that the core structure varies with size without any clear trend in the observed size range from ~100 to ~300 atoms. For the type of surface protection, these nanoclusters do not possess staple motifs with more than three Au(I) atoms. In other words, the observed unique large structures are preferably protected by shorter staple motifs, including monomers, dimers, and bridging thiolates. The optical properties of these nanoclusters are discussed in Chapter 6, which exhibit fundamental differences from those of regular nanoparticles [44–46].

Finally, we provide the following perspectives for future direction of the structure determination of large gold nanoclusters:

(1) The observed structures in the list have a large gap between Au_{146} and Au_{246}, therefore further efforts are needed to solve the structures in the intermediate sizes for understanding the rule of structural evolution with atomic resolution. For example, Negishi et al. experimentally reported Au_{187} [44] and theoretical work by Tlahuice-Flores has predicted the structure of Au_{187} to be a decahedral core [45], but the real structure is yet to be determined by X-ray crystallography.

(2) New sizes over the transition from non-metallic to metallic states (i.e., Au_{246} and Au_{279}) need more investigation as well. The sharp transition with merely 33-atom difference is intriguing and expected to stimulate further fundamental quests on the transition. For example, the smallest number to induce the transition (e.g., single atom augment) is still under investigation, and the question remains unanswered whether or not an increase of smaller than 33 gold atoms can induce the transition.

(3) The effect of detailed structure type (D_h, I_h, or fcc) on the transition is also an interesting direction to pursue, given the currently observed different structures between the critical sizes for the transition (D_h for Au_{246} and fcc for Au_{279}). Structural control might be challenging though.

(4) The structure determination of much larger sizes is of paramount importance to understand the surface

protection pattern by Au–SR. For example, $Au_{333}(SR)_{79}$ has been synthesized/isolated with molecular purity, but its crystal structure remains unsolved [46]. Nevertheless, new properties of $Au_{333}(SR)_{79}$ have been reported in the recent work on femtosecond spectroscopy analysis. Specifically, $Au_{333}(SR)_{79}$ shows unusual electron dynamics unlike plasmonic nanoparticles or excitonic nanoclusters [47]. Deep understanding of the properties has to wait for the future structure determination so that the correlation of unusual electron dynamics with structure can be achieved. The Au–SR pattern on larger nanoclusters will also provide the atomic-level insights into the structure of the self-assembled monolayer [48].

Acknowledgments

We acknowledge the financial support from the National Science Foundation (DMR-1808675) and AFOSR.

References

1. Jin, R., Zeng, C., Zhou, M. and Chen, Y. (2016). Atomically precise colloidal metal nanoclusters and nanoparticles: Fundamentals and opportunities, *Chem. Rev.*, **116**, pp. 10346–10413.
2. Tsukuda, T. and Häkkinen, H. (2015). *Protected Metal Clusters: From Fundamentals to Applications*, 1st ed, Elsevier, The Netherland.
3. Chakraborty, I. and Pradeep, T. (2017). Atomically precise clusters of noble metals: Emerging link between atoms and nanoparticles, *Chem. Rev.*, **117**, pp. 8208–8271.
4. Kurashige, W., Niihori, Y., Sharma, S. and Negishi, Y. (2016). Precise synthesis, functionalization and application of thiolate-protected gold clusters, *Coord. Chem. Rev.*, **320–321**, pp. 238–250.
5. Konishi, K., Iwasaki, M., Sugiuchi, M. and Shichibu, Y. (2016). Ligand-based toolboxes for tuning of the optical properties of subnanometer gold clusters, *J. Phys. Chem. Lett.*, **7**, pp. 4267–4274.
6. Fernando, A., Weerawardene, K. L. D. M., Karimova, N. V. and Aikens, C. M. (2015). Quantum mechanical studies of large metal, metal oxide, and metal chalcogenide nanoparticles and clusters, *Chem. Rev.*, **115**, pp. 6112–6216.

7. Pei, Y. and Zeng, X. C. (2012). Investigating the structural evolution of thiolate protected gold clusters from first-principles, *Nanoscale*, **4**, pp. 4054–4072.

8. Kwak, K. and Lee, D. (2019). Electrochemistry of atomically precise metal nanoclusters, *Acc. Chem. Res.*, **52**, pp. 12–22.

9. Knoppe, S. and Bürgi, T. (2014). Chirality in thiolate-protected gold clusters, *Acc. Chem. Res.*, **47**, pp. 1318–1326.

10. Higaki, T., Liu, C., Zeng, C., Jin, R., Chen, Y., Rosi, N. L. and Jin, R. (2016). Controlling the atomic structure of Au_{30} nanoclusters by a ligand-based strategy, *Angew. Chem. Int. Ed.*, **55**, pp. 6694–6697.

11. Liu, C., Li, T., Li, G., Nobusada, K., Zeng, C., Pang, G., Rosi, N. L. and Jin, R. (2015). Observation of body-centered cubic gold nanocluster, *Angew. Chem. Int. Ed.*, **54**, pp. 9826–9829.

12. Zhou, M., Higaki, T., Hu, G., Sfeir, M. Y., Chen, Y., Jiang, D. and Jin, R. (2019). Three-orders-of-magnitude variation of carrier lifetimes with crystal phase of gold nanoclusters, *Science*, **364**, pp. 279–282.

13. Marks, L. D. and Peng, L. (2016). Nanoparticle shape, thermodynamics and kinetics, *J. Phys. Condens. Matter*, **28**, p. 053001.

14. Baletto, F. and Ferrando, R. (2005). Structural properties of nanoclusters: Energetic, thermodynamic, and kinetic effects, *Rev. Mod. Phys.*, **77**, pp. 371–423.

15. Zeng, C., Chen, Y., Kirschbaum, K., Lambright, K. J. and Jin, R. (2016). Emergence of hierarchical structural complexities in nanoparticles and their assembly, *Science*, **354**, pp. 1580–1584.

16. Higaki, T., Zhou, M., Lambright, K. J., Kirschbaum, K., Sfeir, M. Y. and Jin, R. (2018). Sharp transition from nonmetallic Au_{246} to metallic Au_{279} with nascent surface plasmon resonance, *J. Am. Chem. Soc.*, **140**, pp. 5691–5695.

17. Zhou, M., Zeng, C., Song, Y., Padelford, J. W., Wang, G., Sfeir, M. Y., Higaki, T., Jin, R. (2017). On the non-metallicity of 2.2 nm $Au_{246}(SR)_{80}$ nanoclusters, *Angew. Chem. Int. Ed.*, **56**, pp. 16257–16261.

18. Zeng, C., Liu, C., Chen, Y., Rosi, N. L. and Jin, R. (2014). Gold-thiolate ring as a protecting motif in the $Au_{20}(SR)_{16}$ nanocluster and implications, *J. Am. Chem. Soc.*, **136**, pp. 11922–11925.

19. Jin, R., Qian, H., Wu, Z., Zhu, Y., Zhu, M., Mohanty, A. and Garg, N. (2010). Size focusing: A methodology for synthesizing atomically precise gold nanoclusters, *J. Phys. Chem. Lett.*, **1**, pp. 2903–2910.

20. Qian, H., Zhu, Y. and Jin, R. (2009). Size-focusing synthesis, optical and electrochemical properties of monodisperse $Au_{38}(SC_2H_4Ph)_{24}$ nanoclusters, *ACS Nano*, **3**, pp. 3795–3803.

21. Das, A., Li, T., Nobusada, K., Zeng, C., Rosi, N. L. and Jin, R. (2013). Nonsuperatomic $[Au_{23}(SC_6H_{11})_{16}]^-$ nanocluster featuring bipyramidal Au_{15} kernel and trimeric $Au_3(SR)_4$ motif, *J. Am. Chem. Soc.*, **135**, pp. 18264–18627.

22. Zeng, C., Chen, Y., Das, A. and Jin, R. (2015). Transformation chemistry of gold nanoclusters: From one stable size to another, *J. Phys. Chem. Lett.*, **6**, pp. 2976–2986.

23. Zhu, M., Aikens, C. M., Hollander, F. J., Schatz, G. C. and Jin, R. (2008). Correlating the crystal structure of a thiol-protected Au_{25} cluster and optical properties, *J. Am. Chem. Soc.*, **130**, pp. 5883–5885.

24. Heaven, M. W., Dass, A., White, P. S., Holt, K. M. and Murray, R. W. (2008). Crystal structure of the gold nanoparticle $[N(C_8H_{17})_4]$ $[Au_{25}(SCH_2CH_2Ph)_{18}]$, *J. Am. Chem. Soc.*, **130**, pp. 3754–3755.

25. Qian, H., Eckenhoff, W. T., Zhu, Y., Pintauer, T. and Jin, R. (2010). Total structure determination of thiolate-protected Au_{38} nanoparticles, *J. Am. Chem. Soc.*, **132**, pp. 8280–8281.

26. Qian, H. and Jin, R. (2009). Controlling nanoparticles with atomic precision: The case of $Au_{144}(SCH_2CH_2Ph)_{60}$, *Nano Lett.*, **9**, pp. 4083–4087.

27. Zeng, C., Li, T., Das, A., Rosi, N. L. and Jin, R. (2013). Chiral structure of thiolate-protected 28-gold-atom nanocluster determined by X-ray crystallography, *J. Am. Chem. Soc.*, **135**, pp. 10011–10013.

28. Zeng, C., Qian, H., Li, T., Li, G., Rosi, N. L., Yoon, B., Barnett, R. N., Whetten, R. L., Landman, U., Jin, R. (2012). Total structure and electronic properties of the gold nanocrystal $Au_{36}(SR)_{24}$, *Angew. Chem. Int. Ed.*, **51**, pp. 13114–13118.

29. Zeng, C., Chen, Y., Kirschbaum, K., Appavoo, K., Sfeir, M. Y. and Jin, R. (2015). Structural patterns at all scales in a nonmetallic chiral $Au_{133}(SR)_{52}$ nanoparticle, *Sci. Adv*, **1**, p. e1500045.

30. Mackay, A. L. (1962). A dense non-crystallographic packing of equal spheres, *Acta Cryst.*, **15**, pp. 916–918.

31. Chaki, N. K., Negishi, Y., Tsunoyama, H., Shichibu, Y. and Tsukuda, T. (2008). Ubiquitous 8 and 29 kDa gold: Alkanethiolate cluster compounds: Mass-spectrometric determination of molecular formulas and structural implications, *J. Am. Chem. Soc.*, **130**, pp. 8608–8610.

32. Yan, N., Xia, N., Liao, L., Zhu, M., Jin, F., Jin, R. and Wu, Z. (2018). Unraveling the long-pursued Au_{144} structure by X-ray crystallography, *Sci. Adv.*, **4**, p. eaat7259.

33. Jadzinsky, P. D., Calero, G., Ackerson, C. J., Bushnell, D. A. and Kornberg, R. D. (2007). Structure of a thiol monolayer-protected gold nanoparticle at 1.1 Å resolution, *Science*, **318**, pp. 430–433.

34. Walter, M., Akola, J., Lopez-Acevedo, O., Jadzinsky, P. D., Calero, G., Ackerson, C. J., Whetten, R. L., Grönbeck, H. and Häkkinen, H. (2008). A unified view of ligand-protected gold clusters as superatom complexes, *Proc. Natl. Acad. Sci. U.S.A.*, **105**, pp. 9157–9162.

35. Levi-Kalisman, Y., Jadzinsky, P. D., Kalisman, N., Tsunoyama, H., Tsukuda, T., Bushnell, D. A. and Kornberg, R. D. (2011). Synthesis and characterization of Au_{102}(*p*-MBA)$_{44}$ nanoparticles, *J. Am. Chem. Soc.*, **133**, pp. 2976–2982.

36. Li, G., Zeng, C. and Jin, R. (2014). Thermally robust $Au_{99}(SPh)_{42}$ nanoclusters for chemoselective hydrogenation of nitrobenzaldehyde derivatives in water, *J. Am. Chem. Soc.*, **136**, pp. 3673–3679.

37. Higaki, T., Liu, C., Zhou, M., Luo, T. Y., Rosi, N. L. and Jin, R. (2017). Tailoring the structure of 58-electron gold nanoclusters: $Au_{103}S_2$(S-Nap)$_{41}$ and its implications, *J. Am. Chem. Soc.*, **139**, pp. 9994–10001.

38. Desiraju, G. R. (1997). Designer crystals: Intermolecular interactions, network structures and supramolecular synthons, *Chem. Commun.*, pp. 1475–1482.

39. Chen, Y., Zeng, C., Kauffman, D. R. and Jin, R. (2015). Tuning the magic size of atomically precise gold nanoclusters via isomeric methylbenzenethiols, *Nano Lett.*, **15**, pp. 3603–3609.

40. Chen, Y., Zeng, C., Liu, C., Kirschbaum, K., Gayathri, C., Gil, R. R., Rosi, N. L. and Jin, R. (2015). Crystal structure of barrel-shaped chiral Au_{130}(*p*-MBT)$_{50}$ nanocluster, *J. Am. Chem. Soc.*, **137**, pp. 10076–10079.

41. Ino, S. (1969). Stability of multiply-twinned particles, *J. Phys. Soc. Jpn.*, **27**, pp. 941–953.

42. Vergara, S., Lukes, D. A., Martynowycz, M. W., Santiago, U., Plascencia-Villa, G., Weiss, S. C., De La Cruz, M. J., Black, D. M., Alvarez, M. M., López-Lozano, X., Barnes, C. O., Lin, G., Weissker, H.-C., Whetten, R. L., Gonen, T., Yacaman, M. J. and Calero, G. (2017). MicroED structure of Au_{146}(*p*-MBA)$_{57}$ at subatomic resolution reveals a twinned FCC cluster, *J. Phys. Chem. Lett.*, **8**, pp. 5523–5530.

43. Sakthivel, N. A., Theivendran, S., Ganeshraj, V., Oliver, A. G. and Dass, A. (2017). Crystal structure of faradaurate-279: Au_{279}(SPh-*t*Bu)$_{84}$ plasmonic nanocrystal molecules, *J. Am. Chem. Soc.*, **139**, pp. 15450–15459.

44. Negishi, Y., Sakamoto, C., Ohyama, T. and Tsukuda, T. (2012). Synthesis and the origin of the stability of thiolate-protected Au_{130} and Au_{187} clusters, *J. Phys. Chem. Lett.*, **3**, pp. 1624–1628.

45. Tlahuice-Flores, A. (2015). New insight into the structure of thiolated gold clusters: A structural prediction of the $Au_{187}(SR)_{68}$ cluster, *Phys. Chem. Chem. Phys.*, **17**, pp. 5551–5555.

46. Qian, H., Zhu, Y. and Jin, R. (2012). Atomically precise gold nanocrystal molecules with surface plasmon resonance, *Proc. Natl. Acad. Sci. U.S.A.*, **109**, pp. 696–700.

47. Higaki, T., Zhou, M., He, G., House, S. D., Sfeir, M. Y., Yang, J. C. and Jin, R. (2019). Anomalous phonon relaxation in $Au_{333}(SR)_{79}$ nanoparticles with nascent plasmons, *Proc. Natl. Acad. Sci. U.S.A.*, **116**, pp. 13215–13220.

48. Zeng, C., Liu, C., Chen, Y., Rosi, N. L. and Jin, R. (2016). Atomic structure of self-assembled monolayer of thiolates on a tetragonal Au_{92} nanocrystal, *J. Am. Chem. Soc.*, **138**, pp. 8710–8713.

Chapter 3

Synthesis and Structure of Selenolate-Protected Metal Nanoclusters

Yongbo Song and Manzhou Zhu
Department of Chemistry and Centre for Atomic Engineering of Advanced Materials, Anhui Province Key Laboratory of Chemistry for Inorganic/Organic Hybrid Functionalized Materials, Anhui University, Hefei, Anhui 230601, P. R. China
Key Laboratory of Structure and Functional Regulation of Hybrid Materials (Anhui University), Ministry of Education, Hefei 230601, P. R. China
zmz@ahu.edu.cn

Surface ligands (halogen, phosphine, thiolate, and alkyne) not only stabilize metal nanoclusters (NCs) from aggregation, but also play a crucial role in affecting their geometric structure, physicochemical properties, and applications. In previous works, the thiolate-capped metal nanoclusters, denoted as $Au_n(SR)_m$, have attracted most attention on both their structure and properties. Because of the same periodic group, researchers have been motivated to pursue the selenolate-capped metal nanoclusters ($Au_n(SeR)_m$) by replacing the thiolate (SR) ligands with selenolate (SeR). In this chapter, we focus on the synthesis and structural characterization of $Au_n(SeR)_m$

Atomically Precise Nanoclusters
Edited by Yan Zhu and Rongchao Jin

ISBN 978-981-4877-08-4 (Hardcover), 978-1-003-11990-6 (eBook)
www.jennystanford.com

with precise atoms. The comparison of the precise structures of $Au_n(SeR)_m$ and $Au_n(SR)_m$ will clearly offer new insights into the differences between selenolate and thiolate ligands (e.g., bond length, coordination ability), which will help us further understand how ligands modulate the structure and property of metal nanoclusters. Finally, the challenges and problems associated with the selenolate-capped metal nanoclusters will be discussed.

3.1 Introduction

Owing to their ultrasmall size (<2 nm), metal nanoclusters exhibit strong quantum size effects, which endow them with unique physicochemical properties [1–5], such as multiple absorption bands [6], strong photoluminescence [7, 8], magnetism [9], and outstanding catalytic reactivity [10–12], which are different from those of the large nanoparticles with surface plasmon resonance (SPR). Metal nanoclusters have attracted wide interest since they have a significant potential application as functional materials. Among these metal nanoclusters, ligand-protected metal nanoclusters are most widely studied, and they are generally composed of metal kernels and surface protecting ligands [4, 5, 13–16]. It is known that metal kernels show high chemical and physical activities [17, 18], and the physicochemical properties of metal nanoclusters highly depend on the size, composition, and atom-packing mode of the metal kernel [19–21]. So, the kernel of metal nanoclusters has received relatively more attention.

Recent studies have demonstrated that surface protecting ligands not only can stabilize metal nanoclusters from aggregation [22–24], but also play a significant role in regulating the construction and properties of metal nanoclusters [25–28]. Therefore, ligand engineering has been proposed, which is another good strategy to tailor the physicochemical properties of metal nanoclusters protected by ligands. For example, changing the type (e.g., the donor/withdrawing ability, the size, or bulkiness) of ligands can efficiently modulate the stability, optical properties, electronic structure, and atom-packing mode of metal nanoclusters [29–31]. There are many typical examples, which have been discussed in previous chapters. Among these organic compounds, phosphine, alkyne, and thiolate are most commonly used as the protecting ligands [4, 5, 32–35], and experimental results demonstrate that they have different coordination modes with metal atoms, which have a significant effect on the atom-packing mode of metal nanoclusters.

In recent years, it has been reported that more stable gold clusters can be produced by changing the ligand of these gold clusters from thiolate to selenolate [36–41]. Both S (in thiolates) and Se (in selenolate) belong to the same group in the periodic table and exhibit many similarities. Compared to S, the electronegativity and atomic radius of Se are closer to those of gold [41], which will endow the Au–SeR bond with more covalent and higher bond energy than Au–SR. This is expected to make $Au_n(SeR)_m$ more stable than $Au_n(SR)_m$. Furthermore, the Se atom exhibits more outer orbitals than that of the S atom, and so the Se atom has higher coordination ability with metal atoms. This will significantly alter the arrangement of atoms and thus result in the preparation of gold nanoclusters with novel structures. Finally, in comparison to the thiolate-capped counterparts, we can further understand the relationship between surface protecting ligands and metal core.

To date, some selenolate-protected nanoclusters have been synthesized and studied [42–62], and some of their crystal structures have been detected by X-ray single crystal diffractometer. Some synthetic methods and the structural characterization for selenolate-protected metal nanoclusters with free valence electrons are discussed in the following section.

3.2 Synthetic Methods

Preparing more metal nanoclusters with diverse sizes and structures is very significant to promote the development of nanocluster science. Therefore, the synthesis of selenolate-protected metal nanoclusters plays a significant role in the development of this area. As discussed in the previous chapters, recent research has made great breakthroughs in the synthesis of atomically precise metal nanoclusters protected by thiolate and isolation from a mixture, which offer a good basis for preparing selenolate-capped metal nanoclusters [4, 5]. However, due to the difference between Se and S atoms [41], the synthesis of metal nanoclusters protected by selenolate possesses new challenges and becomes more difficult. Thus, the reaction conditions (e.g., the concentration of the metal precursor, the type and concentration of ligands and the reducing agent, the type of solvent used as the reaction medium, reaction temperature, stirring speed, the mixing of metal salt and ligand, and the addition speed of the reductant) need to be improved and optimized. In terms of synthetic methodologies, three typical routes for preparing selenolate-capped metal nanoclusters are usually adopted: direct synthesis, ligand exchange, and size focusing.

3.2.1 Direct Synthesis

Generally, in this one-pot synthetic method, a mixture of the metal precursor and ligands was directly reduced by the well-established two-phase procedure pioneered by Brust and co-workers [63]. For the selenolate-protected metal nanoclusters, a slight modification has been made in the reaction process. For example, the metal precursor is first reduced in the absence of ligands. After 1 to 2 s, the ligands are rapidly added to the solution. Using this strategy, Negishi et al. synthesized $[Au_{25}(SeC_8H_{17})_{18}]^-$ nanoclusters [43]. Furthermore, a co-reduction synthetic method was also proposed by Song et al. in which ligands (PhSeH) and the reductant ($NaBH_4$) are added to the solution of the metal precursor simultaneously to convert Au(III) into Au(I) or Au(0). With this, $[Au_{25}(SePh)_{18}]^-$ and $[Au_{24}(SePh)_{20}]$ nanoclusters were prepared in high yield [44–46].

3.2.2 Ligand Exchange

In this strategy, as-prepared metal nanoclusters with precise formula are used as precursors to react with another type of ligands (selenolate). With this method, $[Au_{25}(SePh)_{18}]^-$ nanoclusters were first prepared [51]. Interestingly, if the concentration of the precursor ($[Au_{25}(SCH_2CH_2Ph)_{18}]^-$) is reduced, $Au_{18}(SePh)_{14}$ and $Au_{20}(SePh)_{16}$ nanoclusters will be obtained [52], which are separated by HPLC (high-performance liquid chromatography) and characterized by ESI (electrospray ionization) mass spectrometry, indicating the product is highly sensitive to the ratio of precursor and selenolate (e.g., PhSeH). Furthermore, with this method, $[Au_{38}(SeC_8H_{17})_{24}]$,

$[Au_{36}(SePh)_{24}]$, $[Ag_{20}\{Se_2P(OEt)_2\}_{12}]$, and $[MAg_{20}\{Se_2P(OEt)_2\}_{12}]^+$ (M = Ag and Au) nanoclusters were also obtained [53–55]. However, selenolate-capped metal nanoclusters with other sizes have not been prepared with this method, even though thiolate-capped metal nanoclusters with different sizes have been widely reported with ligand exchange.

3.2.3 Size Focusing

In this method, there are two primary steps. In step I, polydisperse metal nanoclusters capped by phosphine are obtained through moderating the reaction conditions (i.e., the static and dynamic factors). In step II, as-prepared polydisperse metal nanoclusters are focused into the monodisperse product by etching or aging under a severe condition. When this condition is applied to the as-prepared polydisperse metal nanoclusters, only the metal nanoclusters with most robustness can survive, while the other products are decomposed or converted to the most stable size [64, 65]. The 'survival of the most robust' principle somewhat resembles nature's law 'survival of the fittest' [66]. With this method, $[Au_{11}(L^5)_4(SePh)_2]^+$, $[Au_{25}(PPh_3)_{10}(SePh)_5Cl_2]^{+/2+}$, $[Au_{60}Se_2(PPh_3)_{10}(SePh)_{15}]^+$, and $[Au_{13}Cu_4(dppy)_4(SePh)_9]$ are obtained through controlling the reaction conditions [56–59].

3.3 Structure of Selenolate-Capped Metal Clusters

Revealing of the total structure of atomically precise metal nanoclusters can be viewed as a holy grail in nanoscience [4], which has helped us not only to understand the core structure (i.e., the arrangements of metal atoms) and the surface structure (i.e., the arrangements of ligands and the bonding between the ligands and the metal core), but also to further understand their novel physicochemical properties (e.g., electronic, optical, catalytic) at the atomic level by correlation with the precise structures. In past decades, thiolate-capped metal nanoclusters have made a great breakthrough due to their extraordinary robustness [4, 5]. In the previous chapter, the structural analysis of thiolate–metal system was discussed in detail. Compared to S atom, the Se atom possesses longer atomic radius and lower electronegativity [41], which endow selenolate ligands with unique coordination modes with metal atoms. In this chapter, we first provide a brief overview of some early works, in which some selenolate-capped metal nanoclusters were structurally determined. Based on the protecting ligands, they can be divided into two categories: (i) the metal nanocluster protected by full selenolate ligands and (ii) the metal nanocluster co-protected by selenolate and phosphine ligands. Furthermore, the comparison of selenolate-capped metal nanoclusters with their thiolate (or phosphine) counterparts is also summarized for further understanding the effect of coordination modes on the atom-packing mode.

3.3.1 Metal Nanocluster Protected by Full Selenolate Ligands

3.3.1.1 Case of $[Au_{24}(SePh)_{20}]$ nanocluster

The structure of $[Au_{24}(SePh)_{20}]$ exhibits a prolate-shaped construction [45]. As shown in Fig. 3.1, the prolate Au_8 kernel can be viewed as two tetrahedral Au_4 units cross-joined together without sharing any Au atoms. In either unit of the dimeric Au_8 kernel, bond lengths range from 2.701 to 2.823 Å (average 2.735 Å). The Au_4 units are joined together via either pair of Au–Au, which could increase the stability of the Au_8 kernel. It is worth noting that the structures of many metal nanoclusters with small size usually contain the Au_4 tetrahedra, and thus the Au_4 unit can be viewed as a basic building block. In previous theoretical works [67, 68], Au_4 units were predicted in several nanocluster structures. However, Au_8 kernel containing two Au_4 units is rarely reported in selenolate or thiolate-capped metal nanoclusters.

Except for the Au_8 kernel, two Au_5Se_6 staples (Fig. 3.1), which are linked with the middle of the Au_8 kernel, were first observed in the surface protecting structure of $Au_{24}(SeC_6H_5)_{20}$. It is worth noting that such Au_5Se_6 staples were indeed observed experimentally for the first time. Furthermore, two Au_3Se_4 staple motifs are linked with the ends of the Au_8 kernel. In comparison to the Au_3S_4 observed in $Au_{23}(SR)_{16}$ nanocluster [69], the bond length of Au–Se (average 2.427 Å) is longer than that of Au–S (average 2.301 Å), which is due to the larger Se atom than the S atom (covalent radius of Se $r = 1.20$ Å versus S atom $r = 1.05$ Å). Interestingly, the two Au_5Se_6 staple motifs are combined by two pairs of Au–Au bonds, and the Au–Au bonds are 3.091 and 3.077 Å. In addition, the Au_3Se_4 staple motifs are closely linked to the Au_5Se_6 staple through the Au–Au bond, with the average Au–Au bond length being 2.986 Å. Considering this, the framework of $Au_{24}(SeC_6H_5)_{20}$ can be viewed as an interlocked catenane-like construction, which is same as the theoretical structure of $Au_{24}(SCH_3)_{20}$ modeled by Pei et al. [70].

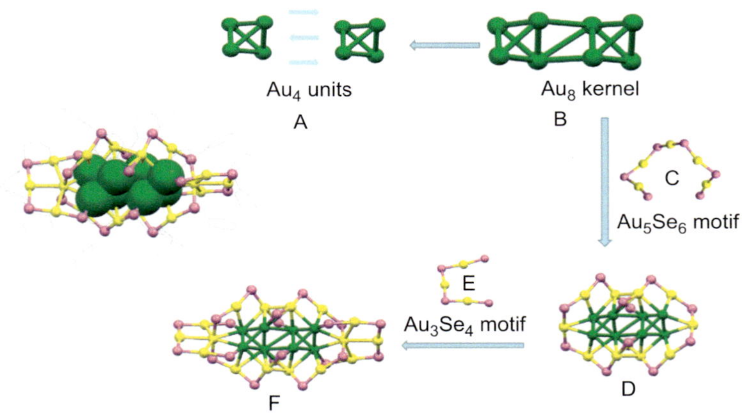

Figure 3.1 The crystal structure of $[Au_{24}(SePh)_{20}]$: (A, B) the prolate Au_8 kernel, (C) Au_5Se_6 staple, and (E) Au_3Se_4 staple (color labels: green/yellow = kernel/surface Au atoms, violet = Se) [45].

Meanwhile, the $[Au_{24}(SCH_2Ph\text{-}^tBu)_{20}]$ nanocluster was structurally determined by Jin and co-workers [71]. It shows a different construction from that of selenolate-capped Au_{24} in both surface staple motifs and Au_8 kernel. Their structures are shown in detail in Fig. 3.2. The $[Au_{24}(SCH_2Ph\text{-}^tBu)_{20}]$ nanocluster also exhibits a Au_8 kernel containing two Au_4 units, but they involve face joining in an antiprismatic manner, which is different from that of $[Au_{24}(SePh)_{20}]$. Furthermore, the Au_8 core is protected by four tetrameric Au_4S_4 staple motifs in $[Au_{24}(SCH_2Ph\text{-}^tBu)_{20}]$ rather than two Au_5S_6 and two Au_3S_4 motifs. The four Au_4S_5 motifs in $[Au_{24}(SCH_2Ph\text{-}^tBu)_{20}]$ are independent and not too correlative. These differences induced by different ligands arouse the interest of some researchers. Based on the precise structures, the electronic structures of $[Au_{24}(SCH_2Ph\text{-}^tBu)_{20}]$ and $[Au_{24}(SePh)_{20}]$ were studied by Jiang and Takagi et al., respectively [60, 72]. Jiang and co-workers think that the higher stability of $[Au_{24}(SCH_2Ph\text{-}^tBu)_{20}]$ is due to its better capacity to respond to the steric effect of the larger $-CH_2Ph\text{-}^tBu$ groups [72]. Takagi thinks that the coordination energy and the Au–Au attractive interaction between the staple motifs and Au_8 core endow the $[Au_{24}(SePh)_{20}]$ with good stability [60]. However, the atom-packing mode can be significantly affected by many factors, and the natural factor resulting in the different construction is still unclear.

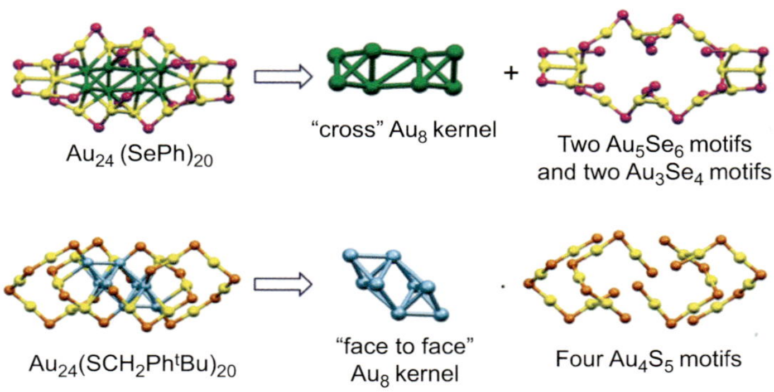

Figure 3.2 The structural comparison of $[Au_{24}(SePh)_{20}]$ and $[Au_{24}(SCH_2Ph^tBu)_{20}]$ nanoclusters (color labels: green/yellow/light blue = Au, violet = Se, orange = S).

3.3.1.2 Case of $[Au_{25}(SePh)_{18}]^-$ nanocluster

X-ray crystallography shows that the $[Au_{25}(SePh)_{18}]^-$ nanocluster exhibits a similar structure to that of $[Au_{25}(SC_2H_4Ph)_{18}]^-$ [46, 73], which contains a centered icosahedral Au_{13} core capped by six $Au_2(SePh)_3$ staple motifs (Fig. 3.3, left). Despite the close resemblance of $[Au_{25}(SePh)_{18}]^-$ to $[Au_{25}(SC_2H_4Ph)_{18}]^-$, there are also some structural differences that are worthy of comments. As shown in Fig. 3.3 (right), compared with Au_2S_3 in the σ_h plane along x–y, the distortion of Se atoms in Au_2Se_3 within the σ_h plane is more obvious, and the same phenomenon is also observed in the x–z plane. However, in the σ_h plane along y–z, the six Se atoms are nearly on the same plane (the sum of angles of the hexagon is 717.24°), which is different from that of $[Au_{25}(SC_2H_4Ph)_{18}]^-$. In addition, the overall diameter of the $Au_{25}Se_{18}$ framework, which was defined as the distance between two outermost Se atoms (12.074 Å, 12.238 Å, and 12.403 Å along the *z*, *x*, and *y* directions, respectively), is larger than the values of $Au_{25}S_{18}$ (11.995 Å, 12.016 Å, and 12.061 Å, respectively). Besides this, the radial Au–Au distance in the Au_{13} icosahedron was also slightly increased upon the replacement of the thiolate ligand by selenolate (i.e., average 2.797 ± 0.01 Å in $[Au_{25}(SePh)_{18}]^-$ versus 2.775 ± 0.01 Å of $[Au_{25}(SC_2H_4Ph)_{18}]^-$). Hence, not only the exterior ligand shell but also the Au_{13} core become more expanded, induced by the ligand change from thiolate to selenolate, which should also affect the distribution of the electron density of the cluster.

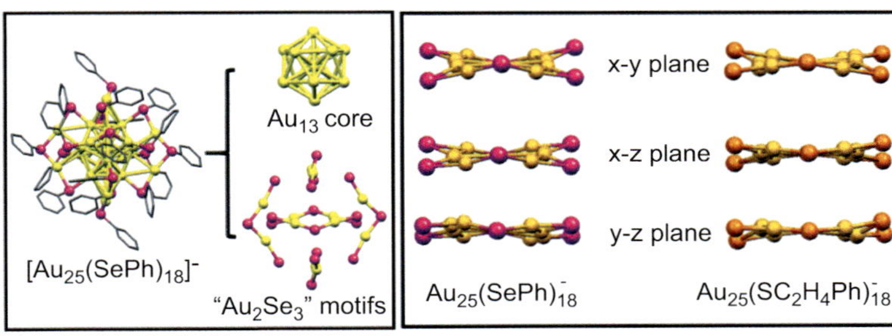

Figure 3.3 The crystal structure of $[Au_{25}(SePh)_{18}]^-$ (left) and the comparison of $Au_2(Se/S)_3$ motiffs between $[Au_{25}(SePh)_{18}]^-$ and $[Au_{25}(SC_2H_4Ph)_{18}]^-$ nanoclusters (color labels: yellow = Au, violet = Se, orange = S) [46].

3.3.1.3 Case of $[Cd_{12}Ag_{32}(SePh)_{36}]$ nanocluster

Recently, a novel $[Cd_{12}Ag_{32}(SePh)_{36}]$ nanocluster was structurally determined [48], which shows an Ag_{28} kernel stabilized by four $Cd_3Ag(SePh)_9$ motifs (Fig. 3.4). As shown in Fig. 3.5A, the structural analysis of Ag_{28} core can start with an Ag_4 tetrahedron at the center with the average Ag–Ag bond distance of 2.84 Å, which is close to that of the bulk Ag [74], indicating the strong Ag–Ag interactions in $Cd_{12}Ag_{32}(SePh)_{36}$ nanoclusters. Furthermore, there are four Ag_6 facets capping the four faces of the Ag_4 tetrahedron (Fig. 3.5B). In each of the Ag_6 facet, there are four triangles with the average Ag–Ag bond distance of 2.958 Å, which is longer than that in Ag_4 tetrahedron.

The four Ag_6 facets interact with each other through Ag–Ag bonding, resulting in an Ag_{24} shell (Fig. 3.5C) that completely encapsulates the Ag_4 tetrahedron, forming a two-shell $Ag_4@Ag_{24}$ kernel (Fig. 3.5D). Notably, the central Ag_3 triangles of Ag_6 facets (placed exactly on top of the triangular faces of the inner Ag_4 tetrahedron) are significantly elongated from the Ag_4 core compared to other Ag–Ag bonds.

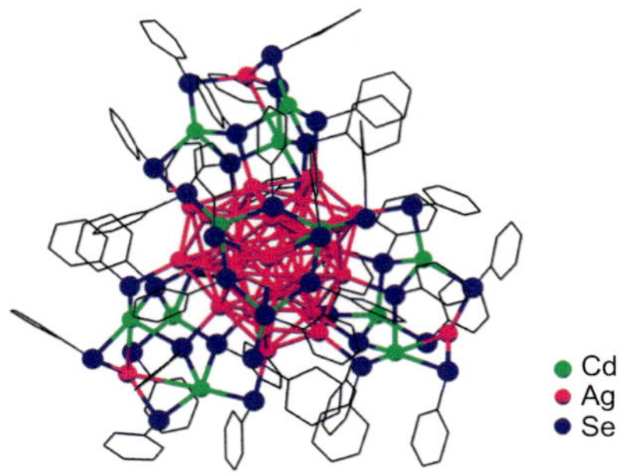

Figure 3.4 The whole structure of $[Cd_{12}Ag_{32}(SePh)_{36}]$ nanocluster (the C atoms are in the mode of wireframe and the H atoms are omitted for clarity). Reprinted with permission from Ref. [48], Copyright 2019, American Chemical Society.

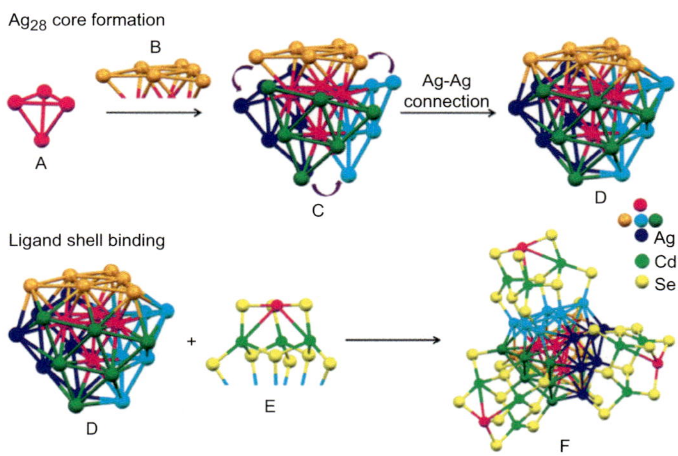

Figure 3.5 Construction of the structure of $[Cd_{12}Ag_{32}(SePh)_{36}]$. (A) Ag_4 inner core and (B) Ag_6 facet. Capping of Ag_4 core with Ag_6 facets and interfacet interactions (purple arrows) result in (C) and (D), respectively. Mounting of $Cd_3Ag(SePh)_9$ motifs (E) on the Ag_{28} core (D) gives the total structure of the cluster (F). The phenyl rings of ligands are omitted for clarity. Reprinted with permission from Ref. [48], Copyright 2019, American Chemical Society.

As for the surface motifs, they contain four $Cd_3Ag(SePh)_9$ motifs (Fig. 3.5E). In each of the $Cd_3Ag(SePh)_9$ motifs, every Cd atom shows a four-coordination mode with four Se atoms, forming a $CdSe_4$ tetrahedron. The Se atoms possess two types of coordination modes: μ_2-Se and μ_3-Se. Based on the construction of $Cd_3Ag(SePh)_9$ motifs, the nine Se atoms can be divided into two layers: (i) three Se atoms at top that bind with one Ag atom to form an $AgSe_3$ cap-like structure and (ii) other six Se atoms at bottom not only coordinate with four Cd atoms, but also attach to the Ag_6 facet through six Se–Ag bonds with the average distance of 2.603 Å. According to the arrangement of the Ag atoms in Ag_6 facet, the six Se atoms form a quasi-triangle. All the four $Cd_3Ag(SePh)_9$ motifs are mounted over four triangular faces of the inner Ag_4 tetrahedron, and therefore it imparts the tetrahedral shape to the final structure (Fig. 3.5F). Owing to the asymmetric Ag_{28} kernel, two $Cd_{12}Ag_{32}(SePh)_{36}$ nanoclusters are observed in a unit cell, which are enantiomers.

3.3.1.4 Case of $[MAg_{20}\{Se_2P(OEt)_2\}_{12}]^+$ nanocluster (M = Au or Ag)

With the ligand-exchange strategy, Liu and co-workers synthesized $[MAg_{20}\{Se_2P(OEt)_2\}_{12}]^+$ (M = Au or Ag) nanoclusters using diselenophosphate as ligands [53]. Their crystal structures were also obtained and they had similar construction. Here, the structure of $[Ag_{21}\{Se_2P(OEt)_2\}_{12}]^+$ nanocluster is discussed in detail. As shown in Fig. 3.6A, $[Ag_{21}\{Se_2P(OEt)_2\}_{12}]^+$ nanocluster exhibits a T symmetry, which was predicted in the previous work. The metal core of $[Ag_{21}\{Se_2P(OEt)_2\}_{12}]^+$ nanocluster can be described as an icosahedral Ag_{13} kernel (Fig. 3.6B) inscribed in a large cube formed by eight capping Ag atoms (Fig. 3.6C), which possesses four threefold rotational axes. In the silver-centered Ag_{13} icosahedron, the radial and peripheral Ag–Ag distances are in the range of 2.7513–2.8296 Å and 2.8514–3.0103 Å, respectively, which are shorter than that of Ag_{ico}–Ag_{cap} (2.8853–3.0586 Å). This construction is totally different from that of $[Ag_{20}\{Se_2P(O^iPr)_2\}_{12}]$ nanocluster, which also contains an icosahedral Ag_{13} kernel capped by seven capping silver atoms, but shows a C_3 symmetry. This demonstrates that adding one silver atom will significantly modulate the arrangement of the capping silver atoms, endowing the total structure with a different symmetry.

Owing to the construction of Se–P–Se in the ligands, the 12 ligands are bridged together through a near-planar $AgSe_3$ coordination mode with the average distance of 2.620(2) Å (Fig. 3.6D). On the basis of the coordination mode, Se atoms can be divided into two categories: R–Se–Ag_2 (Ag_{ico} and Ag_{cap}) and R–Se–Ag_{cap}.

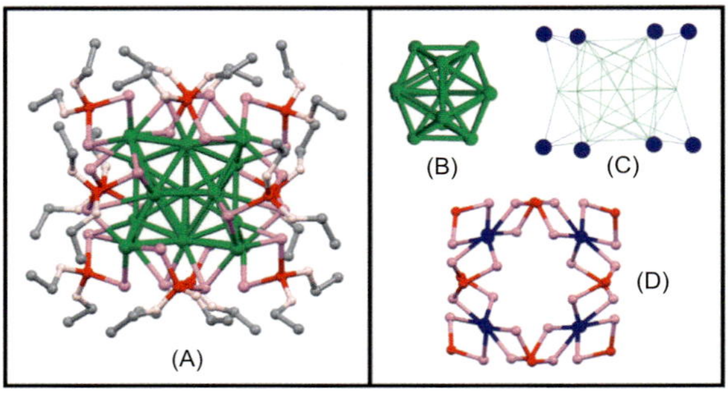

Figure 3.6 (A) The crystal structure of $[Ag_{21}\{Se_2P(OEt)_2\}_{12}]^+$ nanocluster, (B) the icosahedral Ag_{13} kernel, (C) the capping silver atoms, and (D) the $Ag_8(PSe_2)_{12}$ shell (color labels: green/blue = Ag, violet = Se, red = P, gray = C, pink = O; the H atoms are omitted for clarity).

3.3.2 Metal Nanocluster Co-capped by Selenolate and Phosphine

3.3.2.1 Case of $[Au_{11}(L^5)_4(SePh)_2]^+$ nanocluster

With the size-focusing method, $[Au_{11}(L^5)_4(SePh)_2]^+$ nanocluster (L^5 = 1,5-bis(diphenylphosphino)pentane) is synthesized [56] and its total structure is characterized by the X-ray crystallography, which is a undecagon (Fig. 3.7A), similar to $[Au_{11}(PPh_3)_8Cl_2]^+$ (Au_{11}-8) reported by Hutchison et al. [75]. Compared with $[Au_{11}(PPh_3)_8Cl_2]^+$, the eight phosphine ligands and two Cl are replaced by four L^5 ligands and two PhSe ligands in $[Au_{11}(L^5)_4(SePh)_2]^+$ nanocluster, respectively. Different from the monodentate phosphine ligands, the L^5 ligands are in the form of "Au–$PPh_2(CH_2)_5Ph_2P$–Au," which looks like the "staple motif" in the thiolate-capped gold nanoclusters, bridging the two Au atoms together. Due to the confined space of this configuration, some structural differences between $[Au_{11}(L^5)_4(SePh)_2]^+$ and $[Au_{11}(PPh_3)_8Cl_2]^+$ nanoclusters are observed (Fig. 3.7B,C). In $[Au_{11}(L^5)_4(SePh)_2]^+$, the average Au–P distance is 2.31 Å, which is longer than that in the $[Au_{11}(PPh_3)_8Cl_2]^+$ nanocluster (2.29 Å). For comparison, the 11-atom core in these two nanoclusters is divided into three layers: a, b, c (Fig. 3.7B,C). Compared with that of the $[Au_{11}(PPh_3)_8Cl_2]^+$ nanocluster, the Au_{11} core in the $[Au_{11}(L^5)_4(SePh)_2]^+$ nanocluster has great distortion. We propose that this can be explained by the confined space induced by L^5 ligands.

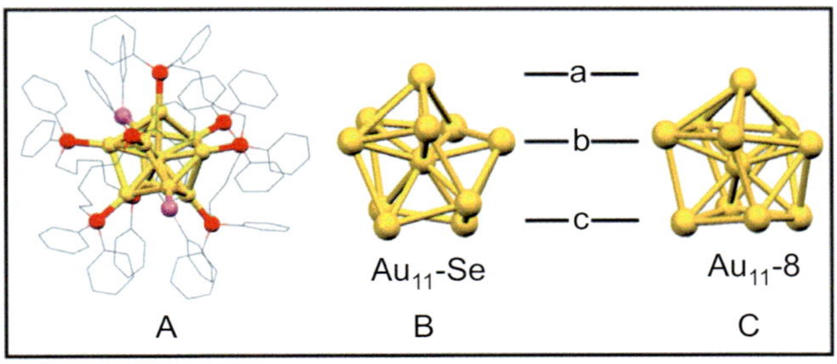

Figure 3.7 (A) The total construction of $[Au_{11}(L^5)_4(SePh)_2]^+$ nanocluster; the gold core of $[Au_{11}(L^5)_4(SePh)_2]^+$ (B) and $[Au_{11}(PPh_3)_8Cl_2]^+$ (C) nanoclusters (color labels: yellow = Au, red = P, violet = Se; the C atoms are in the form of wireframe) [56].

3.3.2.2 Case of rod-like $[Au_{25}(SePh)_5(TPP)_{10}Cl_2]^{+/2+}$ nanoclusters

By changing the ligand from L^5 to triphenylphosphine (TPP), two gold nanoclusters co-protected by PhSeH and TPP are prepared [57]. The X-ray crystallography reveals that these two products possess a similar construction (Fig. 3.8), which contains two icosahedral Au_{13} units sharing one vertex Au atom bridged by five "Au–Se–Au" motifs. Interestingly, the structural analysis demonstrates that these two products possess a different electronic charge, $[Au_{25}(SePh)_5(TPP)_{10}Cl_2]^+$ and $[Au_{25}(SePh)_5(TPP)_{10}Cl_2]^{2+}$, which endows them with different free valence electrons: 16e and 17e. Considering $[Au_{25}(SC_2H_4Ph)_5(TPP)_{10}Cl_2]^{2+}$ [76], these three nanoclusters compose a comparable system, in which the effect of both ligands and the free valence electron on their construction can be discussed in detail.

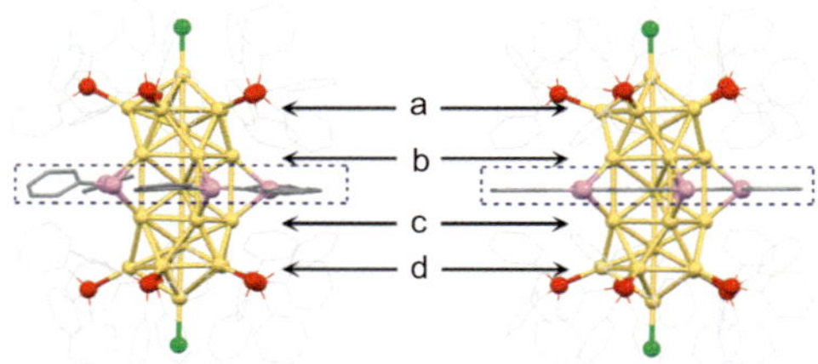

Figure 3.8 Structure of $[Au_{25}(SePh)_5(TPP)_{10}Cl_2]^+$ and $[Au_{25}(SePh)_5(TPP)_{10}Cl_2]^{2+}$ nanoclusters (color labels: Au = yellow, P = red, Cl = green, Se = violet; the C atoms are in wireframe for clarity) [57].

For a better view, the rod-like 25-atom core is divided into four planes: a, b, c, and d (Fig. 3.8). The ligand's effect (selenolate versus thiol) on the structure is analyzed. Except for the difference induced by the longer atomic radius of Se atoms, the distance between planes b and c in $[Au_{25}(SePh)_5(TPP)_{10}Cl_2]^{2+}$ nanocluster (average 3.185 Å) is longer than that in $[Au_{25}(SC_2H_4Ph)_5(TPP)_{10}Cl_2]^{2+}$ nanocluster (average 3.014 Å), indicating that the icosahedral Au_{13} units are stretched due to PhSe ligands. Accordingly, the distance between the apical Au atoms of the $[Au_{25}(SePh)_5(TPP)_{10}Cl_2]^{2+}$ nanocluster (11.221 Å) is longer than that of the $[Au_{25}(SC_2H_4Ph)_5(TPP)_{10}Cl_2]^{2+}$ nanocluster (11.042 Å). These results demonstrate that changing the ligands from thiolate to selenolate will stretch the whole structure. The same phenomenon was also observed in $[Au_{25}(SePh)_{18}]^-$ nanocluster. Next, the influence of the free valence electrons on the atom-packing mode was explored by comparing the structures of $[Au_{25}(SePh)_5(TPP)_{10}Cl_2]^+$ and $[Au_{25}(SePh)_5(TPP)_{10}Cl_2]^{2+}$ nanoclusters. Since they have the same surface protecting ligands, the distance of Au–Se, Au–P, and the "Au–Se–Au" angle in these two nanoclusters is the same to each other. Nonetheless, structural difference induced by free valence electrons still exist. First, $[Au_{25}(SePh)_5(TPP)_{10}Cl_2]^+$ nanocluster exhibits significantly higher symmetry than that of $[Au_{25}(SePh)_5(TPP)_{10}Cl_2]^{2+}$ nanocluster. Second, the Ph of PhSe ligands all lie in the same plane in $[Au_{25}(SePh)_5(TPP)_{10}Cl_2]^+$ nanocluster (Fig. 3.8, highlighted with blue dashed line), but not in $[Au_{25}(SePh)_5(TPP)_{10}Cl_2]^{2+}$ nanocluster.

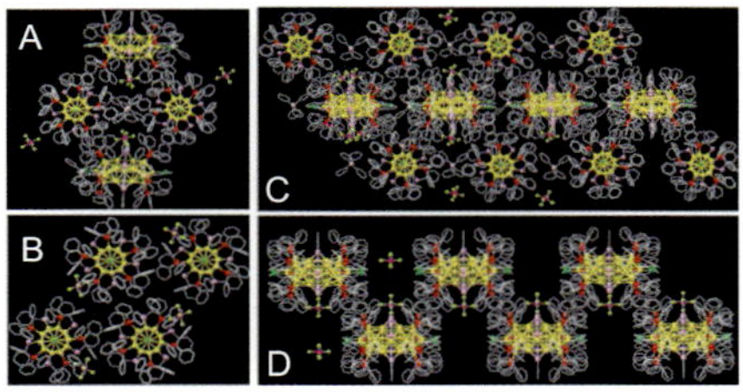

Figure 3.9 Side view of the assemble packing of (A) $[Au_{25}(SePh)_5(TPP)_{10}Cl_2]^{2+}$ and (B) $[Au_{25}(SePh)_5(TPP)_{10}Cl_2]^{+}$ nanoclusters and top view of the assemble packing of (C) $[Au_{25}(SePh)_5(TPP)_{10}Cl_2]^{2+}$ and (D) $[Au_{25}(SePh)_5(TPP)_{10}Cl_2]^{+}$ nanoclusters (color labels: Au = yellow, P = red, Cl = green, Se = violet; the C atoms are in wireframe for clarity) [57].

Furthermore, the two nanoclusters possess different space groups with the same crystal condition. The $[Au_{25}(SePh)_5(TPP)_{10}Cl_2]^{2+}$ nanoclusters have a monoclinic space group P2(1)/c, while the $[Au_{25}(SePh)_5(TPP)_{10}Cl_2]^{+}$ nanoclusters possess an orthorhombic space group Pnma. As shown in Fig. 3.9B (side view), the four $[Au_{25}(SePh)_5(TPP)_{10}Cl_2]^{+}$ nanoclusters are symmetrically arranged. However, in the crystal unit cell of $[Au_{25}(SePh)_5(TPP)_{10}Cl_2]^{2+}$ nanocluster, there are two types of orientations: two molecules located along the vertical flat face direction and the other two molecules located parallel to the plane (Fig. 3.9A, side view). The top view of the assembled packing spread along one direction based on the crystal unit cell in Fig. 3.9C,D better illustrates these rearrangements. Difference between $[Au_{25}(SePh)_5(TPP)_{10}Cl_2]^{+}$ and $[Au_{25}(SePh)_5(TPP)_{10}Cl_2]^{2+}$ nanoclusters might be mainly caused by the presence of the second counterion (BPh_4^-) in the $[Au_{25}(SePh)_5(TPP)_{10}Cl_2]^{2+}$ nanocluster. Without BPh_4^-, the other rod-like $[Au_{25}(SC_2H_4Ph)_5(TPP)_{10}Cl_2]^{2+}$ nanocluster with sole SbF_6^- counterion shows the similar arrangement as that of $[Au_{25}(SePh)_5(TPP)_{10}Cl_2]^{+}$ nanocluster. Nonetheless, some other factors, such as the different steric interactions and different chemical environments of external ligands in these two systems, might also contribute to the arrangement of $[Au_{25}(SePh)_5(TPP)_{10}Cl_2]^{+}$ and $[Au_{25}(SePh)_5(TPP)_{10}Cl_2]^{2+}$ nanoclusters. In one words, the different arrangements induced by different charge states will offer a new insight into how to assemble nanoclusters into macroscopic materials.

3.3.2.3 Case of $[Au_{60}Se_2(SePh)_{15}(TPP)_{10}]^+$ nanocluster

With the similar method as that of $[Au_{25}(SePh)_5(TPP)_{10}Cl_2]^{+/2+}$, a $[Au_{60}Se_2(SePh)_{15}(TPP)_{10}]^+$ nanocluster is synthesized by altering the reaction condition [58]. As shown in Fig. 3.10, the $[Au_{60}Se_2(SePh)_{15}(TPP)_{10}]^+$ nanocluster shows a ring-like construction with the outside diameter of 14.00 Å and the inside diameter of 4.20 Å, in which each of the adjacent two Au_{13} units shares a vertex gold atom in a cyclic fashion ($13 \times 5 - 5 = 60$).

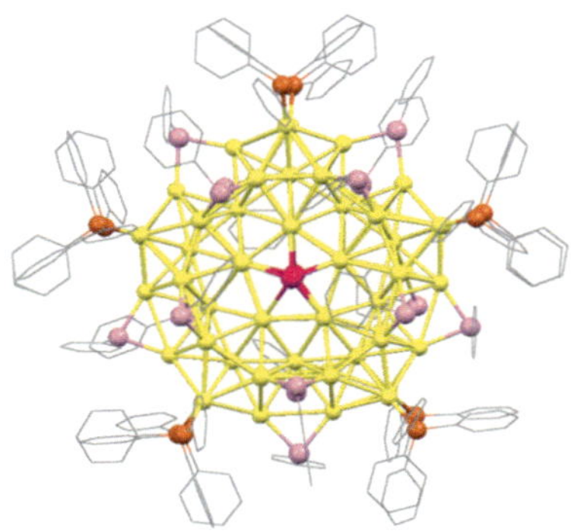

Figure 3.10 The crystal structure of $[Au_{60}Se_2(SePh)_{15}(TPP)_{10}]^+$ nanocluster (color labels: Au = yellow, P = red, Se = violet/magenta; the C atoms are in wireframe for clarity) [58].

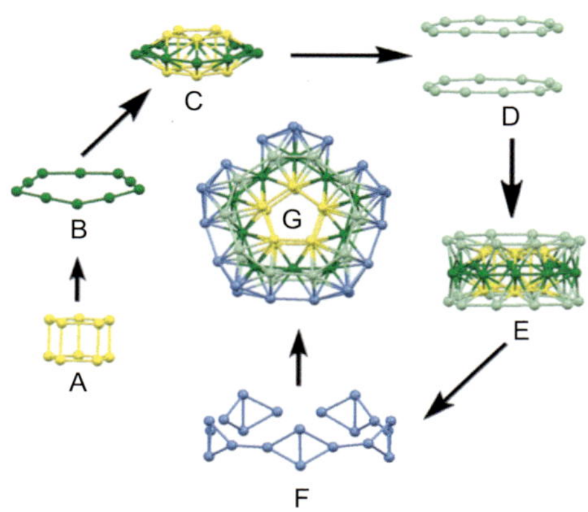

Figure 3.11 The anatomy of the Au_{60} core in $[Au_{60}Se_2(SePh)_{15}(TPP)_{10}]^+$ nanocluster [58].

Due to the ring-like structure, $[Au_{60}Se_2(SePh)_{15}(TPP)_{10}]^+$ shows four-layer construction from inside to outside (Fig. 3.11 A,B,D,F). As shown in Fig. 3.11A, the innermost layer looks like a pentagonal prism: both the top and the bottom are pentagons with the average Au–Au distance of 2.741–2.759. In addition, the four Au atoms in each side face are almost in the same plane (the angle sum of quadrangle: 359.928). Interestingly, layer II (Fig. 3.11B) is a closed decagon constituted by 10 Au(0) atoms—the Au atoms are all connected to the neighboring Au atoms, and this observation has been noted in other gold NCs. The average Au–Au bond length is 2.803 Å. Layers I and II together give rise to an incomplete bipyramid that looks like a flying saucer (Fig. 3.11C). Layer III contains two circular rings, and each of them consists of 10 Au atoms (Fig. 3.11D), and the average Au–Au bond length is 3.011 Å. As shown in Fig. 3.11E, the two circular rings are just like a cylindrical barrel and hold the "flying saucer" inside it. Finally, the remaining 20 Au atoms can be seen as five rhombuses connected by Au–Au bonds, which looks like a crown (Fig. 3.11F). In each rhombus, the average Au–Au bond length is 2.974 Å and the Au_{60} core was constructed (Fig. 3.11G).

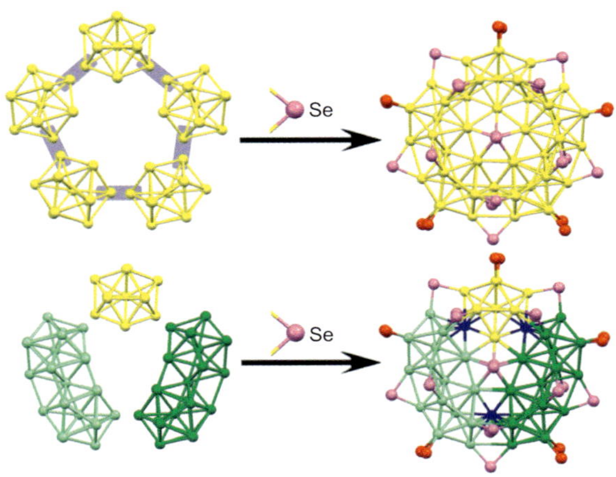

Figure 3.12 Two possible ways of the self-assembly of $[Au_{60}Se_2(SePh)_{15}(TPP)_{10}]^+$ nanocluster [58].

Alternatively, $[Au_{60}Se_2(SePh)_{15}(TPP)_{10}]^+$ can also be viewed as a cluster-assembled material and can be divided into five icosahedral Au_{13} building blocks clipped together by five Au–Se–Au linkages

(Fig. 3.12, above). In each Au_{13} unit, the Au–Au distances are in the range of 2.60–3.12 Å. The bond distances are similar to those in $[Au_{13}(PMe_2Ph)_{10}Cl_2]^{3+}$ and $[Au_{25}(PPh_3)_{10}(SR)_5X_2]^{2+}$ NCs [76, 77]. However, compared with the rod-like Au_{25}, the Au–Au bond distances between two adjacent Au_{13} units are significantly different and can be generally classified into two categories: three long bonds (3.02–3.28 Å) and two short bonds (2.73–2.76 Å). The Au-P bond lengths in $[Au_{60}Se_2(SePh)_{15}(TPP)_{10}]^+$ are in the range of 2.24–2.32 Å, which are typical Au–P bond distances. The Au–Se bonds can also be divided into two categories: (i) the normal format, that is, RSe–Au_2 (bond length 2.46–2.50 Å) and (ii) the bare Se atoms without the phenyl tail: Se–Au_5 with the Au–Se bond length 2.65 Å (significantly longer than that of Se–M_3). The two sets of Se–Au_5 seem like two open hands, tightly connecting the five Au_{13} building blocks through the radial Au–Se bonds. This bonding mode (Se–Au_5) is the first example in the selenolate-capped gold NCs. From the other side, $[Au_{60}Se_2(SePh)_{15}(TPP)_{10}]^+$ also can be divided into two rod-like Au_{25} units and an Au_{13} unit (Fig. 3.12, below).

3.3.2.4 Case of $[Au_{13}Cu_4(PPyPh_2)_3(SePh)_9]$ nanocluster

With the similar method, an Au–Cu nanocluster was synthesized, which was determined to be $[Au_{13}Cu_4(PPyPh_2)_3(SePh)_9]$ [59]. X-ray diffractometer reveals that $[Au_{13}Cu_4(PPyPh_2)_3(SePh)_9]$ possesses an icosahedral Au_{13} core with four capping Cu atoms (Fig. 3.13A), which is similar to that of $[Au_{13}Cu_4(PPh_2Py)_4(SPh^tBu)_8]^+$ [78]. Compared with $[Au_{13}Cu_4(PPh_2Py)_4(SPh^tBu)_8]^+$ nanocluster, not only the eight RS ligands are now replaced by PhSe ligands, but also one PPh_2Py ligand is replaced by one PhSe ligand in the new $[Au_{13}Cu_4(PPyPh_2)_3(SePh)_9]$ nanocluster, and hence one "N–Cu–S_2" motif is changed to a new "Cu–Se_3" motif. The "N–Cu–Se_2" motif can be clearly seen in Fig. 3.13A (the Cu atoms are marked in blue). The length of Cu–Se in "N–Cu–Se_2" motifs is longer (average 2.41 Å) than that of Cu–S in the "N–Cu–S_2" motifs (average 2.28 Å). Furthermore, the average Cu-Se length is 2.39 Å in the Cu–Se_3 motif, which is also longer than that of Cu–S but shorter than that of Cu–Se in the construction of N–Cu–Se_2 motifs.

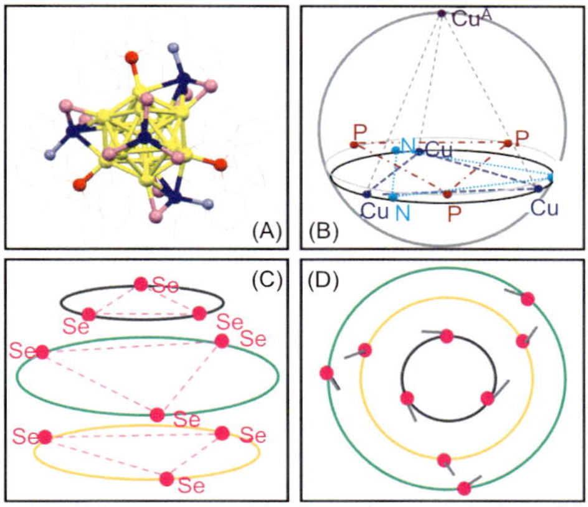

Figure 3.13 Structural analysis of the $[Au_{13}Cu_4(PPyPh_2)_3(SePh)_9]$ nanocluster: the whole structure (A), the cartoon showing the arrangement of the Cu, N, and P atoms (B), the Se atoms (C), and the Ph in PhSe-ligands (D). Color labels: yellow = Au, blue = Cu, light blue = N, violet = Se, red = P, gray = C; the H atoms are all omitted for clarity [59].

Compared to the $[Au_{13}Cu_4(PPh_2Py)_4(SPh^tBu)_8]^+$ counterpart, the single-ligand exchange (phosphine to selenolate) leads to high symmetry in the $[Au_{13}Cu_4(PPyPh_2)_3(SePh)_9]$ nanocluster; see schematic diagrams for the arrangements in the $[Au_{13}Cu_4(PPyPh_2)_3(SePh)_9]$ nanocluster (Fig. 3.13B,C). As one can see in Fig. 3.13B, Cu, N, and P atoms form three equilateral triangles that are parallel to each other. For comparison, the Se atoms in $[Au_{13}Cu_4(PPyPh_2)_3(SePh)_9]$ can also be divided into three groups (Fig. 3.13C), and the Se atoms in each group can also form an equilateral triangle. In addition to these, the Ph– in the selenolate ligand also abides by the rule: as shown in Fig. 3.13D (the gray bar represents the Ph), all the Ph are arranged in a counterclockwise direction, which is consistent with the direction of N–Cu bonds. All these results demonstrate that the single-ligand exchange from $PPhPy_2$ to PhSe has induced the high symmetry.

3.4 Summary

From the above discussion, we can conclude that changing the ligands from thiolates to selenolate can significantly modulate the atom-packing mode of metal nanoclusters, and novel metal–ligand coordination has been observed. It is still unclear how surface protecting ligands control the whole structure of metal nanoclusters due to the insufficiency of the precise structure of selenolate-capped metal nanoclusters. Some interesting questions need to be further explored: What size of the metal nanoclusters can be obtained using selenolate as ligands? How about the construction of surface protecting shell in the selenolate-capped metal nanocluster with large size? Considering these, the structural characterization of metal nanoclusters protected by selenolate is still a great challenge.

The biggest challenge in this field is that the types of selenolate ligands are too less, which seriously hinders the synthesis of selenolate-capped metal nanocluster. On the basis of thiolate-capped metal nanoclusters, the natural features of ligands (e.g., electronegativity, the size, or bulkiness) can efficiently control the construction of metal nanoclusters. Thus, preparing some novel selenolate ligands is beneficial to obtaining more metal nanoclusters with different structures. Future work will shed more light on this and other major scientific questions of selenolate-capped metal nanoclusters.

Acknowledgments

Y. S. acknowledges the financial support by the NSFC (Grant 21801001). M. Z. acknowledges the financial support by the NSFC (Grants 21871001 and 21631001), the Ministry of Education, and the Education Department of Anhui Province.

References

1. Qian, H., Zhu, M., Wu, Z., Jin, R. (2012). Quantum sized gold nanoclusters with atomic precision, *Acc. Chem. Res.*, **45**, pp. 1470–1479.
2. Schmid, G. (2008). The relevance of shape and size of Au_{55} clusters, *Chem. Soc. Rev.*, **37**, pp. 1909–1930.
3. Liu, L., Corma, A. (2018). Metal catalysts for heterogeneous catalysis: From single atoms to nanoclusters and nanoparticles, *Chem. Rev.*, **118**, pp. 4981–5079.
4. Jin, R., Zeng, C., Zhou, M., Chen, Y. (2016). Atomically precise colloidal metal nanoclusters and nanoparticles: Fundamentals and opportunities, *Chem. Rev.*, **116**, pp. 10346.

5. Chakraborty, I., Pradeep, T. (2017). Atomically precise clusters of noble metals: Emerging link between atoms and nanoparticles, *Chem. Rev.*, **117**, pp. 8208.
6. Jin, R. (2010). Quantum sized, thiolate-protected gold nanoclusters, *Nanoscale*, **2**, pp. 343–362.
7. Shang, L., Nienhaus, G. U. (2017). In situ characterization of protein adsorption onto nanoparticles by fluorescence correlation spectroscopy, *Acc. Chem. Res.*, **50**, pp. 387–395.
8. Kang, X., Zhu, M. (2019). Tailoring the photoluminescence of atomically precise nanoclusters, *Chem. Soc. Rev.*, **48**, pp. 2422–2457.
9. Antonello, S., Perera, N. V., Ruzzi, M., Gascón, J. A., Maran, F. (2013). Interplay of charge state, lability, and magnetism in the molecule-like $Au_{25}(SR)_{18}$ cluster, *J. Am. Chem. Soc.*, **135**, pp. 15585–15594.
10. Zhao, J., Jin, R. (2018). Heterogeneous catalysis by gold and gold-based bimetal nanoclusters, *Nano Today*, **18**, pp. 86–102.
11. Du, Y., Sheng, H., Astruc, D., Zhu, M. (2020). Atomically precise noble metal nanoclusters as efficient catalysts: A bridge between structure and properties, *Chem. Rev.*, 120, pp. 526–622.
12. Yamazoe, S., Koyasu, K., Tsukuda, T. (2014). Nonscalable oxidation catalysis of gold clusters, *Acc. Chem. Res.*, **47**, pp. 816–824.
13. Sakthivel, N. A., Dass, A. (2018). Aromatic thiolate-protected series of gold nanomolecules and a contrary structural trend in size evolution. *Acc. Chem. Res.*, **51**, pp. 1774.
14. Yan, J., Teo, B. K., Zheng, N. (2018). Surface chemistry of atomically precise coinage–metal nanoclusters: From structural control to surface reactivity and catalysis, *Acc. Chem. Res.*, **51**, pp. 3084–3093.
15. Wang, S., Li, Q., Kang, X., Zhu, M. (2018). Customizing the structure, composition, and properties of alloy nanoclusters by metal exchange, *Acc. Chem. Res.*, **51**, pp. 2784–2792.
16. Pei, Y., Wang, P., Ma, Z., Xiong, L. (2019). Growth-rule-guided structural exploration of thiolate-protected gold nanoclusters, *Acc. Chem. Res.*, **52**, pp. 23–33.
17. Zhuang, S., Liao, L., Yuan, J., Wang, C., Zhao, Y., Xia, N., Gan, Z., Gu, W., Li, J., Deng, H., Yang, J., Wu, Z. (2018). Kernel homology in gold nanoclusters, *Angew. Chem. Int. Ed.*, **47**, pp. 15676–15680.
18. Hossain, S., Niihori, Y., Nair, L. V., Kumar, B., Kurashige, W., Negishi, Y. (2018). Alloy clusters: Precise synthesis and mixing effects, *Acc. Chem. Res.*, **51**, pp. 3114–3124.

19. Ghosh, A., Mohammed, O. F., Bakr, O. M. (2018). Atomic-level doping of metal clusters, *Acc. Chem. Res.*, **51**, pp. 3094–3103.

20. Liu, Y., Chai, X., Cai, X., Chen, M., Jin, R., Ding, W., Zhu, Y. (2018). Central doping of a foreign atom into the silver cluster for catalytic conversion of CO_2 toward C–C bond formation, *Angew. Chem. Int. Ed.*, **57**, pp. 9775–9779.

21. Wang, S., Meng, X., Das, A., Li, T., Song, Y., Cao, T., Zhu, X., Zhu, M., Jin, R. (2014). A 200-fold quantum yield boost in the photoluminescence of silver-doped Ag_xAu_{25-x} nanoclusters: The 13th silver atom matters, *Angew. Chem. Int. Ed.*, **53**, pp. 2376–2380.

22. Jung, J., Kang, S., Han, Y. (2012). Ligand effects on the stability of thiol-stabilized gold nanoclusters: $Au_{25}(SR)_{18}^-$, $Au_{38}(SR)_{24}$, and $Au_{102}(SR)_{44}$, *Nanoscale*, **4**, pp. 4206–4210.

23. Chevrier, D. M., Raich, L., Rovira, C., Das, A., Luo, Z., Yao, Q., Chatt, A., Xie, J., Jin, R., Akola, J., Zhang, P. (2018). Molecular-scale ligand effects in small gold–thiolate nanoclusters, *J. Am. Chem. Soc.*, **140**, pp. 15430–15436.

24. Tang, Q., Hu, G., Fung, V., Jiang, D. (2018). Insights into interfaces, stability, electronic properties, and catalytic activities of atomically precise metal nanoclusters from first principles, *Acc. Chem. Res.*, **51**, pp. 2793–2802.

25. Fang, J., Zhang, B., Yao, Q., Yang, Y., Xie, J., Yan, N. (2016). Recent advances in the synthesis and catalytic applications of ligand-protected, atomically precise metal nanoclusters, *Coord. Chem. Rev.*, **322**, pp. 1–29.

26. Zeng, C., Chen, Y., Das, A., Jin, R. (2015). Transformation chemistry of gold nanoclusters: From one stable size to another, *J. Phys. Chem. Lett.*, **6**, pp. 2976–2986.

27. Ren, L., Yuan, P., Su, H., Malola, S., Lin, S., Tang, Z., Teo, B. K., Häkkinen, H., Zheng, L., Zheng, N. (2017). Bulky surface ligands promote surface reactivities of $[Ag_{141}X_{12}(S\text{-}Adm)_{40}]^{3+}$ (X = Cl, Br, I) nanoclusters: Models for multiple-twinned nanoparticles, *J. Am. Chem. Soc.*, **139**, pp. 13288–13291.

28. Nasaruddin, R. R., Chen, T., Yan, N., Xie, J. (2018). Roles of thiolate ligands in the synthesis, properties and catalytic application of gold nanoclusters, *Coord. Chem. Rev.*, **368**, pp. 60–79.

29. Konishi, K., Iwasaki, M., Sugiuchi, M., Shichibu, Y. (2016). Ligand-based toolboxes for tuning of the optical properties of subnanometer gold clusters, *J. Phys. Chem. Lett.*, **7**, pp. 4267–4274.

30. Higaki, T., Liu, C., Zhou, M., Luo, T.-Y., Rosi, N. L., Jin, R. (2017). Tailoring the structure of 58-electron gold nanoclusters: $Au_{103}S_2(S\text{-}Nap)_{41}$ and its implications, *J. Am. Chem. Soc.*, **139**, pp. 9994–10001.

31. Wan, X., Wang, J., Nan, Z., Wang, Q. (2017). Ligand effects in catalysis by atomically precise gold nanoclusters, *Sci. Adv.*, **3**, e1701823.

32. Lei, Z., Wan, X., Yuan, S., Guan, Z., Wang, Q. (2018). Alkynyl approach toward the protection of metal nanoclusters, *Acc. Chem. Res.*, **51**, pp. 2465.

33. Chen, Y., Zeng, C., Kauffman, D. R., Jin, R. (2015). Tuning the magic size of atomically precise gold nanoclusters via isomeric methylbenzenethiols, *Nano Lett.*, **15**, pp. 3603–3609.

34. Zhang, Q., Chen, X., Wang, L. (2018). Toward solution syntheses of the tetrahedral Au_{20} pyramid and atomically precise gold nanoclusters with uncoordinated sites, *Acc. Chem. Res.*, **51**, pp. 2159–2168.

35. Konishi, K., Iwasaki, M., Shichibu, Y. (2018). Phosphine-ligated gold clusters with core+exo geometries: Unique properties and interactions at the ligand–cluster interface, *Acc. Chem. Res.*, **51**, pp. 3125–3133.

36. Zaluzhna, O., Li, Y., Allison, T. C., Tong, Y. J. (2012). Inverse-micelle-encapsulated water-enabled bond breaking of dialkyl diselenide/disulfide: A critical step for synthesizing high-quality gold nanoparticles, *J. Am. Chem. Soc.*, **134**, pp. 17991–17996.

37. Hohman, J. N., Thomas, J. C., Zhao, Y., Auluck, H., Kim, M., Vijselaar, W., Kommeren, S., Terfort, A., Weiss, P. S. (2014). Exchange reactions between alkanethiolates and alkaneselenols on Au{111}, *J. Am. Chem. Soc.*, **136**, pp. 8110–8121.

38. Brust, M., Stuhr-Hansen, N., Nørgaard, K., Christensen, J. B. Nielsen, L. K., Bjørnholm, T. (2001). Langmuir–Blodgett films of alkane chalcogenide (S, Se, Te) stabilized gold nanoparticles, *Nano Lett.*, **1**, pp. 189.

39. Zelakiewicz, B. S., Yonezawa, T., Tong, Y. (2004). Observation of selenium-77 nuclear magnetic resonance in octaneselenol-protected gold nanoparticles, *J. Am. Chem. Soc.*, **126**, pp. 8112–8113.

40. Li, Y., Zaluzhna, O., Xu, B., Gao, Y., Modest, J. M., Tong, Y. J. (2011). Mechanistic insights into the Brust–Schiffrin two-phase synthesis of organo-chalcogenate-protected metal nanoparticles, *J. Am. Chem. Soc.*, **133**, 2092–2095.

41. Yee, C. K., Ulman, A., Ruiz, J. D., Parikh, A., White, H., Rafailovich, M. (2003). Alkyl selenide- and alkyl thiolate-functionalized gold nanoparticles: Chain packing and bond nature, *Langmuir*, **19**, pp. 9450–9458.

42. Muñoz-Castro, A. (2019). On the ligand–core interaction in ligand-protected gold superatoms. Insights from $Au_{25}(XR)_{18}$ (X = S, Se, Te) *via* relativistic DFT calculations. *Phys. Chem. Chem. Phys.*, **21**, pp. 13022–13029.

43. Negishi, Y., Kurashige, W., Kamimura, U. (2011). Isolation and structural characterization of an octaneselenolate-protected Au_{25} cluster, *Langmuir*, **27**, pp. 12289–12292.

44. Song, Y., Cao, T., Deng, H., Zhu, X., Li, P., Zhu, M. (2014). Kinetically controlled, high-yield, direct synthesis of $[Au_{25}(SePh)_{18}]^{-}TOA^{+}$, *Sci. China Chem.*, **57**, pp. 1218–1224.

45. Song, Y., Wang, S., Zhang, J., Kang, X., Chen, S., Li, P., Sheng, H., Zhu, M. (2014). Crystal structure of selenolate-protected $Au_{24}(SeR)_{20}$ nanocluster, *J. Am. Chem. Soc.*, **136**, pp. 2963–2965.

46. Song, Y., Zhong, J., Yang, S., Wang, S., Cao, T., Zhang, J., Li, P., Hu, D., Pei, Y., Zhu, M. (2014). Crystal structure of $Au_{25}(SePh)_{18}$ nanoclusters and insights into their electronic, optical and catalytic properties, *Nanoscale*, **6**, pp. 13977–13985.

47. Chakraborty, I., Kurashige, W., Kanehira, K., Gell, L., Häkkinen, H., Negishi, Y., Pradeep, T. (2013). $Ag_{44}(SeR)_{30}$: A hollow cage silver cluster with selenolate protection, *J. Phys. Chem. Lett.*, **4**, pp. 3351–3355.

48. Bootharaju, M. S., Chang, H., Deng, G., Malola, S., Baek, W., Häkkinen, H., Zheng, N., Hyeon, T. (2019). $Cd_{12}Ag_{32}(SePh)_{36}$: Non-noble metal doped silver nanoclusters, *J. Am. Chem. Soc.*, **141**, pp. 8422–8425.

49. Kurashige, W., Munakata, K., Nobusada, K., Negishi, Y. (2013). Synthesis of stable Cu_nAu_{25-n} nanoclusters (n = 1–9) using selenolate ligands, *Chem. Commun.*, **49**, pp. 5447–5449.

50. Kurashige, W., Yamaguchi, M., Nobusada, K., Negishi, Y. (2012). Ligand-induced stability of gold nanoclusters: Thiolate versus selenolate, *J. Phys. Chem. Lett.*, **3**, pp. 2649–2652.

51. Meng, X., Xu, Q., Wang, S., Zhu, M. (2012). Ligand-exchange synthesis of selenophenolate-capped Au_{25} nanoclusters, *Nanoscale*, **4**, pp. 4161–4165.

52. Xu, Q., Wang, S., Liu, Z., Xu, G., Meng, X., Zhu, M. (2013). Synthesis of selenolate-protected $Au_{18}(SeC_6H_5)_{14}$ nanoclusters, *Nanoscale*, **5**, pp. 1176–1182.

53. Chang, W., Lee, P., Liao, J., Chakrahari, K., Kahlal, S., Liu, Y., Chiang, M., Saillard, J., Liu, C. (2017). Eight-electron silver and mixed gold/silver nanoclusters stabilized by selenium donor ligands, *Angew. Chem. Int. Ed.*, **56**, pp. 10178–10182.

54. Rambukwella, M., Chang, L., Ravishanker, A., Fortunelli, A., Stener, M., Dass, A. (2018). $Au_{36}(SePh)_{24}$ nanomolecules: Synthesis, optical spectroscopy and theoretical analysis, *Phys. Chem. Chem. Phys.*, **20**, pp. 13255–13262.

55. Kurashige, W., Yamazoe, S., Kanehira, K., Tsukuda, T., Negishi, Y. (2013). Selenolate-protected Au_{38} nanoclusters: Isolation and structural characterization, *J. Phys. Chem. Lett.*, **4**, pp. 3181–3185.

56. Kang, X., Song, Y., Deng, H., Zhang, J., Liu, B., Pan, C., Zhu, M. (2015). Ligand-induced change of the crystal structure and enhanced stability of the Au_{11} nanocluster. *RSC Adv.*, **5**, pp. 66879–66885.

57. Song, Y., Jin, S., Kang, X., Xiang, J., Deng, H., Yu, H., Zhu, M. (2016). How a single electron affects the properties of the "non-superatom" Au_{25} nanoclusters, *Chem. Mater.*, **28**, pp. 2609–2617.

58. Song, Y., Fu, F., Zhang, J., Chai, J., Kang, X., Li, P., Li, S., Zhou, H., Zhu, M. (2015). The magic Au_{60} nanocluster: A new cluster-assembled material with five Au_{13} building blocks, *Angew. Chem. Int. Ed.*, **54**, pp. 8430–8434.

59. Song, Y., Lv, Y., Zhou, M., Luo, T.-Y., Zhao, S., Rosi, N. L., Yu, H., Zhu, M., Jin, R. (2018). Single-ligand exchange on an Au–Cu bimetal nanocluster and mechanism, *Nanoscale*, **10**, pp. 12093–12099.

60. Takagi, N., Ishimura, K., Matsui, M., Fukuda, R., Matsui, T., Nakajima, T., Ehara, M., Sakaki, S. (2015). How can we understand Au_8 cores and entangled ligands of selenolate- and thiolate-protected gold nanoclusters $Au_{24}(ER)_{20}$ and $Au_{20}(ER)_{16}$ (E = Se, S, R = Ph, Me)? A theoretical study, *J. Am. Chem. Soc.*, **137**, pp. 8593–8602.

61. Zhong, J., Tang, X., Tang, J., Su, J., Pei, Y. (2015). Density functional theory studies on structure, ligand exchange, and optical properties of ligand-protected gold nanoclusters: Thiolate versus selenolate, *J. Phys. Chem. Lett.*, **119**, pp. 9205–9214.

62. Song, Y., Abroshan, H., Chai, J., Kang, X., Kim, H. J., Zhu, M., Jin, R. (2017). Molecular-like transformation from PhSe-protected Au_{25} to Au_{23} nanocluster and its application, *Chem. Mater.*, **29**, pp. 3055–3061.

63. Brust M., Walker M., Bethell D., Schiffrin D. J., Whyman R. (1994). Synthesis of thiol-derivatised gold nanoparticles in a two-phase liquid-liquid system, *J Chem. Soc. Chem. Commun.*, **7**, pp. 801–802.

64. Wu Z., MacDonald M., Chen J., Zhang P., Jin R. (2011). Kinetic control and thermodynamic selection in the synthesis of atomically precise gold nanoclusters, *J. Am. Chem. Soc.*, **133**, pp. 9670–9673.

65. Qian H., Zhu Y., Jin R. (2009). Size-focusing synthesis, optical and electrochemical properties of monodisperse $Au_{38}(SC_2H_4Ph)_{24}$ nanoclusters, *ACS Nano*, **3**, pp. 3795–3803.

66. Jin R., Qian H., Wu Z., Zhu Y., Zhu M., Mohanty A., Garg N. (2010). Size focusing: A methodology for synthesizing atomically precise gold nanoclusters, *J. Phys. Chem. Lett.*, **1**, pp.2903–2910.

67. Zeng, C., Liu, C., Chen, Y., Rosi, N. L., Jin, R. (2014). Gold–thiolate ring as a protecting motif in the $Au_{20}(SR)_{16}$ nanocluster and implications, *J. Am. Chem. Soc.*, **136**, pp. 11922–11925.

68. Pei, Y., Gao, Y., Shao, N., Zeng, X. C. (2009). Thiolate-protected $Au_{20}(SR)_{16}$ cluster: Prolate Au_8 core with new $[Au_3(SR)_4]$ staple motif, *J. Am. Chem. Soc.*, **131**, pp. 13619–13621.

69. Das, A., Li, T., Nobusada, K., Zeng, C., Rosi, N. L., Jin, R. (2013). Nonsuperatomic $[Au_{23}(SC_6H11)_{16}]^-$ nanocluster featuring bipyramidal Au_{15} kernel and trimeric $Au_3(SR)_4$ motif, *J. Am. Chem. Soc.*, **135**, pp. 18264–18267.

70. Pei, Y., Pal, R., Liu, C., Gao, Y., Zhang, Z., Zeng, X. C. (2012). Interlocked catenane-like structure predicted in $Au_{24}(SR)_{20}$: Implication to structural evolution of thiolated gold clusters from homoleptic gold(I) thiolates to core-stacked nanoparticles, *J. Am. Chem. Soc.*, **134**, pp. 3015–3024.

71. Das, A., Li, T., Li, G., Nobusada, K., Zeng, C., Rosi, N. L., Jin, R. (2014). Crystal structure and electronic properties of a thiolate-protected Au_{24} nanocluster, *Nanoscale*, **6**, pp. 6458–6462.

72. Tang, Q., Ouyang, R., Tian, Z., and Jiang, D.-e. (2015). The ligand effect on the isomer stability of $Au_{24}(SR)_{20}$ clusters, *Nanoscale*, **7**, pp. 2225–2229.

73. Zhu, M., Aikens, C. M., Hollander, F. J., Schatz, G. C., Jin, R. (2008). Correlating the crystal structure of a thiol-protected Au_{25} cluster and optical properties, *J. Am. Chem. Soc.*, **130**, pp. 5883–5885.

74. Jin, R., Nobusada, K. (2014). Doping and alloying in atomically precise gold nanoparticles, *Nano Res.*, **7**, pp. 285–300.

75. McKenzie, L. C., Zaikova, T. O., Hutchison, J. E. (2014). Structurally similar triphenylphosphine-stabilized undecagolds, $Au_{11}(PPh_3)_7Cl_3$ and $[Au_{11}(PPh_3)_8C_{12}]Cl$, exhibit distinct ligand exchange pathways with glutathione, *J. Am. Chem. Soc.*, **136**, pp. 13426–13435.

76. Shichibu, Y., Negishi, Y., Watanabe, T., Chaki, N. K., Kawaguchi, H., and Tsukuda, T. (2007). Biicosahedral gold clusters $[Au_{25}(PPh_3)_{10}(SC_nH_{2n+1})_5Cl_2]^{2+}$ (n = 2–18): A stepping stone to cluster-assembled materials, *J. Phys. Chem. C*, **111**, pp. 7845–7847.

77. Briant, C. E., Theobald, B. R. C., White, J. W., Bell, L. K., Mingos, D. M. P., Welch, A. J. (1981). Synthesis and X-ray structural characterization of the centred icosahedral gold cluster compound $[Au_{13}(PMe_2Ph)_{10}Cl_2](PF_6)_3$, the realization of a theoretical prediction, *J. Chem. Soc. Chem. Commun.*, pp. 201–202.

78. Yang, H., Wang, Y., Lei, J., Shi, L., Wu, X., Mäkinen, V., Lin, S., Tang, Z., He, J., Häkkinen, H., Zheng, L., Zheng, N. (2013). Ligand-stabilized $Au_{13}Cu_x$ ($x = 2, 4, 8$) bimetallic nanoclusters: Ligand engineering to control the exposure of metal sites, *J. Am. Chem. Soc.*, **135**, pp. 9568–9571.

Chapter 4

Strategy for Structural Prediction of Thiolate-Protected Au Nanoclusters Based on Density Functional Theory

Chunyan Liu,[a] Limu Hu,[b] and Jing Ma[b]

[a]*College of Chemistry and Chemical Engineering, Hunan University, Changsha 410082, Hunan, P. R. China*

[b]*Key Laboratory of Mesoscopic Chemistry of Ministry of Education, School of Chemistry and Chemical Engineering, Nanjing University, Nanjing 210023, P. R. China*

majing@nju.edu.cn

With its excellent stability, biocompatibility, and tunability, the thiolate-protected Au nanocluster exhibits potential applications in biosensor, drug delivery, and antimicrobial activity. Determination of the structure of cluster with different sizes and the further size effect are the key issues in the application of Au nanoclusters. In the past two decades, a mass of clusters has been resolved experimentally and theoretically, which reveals valuable structural information about the thiolate-protected Au cluster. In this chapter, we focus on the theoretical work on the determination of cluster and summarize two general strategies, that is, unbiased and biased. The unbiased

Atomically Precise Nanoclusters
Edited by Yan Zhu and Rongchao Jin

ISBN 978-981-4877-08-4 (Hardcover), 978-1-003-11990-6 (eBook)
www.jennystanford.com

global search works well for the small-sized cluster and cluster without the available structural feature, while the biased approach is more suitable for the large-sized clusters with diverse structural patterns. We hope that the current work will be helpful in providing a basic understanding of the theoretical prediction strategy and promote a more effective strategy.

4.1 Introduction

With unique structures and excellent properties, such as high catalysis activity and antibacterial performance, the thiolate-protected Au nanoparticles (abbreviated as RS–AuNPs) demonstrate extensive potential applications in many fields [1–5]. However, the nanoparticle is highly active and easily aggregates with neighboring particles due to its high surface atom ratio, giving rise to the difficulty in the precise characterization of the chemical formula and geometry of RS–AuNPs in solution. The understanding and tuning of diverse properties of RS–AuNPs are urgently demanding for a new strategy that is capable to generate particles with precise size, formula, and structure.

The newly developed size-focusing [6] and ligand exchange-induced size/structure transformation [7] are presented as an efficient method to produce monodisperse RS–AuNPs (also denoted as thiolate-protected Au nanocluster if the diameter of nanoparticle is smaller than 3 nm), providing a perfect platform to systematically explore diverse properties and applications of Au cluster. It is well recognized that the topological geometry, the HOMO–LUMO (highest occupied and lowest unoccupied molecular orbitals) gap, and the catalytic activity of RS–AuNPs are size dependent [1], among which a precise structure of RS–AuNPs is the precondition to understand the other properties and further establish the definite size–geometry–property correlation. Hence, the structure determination becomes the key issue for both experimentalists and theorists. Experimentally, a lot of efforts have been dedicated to the synthesis of the atomic monodisperse thiolate-protected cluster whose structure could be precisely resolved. However, the synthesis process is highly sensitive to experimental conditions [8–13]. The final product is either changed with mildly tuned experimental conditions or found to be a mixture of several different sized clusters, posing dilemma for the subsequent structure as well as the property characterization step and further applications. Under this circumstance, the density functional theory (DFT) computation acts as a pivotal part in many areas of RS–AuNPs, such as structural and electronic structures, ligand effect, and optical properties.

This chapter focuses mainly on the structure determination strategies of RS–AuNPs based on the DFT method that are summarized and categorized into two kinds: unbiased strategies with global search algorithm and biased strategies with the help of available structural knowledge. In the following sections, the principle and example of the two strategies will be elaborated. For other theoretical research fields, such as electronic structure [14] and ligand effect [15], the relevant reviews are recommended. We hope that this chapter will be helpful to advance more effective structure prediction strategies and further establish a comprehensive size effect of RS–AuNPs.

4.2 Structural Predictions of RS–AuNPs

To find the most stable isomer of cluster means to locate the global minimum on the highly sophisticated potential energy surface (PES) using the unbiased or biased method. Considering that the number of local minima increases exponentially with the size of cluster, the unbiased strategy, that is, the heuristic global search method, is believed to be more suitable for small-sized RS–AuNPs. The biased strategy works well for RS–AuNPs with relatively larger size and available structure knowledge from experimental and theoretical research.

4.2.1 Unbiased Prediction Method

The unbiased method refers in particular to the global search from scratch, which can be roughly categorized into two strategies. The first one is random sampling search based on the Monte Carlo move operating on the single individual, for example, basin hopping (BH) [16] and simulated annealing [17]. Another popular strategy is the algorithm guided under specific evolution rules, typified by genetic algorithm (GA) [18] and particle swarm optimization [19]. We will lay emphasis on two popular global search methods, that is, BH and GA. Considering that unbiased global search is also the initial and important step of the biased prediction method, the principle, advantages, and disadvantages are presented in this section.

BH algorithm: The rather complicated PES with many local minima needs to be explored with the combination of random move and optimization operation on a single individual [16, 20]. In this algorithm, the trial structure is normally generated after a random deformation on the structure of the last step within a set step length, that is, the hopping behavior illustrated in Scheme 4.1. The essence of this algorithm is performing optimization for every

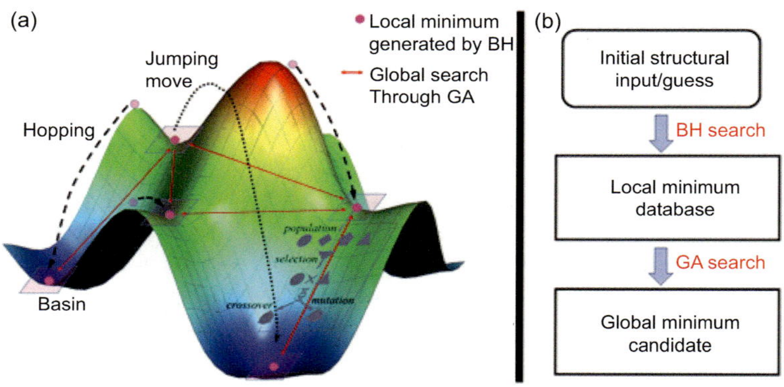

Scheme 4.1 (a) Schematic representation of BH and GA and (b) the flowchart of a hybrid strategy combining BH and GA. The pink dots represent the local minima on PES through the combination of random move and optimization operation in BH. The red lines with double arrowhead represent the evolution paths between local minima through the crossover and mutation operation under the GA.

structure after a random sample, which transforms the original PES into the combination of a series of basins (the square plane in Scheme 4.1) representing local minima. Given that the goal is to find the lowest lying isomer, the Metropolis criterion is utilized to decide whether the current generated geometry is accepted. Specifically, according to the energy difference ΔE between the current structure and the structure of the last step, the current structure is accepted as a favored isomer if the $\Delta E < 0$ or the Boltzmann factor $\exp[-\Delta E/k_B T]$ is larger than a generated random number in case that $\Delta E > 0$, where k_B is the Boltzmann constant and T is the set temperature. Subsequently, a new structure will be generated for either the next step or another trial structure of the current step. The procedure is repeated until the search process terminates or converges. With the aid of optimization operation, the transition states on the PES are eliminated, which effectively accelerates the search process. On the other hand, due to the fact that the operating object remains to be an individual, the search process will be easily trapped when locating on a deep potential funnel.

GA: In contrast to the BH algorithm, GA represents another typical evolution search where the operating object is a population of individuals instead of one single individual [18, 20]. Being similar to the BH algorithm, a local optimization is performed for each individual within the search cycling. The fitness (f) of those optimized structures is evaluated based on their energies. One popular way of fitness calculation is given in Eq. 4.1:

$$f_i = \frac{E_i - E_{\min}}{E_{\max -} E_{\min}} \tag{4.1}$$

Here, f_i and E_i are the fitness and energy of the ith individual, respectively, and $E_{\min}$ and $E_{\max}$ represent the lowest and the highest energy of the current population, respectively. In this way, the isomer with lower energy is evaluated to possess higher fitness. Then, on the principle of the survival of the fittest, a subset of individuals with higher stability are chosen and grouped to constitute the parent generation. Through the crossover and mutation operation as in Scheme 4.1a, the structures of the parent individuals are randomly fragmented, packed together, and then deformed to produce the offspring generation. Next, the combinations of local minimization, fitness evaluation, selection, crossover, and mutation are repeatedly operated until the energy/structure converges or the search cycle is completed. With the pivotal selection, crossover, and mutation operation mimicking evolution, the favored structure genes are kept and assembled to generate more fitted offspring structures during the search process. Although the operation on the population is beneficial to globally explore the sophisticated PES, it also gives rise to a problem that the search is dependent on the initial population.

Based on the above description, there is both evident discrepancy and complementarity between two algorithms. Through the Monte Carlo move and geometry optimization on the single individual, the BH algorithm possesses the excellent local search ability. For GA, since the operation object is the isomer population, the global search ability turns to be the superiority. To make full use of both algorithms, the combination is suggested in case that a single algorithm works poorly. As shown in Scheme 4.1, a BH search is first implemented and then it generates diverse local minima that represent different valleys on PES. The local minima are then utilized as the initial population of GA, and the thorough exploration between different local potential energy valleys is completed with the help of evolution operation. This hybridization ensures both local and global search capacities of the search process, enhancing the effect of the search process.

Improvement on the search efficiency: Apart from the choice of the algorithm, one issue of great importance in global search is how to prevent the stagnancy of search process. For BH algorithm, the search process will easily stagnate when a deep potential energy valley is located. As for GA, the evolution operations will keep the favored structure genes and naturally generate population with more and more similar geometry, resulting in invalid search on the restricted potential energy range. The key to prevent stagnancy concerns two parts, that is, an effective geometry similarity characterization and a timely feedback modulation.

Considering the isomers with a minor geometrical difference are normally located on the same potential energy funnel and optimized to the same local minimum, the geometry similarity characterization is needed to prevent repeated local search and enhance the global exploration on PES. The root mean square deviation (RMSD) is frequently adopted as a characterization index. Due to the defect that the RMSD is highly sensitive to the atom order of the structure, the calculation based on the quaternion method is more recommended [21], which minimizes the RMSD deviation originating from the atom sequence with the aid of translation and rotation operation. The quaternion RMSD calculation gives a reasonable evaluation of the radial distance distribution deviation around the mass center, but neglects the topological connectivity between atoms. Instead, a difference quantification between connectivity matrix of varied isomer geometries is more powerful to measure the geometry similarity from the point of the topological connectivity [22]. The incorporation of quaternion RMSD and topological connectivity constitutes an effective geometry similarity characterization.

With the introduction of geometry similarity characterization, the search program can automatically recognize the isomers with similar geometry and, correspondingly, define explored PES zones. Once the invalid revisits on the marked potential energy funnel occur frequently, the search process should be able to identify the timely feedback and tune the next search orientation. Many strategies have been developed to resolve this issue, in which adopting the adaptive step length or probability is a common way [23, 24]. Exampled as "the BH with occasional jumping" strategy [23], both step length and accept probability of the next trial moves will be tentatively increased until the current energy valley is escaped, illustrated as the "jumping move" in Scheme 4.1. Together, a heuristic global search algorithm embedding geometry similarity characterization and feedback modulation system is capable of self-optimization and guiding the search orientation via adaptive step length and probability adjustment, ensuring both the effect and the reliability of the search process.

Nucleation and growth mechanism of intermediate RS–AuNPs based on global search

The controllable synthesis and characterization of monodisperse RS–AuNPs is a challenging task for researchers, stemming from the lack of size evolution information on intermediates with different sizes. In a synthesis experiment of $Au_{25}(SR)_{18}^{-}$, dozens of intermediates have been identified through tracking the synthesis process with mass spectrum [25]. Based on this research, structures of a series of intermediate clusters $Au_m(SR)_n$ with m and n ranging from 5 to 12 have been predicted through the hybrid BH/GA search with PBE/DND level in $DMol^3$ package.

The low-lying structures of different sized clusters are shown in Fig. 4.1, and three evident growth patterns can be concluded, that is, the core growth, core dissolution, and staple-motif growth [26]. Taking core-growth pattern for instance, along with the continuously added Au atom, the inner core evolves from the linear Au_2 core to triangular or linear Au_3, tetrahedral Au_4, bipyramid Au_5, and vertex-sharing bi-tetrahedral Au_7 step by step, while the outer ligands remain unchanged. Notably, this core-growth pattern works for the growth processes starting from diverse precursors, testifying the universality of this nucleation pattern. In case that the number of thiolates is increased but the number of Au atoms remains the same, the added thiolate tends to break a original Au–Au bond in kernel and forms longer ligand, but diminishes the inner kernel. Hence, this pattern is defined as core-dissolution rule. As for the circumstance that the Au atom and thiolate are simultaneously increased, the ligand becomes larger while the inner core remains unchanged, that is, the staple motif–growth rule.

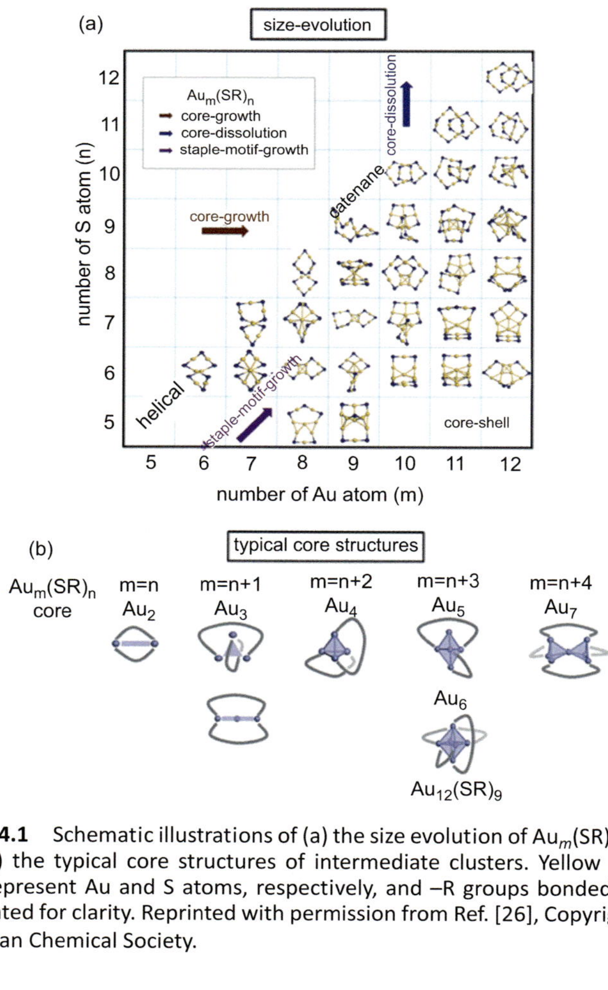

Figure 4.1 Schematic illustrations of (a) the size evolution of $Au_m(SR)_n$ clusters and (b) the typical core structures of intermediate clusters. Yellow and blue balls represent Au and S atoms, respectively, and –R groups bonded to S are eliminated for clarity. Reprinted with permission from Ref. [26], Copyright 2015, American Chemical Society.

The small size of the $Au_6(SR)_6$ cluster ($0e^-$ system) limits the formation of Au core. Notably, along with a successive increase of two Au atoms from $Au_8(SR)_6$ ($2e^-$) to $Au_{14}(SR)_6$ ($8e^-$), an evolution of tetrahedron Au_4 to vertex-sharing bi-tetrahedron Au_7, double vertex-sharing tri-tetrahedron Au_{10}, and quadruple vertex-sharing tetra-tetrahedron Au_{12} is observed in Fig. 4.2 [26]. Every increase of two Au atoms is found to be associated with a newly formed Au_4 tetrahedron in kernel and $2e^-$ growth, revealing the compact connection between topological and electronic structures. Considering the fact that the Au_4 core could be viewed as a superatom with $2e^-$, the kernel of $Au_{25}(SR)_{18}^-$ is divided into four Au_4 tetrahedrons from the perspective of both topological structure and electric structure. The Au_{13} icosahedron core of $Au_{25}(SR)_{18}^-$ is presumed to be the sequential assemble of a successive Au_4 tetrahedron, which matches the $2e^- \rightarrow 4e^- \rightarrow 6e^- \rightarrow 8e^-$ growth rule observed in the experiment [25]. Moreover, based on the calculated average binding energies of intermediate clusters, the intermediates with even valence electrons are found to possess higher stability, which validates the phenomenon that only clusters with even valence electrons are detected in experiment and strengthen the reliability of $2e^-$ growth pattern.

In addition, a further analysis on the average connectivity has been carried out on the low-lying structures of intermediate clusters. From the initial homoleptic clusters, intermediate clusters, to the final product, the average connectivity of three types increases successively, proving that intermediate clusters are highly active and easily participate in the later growth and size-focusing stage. Considering the core similarity between the intermediate cluster and experimentally resolved clusters, it is clear that the intermediate clusters serve as the building blocks and assemble into diverse final kernel structures under diverse experimental conditions. Together, based on global search and DFT, the predicted structures vividly display a preliminary nucleation process and the subsequent size-focusing growth process.

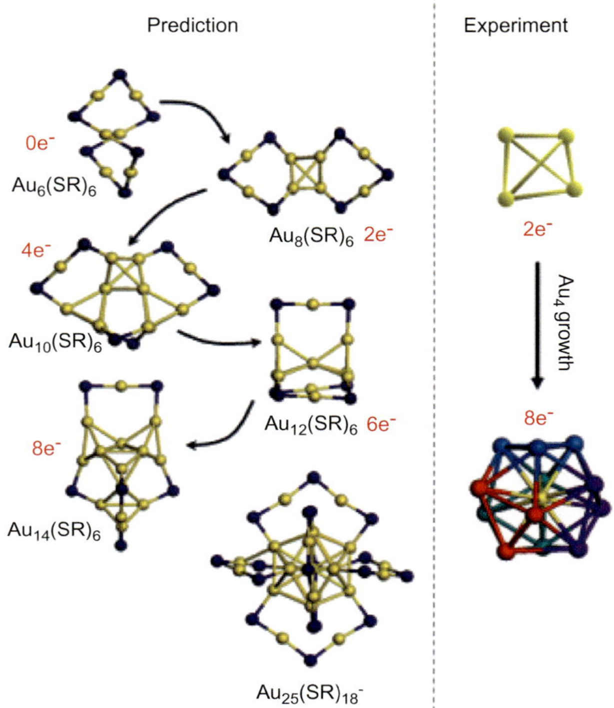

Figure 4.2 Illustration of the successive $2e^-$ growth patterns.

4.2.2 Biased Prediction Strategy for RS–AuNPs

Intrinsic structural rules of RS-AuNPs

If the unbiased prediction strategy means the exhausted search on the sophisticated PES, the biased strategy represents another oriented search based on the available structure acknowledge of clusters whose structures have been resolved. The prediction for RS–AuNPs could be traced back to the structure prediction work of $Au_{38}(SR)_{24}$ at the end of the last century [27]. In the earliest research, all Au atoms were assumed to form an intact kernel, with thiolates forming the outer ligand shell only [27, 28]. Later, the strong Au–S interaction, which may lead to a mix Au–S ligand, was pointed out [29]. Owing to the lack of accurate experimental data, the theoretical assumptions of RS–AuNPs were not validated and the progress made was slow.

In 2007, the successful characterization of $Au_{102}(p\text{-MBA})_{44}$ through single-crystal X-ray diffraction was presented as the first breakthrough of thiolate-protected Au cluster, in which 79 Au atoms constitute the high-symmetry D_{5h} inner core, while the remaining Au atoms and thiolates form the outer ligand shell in the shape of 19 $Au(SR)_2$ and 2 $Au_2(SR)_3$ staple motifs [30]. Notably, this structural feature in which Au atoms participate in both inner core and the outer ligand was theoretically predicted as the "divide and protect" pattern by Häkkinen in 2006 [31]. The unprecedented structural pattern inspired a number of researchers. Soon later, the lowest energy structure of $Au_{25}(SR)_{18}^-$ cluster was theoretically resolved (Fig. 4.3) [32]. The isomer with the highest stability is presumed to contain an icosahedron Au_{13} kernel and six mutually perpendicular $Au_2(SR)_3$ protecting units. The Au_{13} kernel can be taken as a single Au atom centered on Au_{12} polyhedron surface, in which each surficial Au atom is bound to one terminal thiolate of $Au_2(SR)_3$ motifs. Coincidently, this lowest energy theoretical structure has been validated by the experimental crystal structure at the same time and also conforms to the "divide and protect" pattern [33, 34].

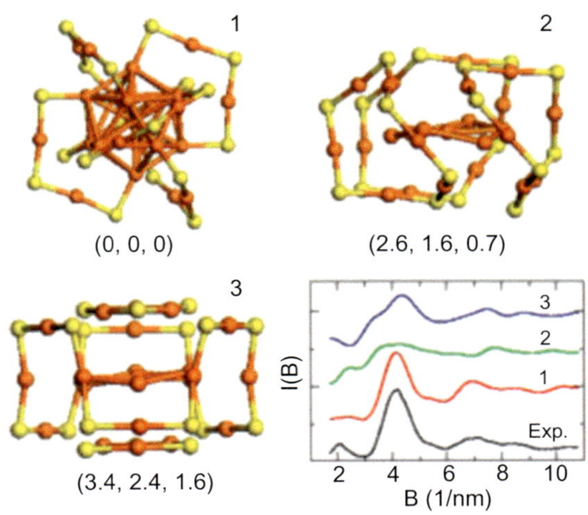

Figure 4.3 Optimized anionic structures and corresponding XRD curves. Orange and yellow represent Au and S, respectively. The R groups are eliminated for simplicity. Reprinted with permission from Ref. [32], Copyright 2008, American Chemical Society.

Inspired by the structure of $Au_{25}(SR)_{18}^-$, in 2008, the structure of $Au_{38}(SR)_{24}$ was proposed to be Au_{23} inner core covered with six $Au_2(SR)_3$ and three $Au(SR)_2$ motifs [35]. As shown in Fig. 4.4, the Au_{23} kernel is assembled by two Au_{13} icosahedra found in $Au_{25}(SR)_{18}^-$ sharing a common Au_3 triangular face. Each side of the bi-icosahedron kernel is protected by three $Au_2(SR)_3$ units, while three $Au(SR)_2$ units located in the middle of two subunits are bonded to both Au_{13} subunits. The high-symmetry kernel and uniform ligand orientation generate a structure with C_{3h} symmetry. This proposed $Au_{23}[Au(SR)_2]_3[Au_2(SR)_3]$ is found to be the lowest energy isomer so far, but there is discrepancy between the experimental and the theoretical circular dichroism spectrum. Two years later, through adjustment on the rotation orientation of $Au_2(SR)_3$ motifs on one side of Au_{13} core [36], this new structure with D_3 symmetry was evaluated to possess higher stability and lately validated experimentally [37]. Based on the theoretical and experimental research so far, several basic structure rules for RS–AuNCs have been established: (1) the cluster contains a high-symmetry inner core and an Au–S outer ligand shell formed by $Au_x(SR)_{x+1}$ staple motifs with different lengths and (2) each Au atom on the core surface is anchored with a thiolate terminal in the staple motifs to achieve high stability [35, 36].

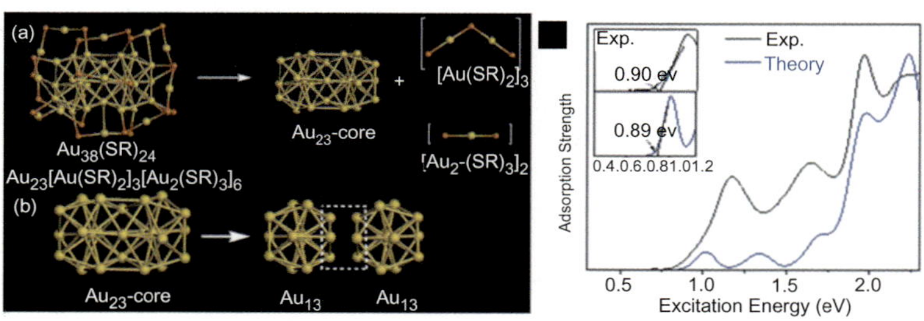

Figure 4.4 The assemble pattern of (a) $Au_{38}(SR)_{24}$ cluster and (b) Au_{23} kernel; (c) the comparison of the experimental and the theoretical optical spectrum. Reprinted with permission from Ref. [35], Copyright 2008, American Chemical Society.

Structure prediction with the force field–based divide-and-protect approach

In an attempt to resolve the structure of $Au_{24}(SR)_{20}$, a new force field–based divide-and-protect strategy was proposed on the basis of the intrinsic structural feature of RS–AuNPs [38]. As shown in

Scheme 4.2, the predicted cluster was first decomposed into $Au_{a+a'}[Au(SR)_2]_b[Au_2(SR)_3]_c$.... In the formula, a and a' represent the number of Au atoms beneath and on the surface of kernel, while b and c represent the number of the first and second ligand motifs, respectively. Therefore, the construction of a reasonable core became the important issue of the strategy. With the high symmetry of inner core, the BH global search under Sutton–Chen potential [38] or fitted non-bond potential based on accurate PES data [39] serves as a reasonable resolution to generate the core database. The generated cores are selected after the consideration of the condition that each Au atom is protected by one terminal thiolate of the staple ligand, that is, $a' = 2(b + c)$. For selected core candidates $Au_{a+a'}$, the outer ligands are assembled on the inner core with possible different orientations to form diverse cluster integrities. The isomers are then optimized using high accurate DFT method and evaluated, in which the lowest energy isomer is chosen as the global minimum and further verified through comparison of other theoretical optical properties with experimental ones.

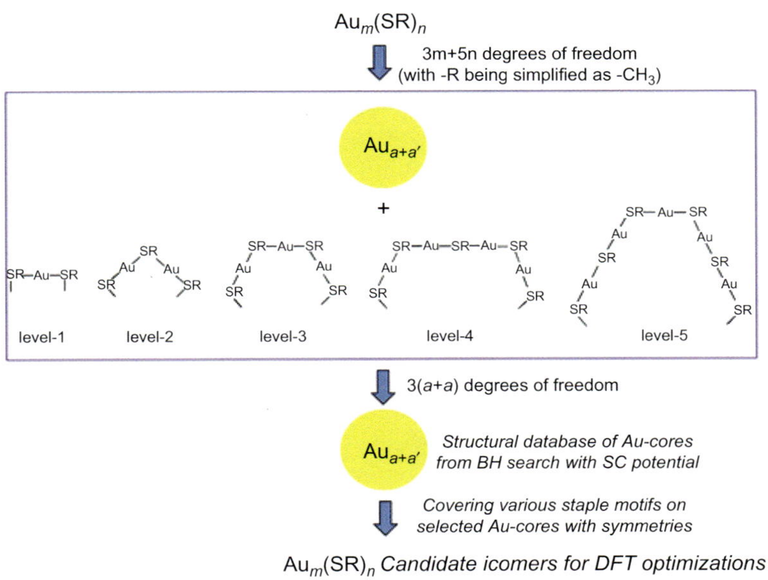

Scheme 4.2 Illustration of force-field based divide-and-protect strategy for thiolate-protected Au nanocluster. Reprinted with permission from Ref. [38]. Copyright 2012, American Chemical Society.

According to this force field–based divide-and-protect approach, the structure of $Au_{24}(SR)_{20}$ was predicted. Given the low ratio of Au/S atoms, the longer $Au_3(SR)_4$, $Au_4(SR)_5$, and $Au_5(SR)_6$ staple motifs were first introduced into the structural formula, giving rise to five possible formulas [38], that is, $Au_8[Au_4(SR)_5]_4$, $Au_8[Au_3(SR)_4]_2[Au_5(SR)_6]_2$, $Au_8[Au_3(SR)_4][Au_4(SR)_5]_2[Au_5(SR)_6]$, $Au_8[Au_2(SR)_3][Au_4(SR)_5][Au_5(SR)_6]_2$, and $Au_8[Au(SR)_2][Au_5(SR)_6]_3$. Subsequently, the diverse assemblance of ligands on the search Au_8 core and optimization operation with DFT method were performed. The three isomers with the highest stability are shown in Fig. 4.5. The lower lying isomers 1 and 2 possess identical formula $Au_8[Au_4(SR)_5]_4$ but with a different ligand orientation. The less stable isomer 3 is presumed to be of $Au_8[Au_3(SR)_4]_2[Au_5(SR)_6]_2$ pattern. All isomers possess the bi-tetrahedron Au_8 inner core, in which each Au_4 subunit is linked with a neighboring staple motif and a motif in the opposite direction. Remarkably, the staple motifs in three isomers exhibit the common interlocked feature, which is also demonstrated in small-sized homoleptic RS–AuNPs. Hence, the $Au_{24}(SR)_{20}$ is presumed to be a transition structure from the homoleptic Au(I) cluster to stable core-stacked RS–AuNPs. Nevertheless, many other properties such as XRD curve and UV–Vis optical spectrum are not matched with the experimental data. The single crystal structure of $Au_{24}(SCH_2Ph\text{-}t\text{-}Bu)_{20}$ resolved 2 years later was reported to possess the same interlocked staple motif pattern as well as structure division formula as the predicted isomer 2 but with a slightly different Au_8 inner core [40]. Notably, the structure of isomer 1 is found to be the same as the crystal structure of $Au_{24}(SeC_6H_5)_{20}$ [41]. This structure difference between thiolate-protected AuNC and selenium-protected AuNC indicates the importance of the ligands, which is revealed in another theoretical work of Jiang's group that varied ligands reverse the stability order of the isomer [42].

Compared to the unbiased global search on the PES of the whole cluster, the new force field–based divide-and-protect strategy mainly focuses on the search of the inner core. Due to the fact that the high-symmetry core structures are normally not the stable minima after DFT optimization, the high-symmetry Au cores are not obtained through direct search under DFT calculation. Instead, the introduction of Sutton–Chen potential or properly fitted non-bond potential according to high-accuracy PES not only guarantees the generation of reasonable high-symmetry kernels, but also drastically decreases the demanding computation [38, 39]. On the basis of the structure rules of RS–AuNPs, the divide-and-protect strategy comprising structure decomposition, the search and selection of the inner core, the integration of cluster through assemblance of outer ligands, and the final DFT calculation is proven to be a powerful tool in the structure prediction of RS–AuNPs, evidenced by the successful prediction of $Au_{25}(SR)_{18}^{-}$, $Au_{38}(SR)_{24}$, $Au_{24}(SR)_{20}$, and so on. To promote the accuracy and efficiency of this force field–based divide-and-protect strategy, the new structural feature is required to be updated timely, for example, the introduction of $Au_6(SR)_6$ ring in the prediction of $Au_{22}(SR)_{18}$ [43] inspired by $Au_8(SR)_8$ in $Au_{20}(SR)_{16}$ [44]. Yet there are still multiple predicted structures of clusters that are found to be incorrect or remain to be validated. The unknown field demands both experimental and theoretical efforts to advance more in-depth understanding of the structure rule of RS–AuNPs that leads to a more effective prediction strategy in return.

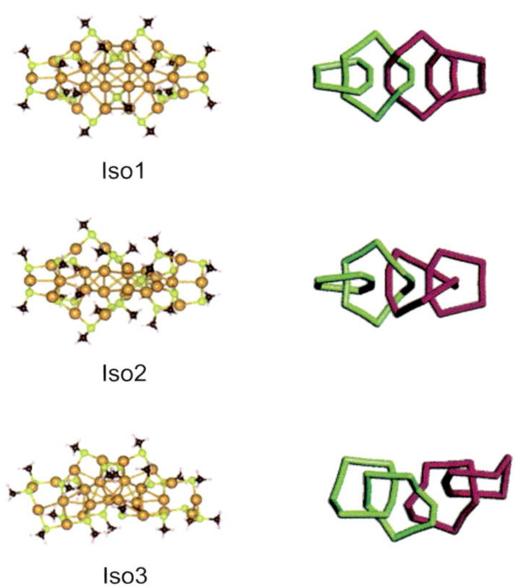

Figure 4.5 The structure of three lowest energy isomers of $Au_{24}(SR)_{20}$. Reprinted with permission from Ref. [38]. Copyright 2012, American Chemical Society.

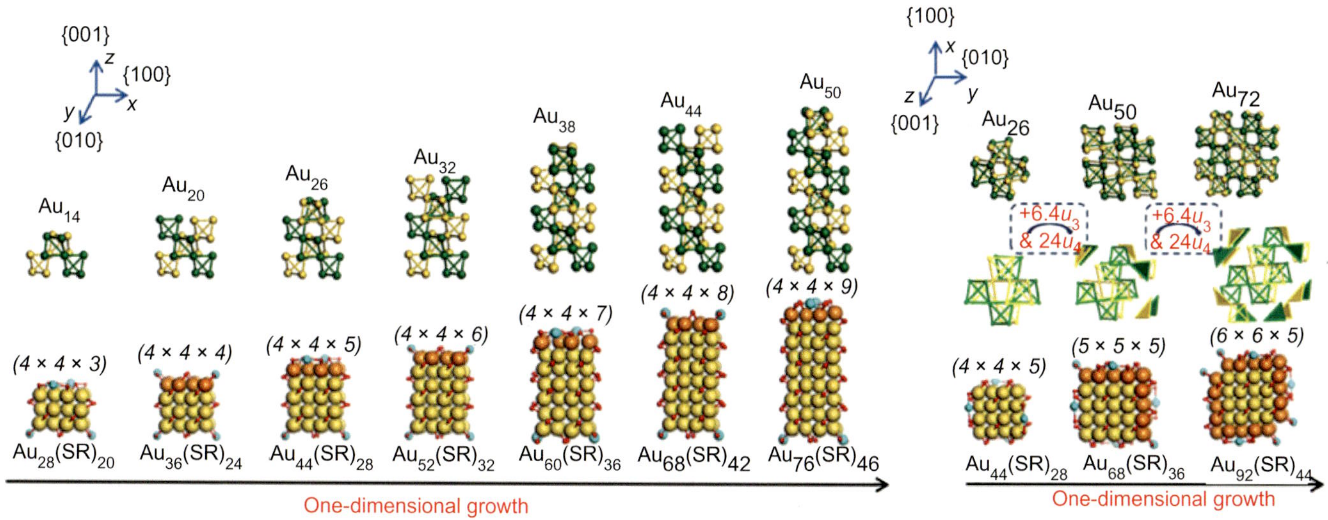

Figure 4.6 The one- and two-dimensional crystal facet growth patterns and metal core evolutions from $Au_{28}(SR)_{20}$ to $Au_{76}(SR)_{46}$ series and from $Au_{44}(SR)_{18}$ to $Au_{92}(SR)_{44}$ series. The red and green represent S and Au, respectively. The R groups are eliminated for simplicity. Reprinted with permission from Ref. [53]. Copyright 2019, American Chemical Society.

Structure prediction for cluster with specific growth pattern

Along with the experimental progress on RS–AuNPs, more and more RS–AuNPs have been gradually synthesized and characterized. In particular, a part of clusters displays evident growth rules from the perspective of both topological geometry and electronic structure. Exampled as $Au_{28}(SR)_{20}$ [45] and $Au_{36}(SR)_{24}$ [46], the Au atoms in kernel of both clusters demonstrate an unprecedented face-centered cubic stacking pattern. Specifically, the kernel of $Au_{28}(SR)_{20}$ and $Au_{36}(SR)_{24}$ could be seen as the 2 × 2 and 2 × 3 two-dimensional matrix where the Au_4 tetrahedron serves as the basic structure unit. Naturally, this discovery poses a question whether there are more clusters following the same growth pattern. Aiming at this question, an unresolved $Au_{44}(SR)_{18}^{2-}$ cluster had been noticed [47]. From $Au_{28}(SR)_{20}$ to $Au_{36}(SR)_{24}$ and $Au_{44}(SR)_{18}^{2-}$, there is a sequential $Au_8(SR)_4$ addition between two neighboring clusters. Based on the kernel growth pattern, a 2 × 4 tetrahedron kernel has been constructed as shown in Fig. 4.6, which is covered with eight $Au_2(SR)_3$ and two $Au(SR)_2$ ligands [48]. The match between theoretical and experimental optical spectra ensures the correctness of the structure. Moreover, the charge state is further revised to be neutral instead of the earlier presumed negative charge state. The AdNDP analysis also points out that Au_4 indeed acts as a $4c–2e^-$ unit participating in the growth process. Fortunately, the structural prediction and growth pattern have been verified by the successful characterization of $Au_{44}(SR)_{28}$ [49] and $Au_{52}(SR)_{32}$ [50] in the recent years. These four face-centered cubic clusters constitute a unique series $Au_{20+8N}(SR)_{16+4N}$, with N ranging from 1 to 4. From the perspective of geometry, the clusters clearly follow a one-dimensional growth along the direction of double helix kernel through sharing the common vertex in neighboring tetrahedron layers, as illustrated in Fig. 4.6. In fact, following this growth pattern, the theoretical prediction of the series cluster has been expanded to larger $Au_{76}(SR)_{46}$ [51] and $Au_{68}(SR)_{26}$ [52] along one-dimensional and two-dimensional growth patterns, respectively. Except the $Au_{20+8N}(SR)_{16+4N}$ series, another magic series from $Au_{22}(SR)_{18}$ to $Au_{28}(SR)_{20}$, $Au_{34}(SR)_{22}$, and $Au_{40}(SR)_{24}$ have been recognized, in which a sequential addition of $Au_6(SR)_2$

results in the sequential formation of a vertex-sharing bi-tetrahedron Au_7 core [53]. For a deeper understanding of the evolution pattern of RS–AuNPs, see Refs. [4, 53, 54].

Compared with the unbiased global search strategy and "divide-and-protect" strategy, the structural prediction based on the certain evolution rule is more straightforward and more reliable on the available structural information. In future, more structural evolution patterns are believed to emerge. An in-depth research on the evolution pattern and the corresponding prediction strategy is crucial to explore the size effect of RS–AuNPs that exists in many fields, such as catalysis and biotherapy.

4.3 Conclusion

In this chapter, we summarize two general strategies for structure prediction of the thiolate-protected Au nanocluster. The unbiased global search is more suitable for small-sized cluster without evident structure feature. For clusters belonging to certain specific evolution pattern, the biased construction following the evolution rule seems to be a straightforward and reasonable approach. The force field–based divide-and-protect strategy serves as another general and biased strategy suitable for most clusters. With these strategies, a number of RS–AuNPs have been resolved theoretically. Together with the experimental progress, valuable information like the intrinsic rule and growth pattern of RS–AuNPs is revealed, which is incorporated into the prediction strategy to revise the strategy in return. In fact, these structure searching strategies are believed to be suitable for other types of clusters after proper modification. To promote the efficiency of the prediction method, it is of importance to timely incorporate those new structural features and rules found in the future research. Except the structure determination, for the purpose of advancing the practical application of RS–AuNPs, the theoretical calculation is in deep need for many other fields, such as the unveiling role of ligand and the explicit size effect mechanism linking geometry and applications. Together with the effort from both experiment and theory, the prosperity of RS–AuNPs chemistry is expected to occur soon.

Acknowledgments

This work was supported by the National Key Research and Development Program of China (2017YFB0702601), the National Natural Science Foundation of China (grant nos. 21673111, 21873045).

References

1. Jin, R., Zeng, C., Zhou, M., Chen, Y. (2016). Atomically precise colloidal metal nanoclusters and nanoparticles: Fundamentals and opportunities, *Chem. Rev.*, **116**, pp. 10346–10413.
2. Dreaden, E. C., Alkilany, A. M., Huang, X., Murphy, C. J., El-Sayed, M. A. (2012). The golden age: Gold nanoparticles for biomedicine, *Chem. Soc. Rev.*, **41**, pp. 2740–2779.
3. Daniel, M.-C., Astruc, D. (2004). Gold nanoparticles: Assembly, supramolecular chemistry, quantum-size-related properties, and application toward biology, catalysis, and nanotechnology, *Chem. Rev.*, **104**, pp. 293–346.
4. Pei, Y., Zeng, X. C. (2012). Investigating the structural evolution of thiolate protected gold clusters from first-principles, *Nanoscale*, **4**, pp. 4054–4072.
5. Häkkinen, H. (2012). The gold-sulfur interface at the nanoscale, *Nat. Chem.*, **4**, pp. 443–455.
6. Jin, R., Qian, H., Wu, Z., Zhu, Y., Zhu, M., Mohanty, A., Garg, N. (2010). Size focusing: A methodology for synthesizing atomically precise gold nanoclusters, *J. Phys. Chem. Lett.*, **1**, pp. 2903–2910.
7. Zeng, C., Chen, Y., Das, A., Jin, R. (2015). Transformation chemistry of gold nanoclusters: From one stable size to another, *J. Phys. Chem. Lett.*, **6**, pp. 2976–2986.
8. Luo, Z., Yuan, X., Yu, Y., Zhang, Q., Leong, D. T., Lee, J. Y., Xie, J. (2012). From aggregation-induced emission of Au(I)–thiolate complexes to ultrabright Au(0)@Au(I)–thiolate core–shell nanoclusters, *J. Am. Chem. Soc.*, **134**, pp. 16662–16670.
9. Wu, Z., Macdonald, M. A., Chen, J., Zhang, P., Jin, R. (2011). Kinetic control and thermodynamic selection in the synthesis of atomically precise gold nanoclusters, *J. Am. Chem. Soc.*, **133**, pp. 9670–9673.
10. Yuan, X., Zhang, B., Luo, Z., Yao, Q., Leong, D. T., Yan, N., Xie, J. (2014). Balancing the rate of cluster growth and etching for gram-scale

synthesis of thiolate-protected Au_{25} nanoclusters with atomic precision, *Angew. Chem. Int. Ed.*, **53**, pp. 4623–4627.

11. Liu, C., Li, G., Pang, G., Jin, R. (2013). Toward understanding the growth mechanism of $Au_n(SR)_m$ nanoclusters: Effect of solvent on cluster size, *RSC Adv.*, **3**, pp. 9778–9784.
12. Yu, Y., Yao, Q., Yuan, X., Lee, J. Y., Xie, J. (2013). Precursor engineering and controlled conversion for the synthesis of monodisperse thiolate-protected metal nanoclusters, *Nanoscale*, **5**, pp. 4606–4620.
13. Yu, Y., Luo, Z., Yu, Y., Lee, J. Y., Xie, J. (2012). Observation of cluster size growth in CO-directed synthesis of $Au_{25}(SR)_{18}$ nanoclusters, *ACS Nano*, **6**, pp. 7920–7927.
14. Xu, W. W., Zeng, X. C., Gao, Y. (2018). Application of electronic counting rules for ligand-protected gold nanoclusters, *Acc. Chem. Res.*, **51**, pp. 2739–2747.
15. Tang, Q., Hu, G., Fung, V., Jiang, D-E. (2018). Insights into interfaces, stability, electronic properties, and catalytic activities of atomically precise metal nanoclusters from first principles, *Acc. Chem. Res.*, **51**, pp. 2793–2802.
16. Wales, D. J., Doye, J. P. K. (1997). Global optimization by basin-hopping and the lowest energy structures of Lennard-Jones clusters containing up to 110 atoms, *J. Phys. Chem. A*, **101**, pp. 5111–5116.
17. Kirkpatrick, S., Gelatt, C. D., Vecchi, M. P. (1983). Optimization by simulated annealing, *Science*, **220**, pp. 671–680.
18. Johnston, R. L. (2003). Evolving better nanoparticles: Genetic algorithms for optimising cluster geometries, *Dalton Trans.*, pp. 4193–4207.
19. Kennedy, J., Eberhart, R. (1995). Particle swarm optimization, *Proc. IEEE Int. Conf.*, pp. 1942–1948 (in Australia).
20. Heiles, S., Johnston, R. L. (2013). Global optimization of clusters using electronic structure methods, *J. Quantum Chem.*, **113**, pp. 2091–2109.
21. Coutsias, E. A., Seok, C., Dill, K. A. (2004). Using quaternions to calculate RMSD, *J. Comput. Chem.*, **25**, pp. 1849–1857.
22. Rodrigue, J.-P., Comtois, C., Slack, B. (2006). Transportation and geography, in *The geography of Transport Systems*, Taylor & Francis, pp. 1–37.
23. Iwamatsu, M., Okabe, Y. (2004). Basin hopping with occasional jumping, *Chem. Phys. Lett.*, **399**, pp. 396–400.

24. Srinivas, M., Patnaik, L. M. (1994). Adaptive probabilities of crossover and mutation in genetic algorithms, *IEEE Trans. Syst. Man Cybern.*, **24**, pp. 656–667.

25. Luo, Z., Nachammai, V., Zhang, B., Yan, N., Leong, D. T., Jiang, D-E., Xie, J. (2014). Toward understanding the growth mechanism: Tracing all stable intermediate species from reduction of Au(I)-thiolate complexes to evolution of Au_{25} nanoclusters, *J. Am. Chem. Soc.*, **136**, pp. 10577–10580.

26. Liu, C., Pei, Y., Sun, H., Ma, J. (2015). The nucleation and growth mechanism of thiolate-protected Au nanoclusters, *J. Am. Chem. Soc.*, **137**, pp. 15809–15816.

27. Luedtke, W. D. and Landman, U. (1996). Structure, dynamics, and thermodynamics of passivated gold nanocrystallites and their assemblies, *J. Phys. Chem.*, **100**, pp. 13323–13329.

28. Häkkinen, H., Barnett, R. N., Landman, U. (1999). Electronic structure of passivated $Au_{38}(SR)_{24}$ nanocrystal, *Phys. Rev. Lett.*, **82**, pp. 3264–3267.

29. Garzón, I. L., Rovira, C., Michaelian, K., Beltrán, M. R., Ordejón, P., Junquera, J., Sánchez-Portal, D., Artacho, E., Soler, J. M. (2000). Do thiols merely passivate gold nanoclusters? *Phys. Rev. Lett.*, **85**, pp. 5250–5251.

30. Jadzinsky, P. D., Calero, G., Ackerson, C. J., Bushnell, D. A., Kornberg, R. D. (2007). Structure of a thiol monolayer-protected gold nanoparticle at 1.1 Å resolution, *Science*, **318**, pp. 430–433.

31. Häkkinen, H., Walter, M., Grönbeck, H. (2006). Divide and protect: Capping gold nanoclusters with molecular gold-thiolate rings, *J. Phys. Chem. B*, **110**, pp. 9927–9931.

32. Akola, J., Walter, M., Whetten, R. L., Häkkinen, H., Grönbeck, H. (2008). On the structure of thiolate-protected Au_{25}, *J. Am. Chem. Soc.*, **130**, pp. 3756–3757

33. Zhu, M., Aikens, C. M., Hollander, F. J., Schatz, G. C., Jin, R. (2008). Correlating the crystal structure of a thiol-protected Au_{25} cluster and optical properties, *J. Am. Chem. Soc.*, **130**, pp. 5883–5885.

34. Heaven, M. W., Dass, A., White, P. S., Holt, K. M., Murray, R. W. (2008). Crystal structure of the gold nanoparticle $[N(C_8H_{17})_4]$-$[Au_{25}(SCH_2CH_2Ph)_{18}]$, *J. Am. Chem. Soc.*, **130**, pp. 3754–3755.

35. Pei, Y., Gao, Y., Zeng, X. C. (2008). Structural prediction of thiolate-protected Au_{38}: A face-fused bi-icosahedral Au core, *J. Am. Chem. Soc.*, **130**, pp. 7830–7832.

36. Lopez-Acevedo, O., Tsunoyama, H., Tsukuda, T., Häkkinen, H., Aikens, C. M. (2010). Chirality and electronic structure of the thiolate-protected Au_{38} nanocluster, *J. Am. Chem. Soc.*, **132**, pp. 8210–8218.

37. Qian, H., Eckenhoff, W. T., Zhu, Y., Pintauer, T., Jin, R. (2010). Total structure determination of thiolate-protected Au_{38} nanoparticles, *J. Am. Chem. Soc.*, **132**, pp. 8280–8281.

38. Pei, Y., Pal, R., Liu, C., Gao, Y., Zhang, Z., Zeng, X. C. (2012). Interlocked catenane-like structure predicted in $Au_{24}(SR)_{20}$: Implication to structural evolution of thiolated gold clusters from homoleptic gold(I) thiolates to core-stacked nanoparticles, *J. Am. Chem. Soc.*, **134**, pp. 3015–3024.

39. Li, Y., Liu, C., Pu, Z., Bao, L., Zhu, Y., Ma, J. (2019). Planar versus nonplanar Pd clusters: stability and CO oxidation activity of Pd clusters with and without $TiO_2(110)$ substrate, *J. Phys. Chem. C*, **123**, pp. 13739–13747.

40. Das, A., Li, T., Li, G., Nobusada, K., Zeng, C., Rosi, N. L., Jin, R. (2014). Crystal structure and electronic properties of a thiolate-protected Au_{24} nanocluster. *Nanoscale*, **6**, pp. 6458–6462.

41. Song, Y., Wang, S., Zhang, J., Kang, X., Chen, S., Li, P., Sheng, H., Zhu, M. (2014). Crystal structure of selenolate-protected $Au_{24}(SeR)_{20}$ nanocluster, *J. Am. Chem. Soc.*, **136**, pp. 2963–2965.

42. Tang, Q., Ouyang, R., Tian, Z., Jiang, D. (2015). The ligand effect on the isomer stability of $Au_{24}(SR)_{20}$ clusters, *Nanoscale*, **7**, pp. 2225–2229.

43. Pei, Y., Tang, J., Tang, X., Huang, Y., Zeng, X. C. (2015). New structure model of $Au_{22}(SR)_{18}$: Bitetrahedron golden kernel enclosed by $[Au_6(SR)_6]$ Au(I) complex, *J. Phys. Chem. Lett.*, **6**, pp. 1390–1395.

44. Zeng, C., Liu, C., Chen, Y., Rosi, N. L., Jin, R. (2014). Gold-thiolate ring as a protecting motif in the $Au_{20}(SR)_{16}$ nanocluster and implications, *J. Am. Chem. Soc.*, **136**, pp. 11922–11925.

45. Zeng, C., Li, T., Das, A., Rosi, N. L. and Jin, R. (2013). Chiral structure of thiolate-protected 28-gold-atom nanocluster determined by X-ray crystallography, *J. Am. Chem. Soc.*, **135**, pp. 10011–10013.

46. Zeng, C., Qian, H., Li, T., Li, G., Rosi, N. L., Yoon, B., Barnett, R. N., Whetten, R. L., Landman, U., Jin, R. (2012). Total structure and electronic properties of the gold nanocrystal $Au_{36}(SR)_{24}$, *Angew. Chem. Int. Ed.*, **51**, pp. 13114–13118.

47. Price, R. C., Whetten, R. L. (2005). All-aromatic, nanometer-scale, gold-cluster thiolate complexes, *J. Am. Chem. Soc.*, **127**, pp. 13750–13751.

48. Pei, Y., Lin, S., Su, J., Liu, C. (2013). Structure prediction of $Au_{44}(SR)_{28}$: A chiral superatom cluster, *J. Am. Chem. Soc.*, **135**, pp. 19060–19063.

49. Zeng, C., Chen, Y., Iida, K., Nobusada, K., Kirschbaum, K., Lambright, K. J., Jin, R. (2016). Gold quantum boxes: On the periodicities and the quantum confinement in the Au_{28}, Au_{36}, Au_{44}, and Au_{52} magic series, *J. Am. Chem. Soc.*, **138**, pp. 3950–3953.

50. Zeng, C., Chen, Y., Liu, C., Nobusada, K., Rosi, N. L., Jin, R. (2015). Gold tetrahedra coil up: Kekule-like and double helical superstructures, *Sci. Adv.*, **1**, pp. e1500425(1–6).

51. Ma, Z., Wang, P., Zhou, G., Tang, J., Li, H., Pei, Y. (2016). Correlating the structure and optical absorption properties of $Au_{76}(SR)_{44}$ cluster, *J. Phys. Chem. C*, **120**, pp. 13739–13748.

52. Wang, P., Sun, X., Liu, X., Xiong, L., Ma, Z., Pei, Y. (2017). Theoretical predictions of a new ~14 kDa core-mass thiolate-protected gold nanoparticle: $Au_{68}(SR)_{36}$, *J. Phys. Chem. Lett.*, **8**, pp. 1248–1252.

53. Pei, Y., Wang, P., Ma, Z., Xiong, L. (2019). Growth-rule-guided structural exploration of thiolate-protected gold nanoclusters, *Acc. Chem. Res.*, **52**, pp. 23–33.

54. Jiang, D-E. (2013). The expanding universe of thiolate gold nanoclusters and beyond, *Nanoscale*, **5**, pp. 7149–7160.

Part II

ELECTRONIC AND OPTICAL PROPERTIES OF NANOCLUSTERS

Chapter 5

Toward Understanding the Structure of Gold Nanoclusters

Endong Wang[a] and Yi Gao[b]
[a]*Shanghai Institute of Applied Physics, Chinese Academy of Sciences, Shanghai 201800, P. R. China*
[b]*Zhangjiang Laboratory, Shanghai Advanced Research Institute, Chinese Academy of Sciences, Shanghai 201210, P. R. China*
gaoyi@zjlab.org.cn

5.1 Introduction

Gold is one of the most amazing elements that has attracted many researchers because of its unique properties. The last several decades have witnessed the blooming of researches on gold, especially gold nanoclusters (AuNCs) [1].

Neutral Au atom possesses [Xe]$4f^{14}5d^{10}6s^1$ electron configuration and shares the same group with Ag and Cu. But Au displays much stronger relativistic effects [2]. One significant change brought by relativity is the contraction of 6s orbital and the expansion of 5d orbitals [3]. On comparing with atoms after Sn (Z = 50), theoretical results show that Au exhibits the most drastic contraction of

Atomically Precise Nanoclusters
Edited by Yan Zhu and Rongchao Jin

ISBN 978-981-4877-08-4 (Hardcover), 978-1-003-11990-6 (eBook)
www.jennystanford.com

6s orbital, which induces greater ionization energies [2]. Also, relativistic effects can cause more obvious d-orbital splitting with elevated orbital energies for Au [4]. Due to the differences in relativity, Cu, Ag, and Au show rather different physical and chemical properties. Taking $[AuPH_3]^+$ and $[AgPH_3]^+$ as examples, Schwerdtfeger et al. compared the influence of relativity through including or excluding the energy-consistent scalar-relativistic small-core pseudopotentials in the calculations [5]. The authors stated that the electron population on the 5s orbital of Ag is 0.156. For non-relativistic calculations of Au, electron population on the 6s orbital is 0.202. However, this value increases to 0.438 if the relativity for Au atom is considered. Thus, the covalent property of Au–P bond is strengthened compared to the Ag–P bond. The dissociation energies for detaching PH_3 from $[AgPH_3]^+$ and $[AuPH_3]^+$ were also calculated. The results were 38.00 and 63.91 kcal/mol, respectively. As a comparison, the dissociation energy is as low as 28.90 kcal/mol for Au when excluding relativity. This is consistent with the analysis of electron population. This example demonstrates that relativity can bring different chemical properties to Au atoms. Besides, the reason that gold exhibits yellow color is also related to relativity [6]. With relativistic effects, the gap between the center of 5d orbitals and that of 6s orbitals was reduced. Thus, the excitation energies lie within the range of visible light. Or it will be in the UV range.

It is also necessary to consider relativity for AuNCs. Bulk gold generally adopts the fcc (face-centered cubic) structure. When the size of this fcc structure narrows, its structural pattern may adopt other non-fcc packing like hcp (hexagonal close packing) [7]. In the meantime, electronic structure properties of AuNCs may change as well. As depicted by Pyykkö, a significant change influenced by size is the alternation of the energy level of electronic states [8]. A typical size limit to determine if it belongs to AuNCs is around 2 nm. By decreasing the size of bulk gold to lower than 2 nm, the energy levels of the orbitals can be altered gradually from continuous to discrete. This discrete energy level makes AuNCs to be molecule-like [9]. It is at this scale that relativity matters. Because when AuNCs become molecule-like, the interactions between frontier molecular orbitals dominate among their interactions with other clusters or molecules [10]. As described above, relativity may bring the contraction of 6s orbitals and the splitting of 5d orbitals. The frontier molecular orbitals of AuNCs are the combinations of the atomic orbitals of Au atoms, which include 5d and 6s orbitals. Thus, relativity is important for AuNCs.

Besides the change from continuous energy levels to discrete energy levels brought by size effects, the macroscopic change of properties is also pronounced. The most obvious one is the increasing ratio between surface and volume. This is an important piece of merit for utilizing AuNCs as the catalyst. Actually, AuNC is found to be useful only in catalysis reactions when it is small sized [11]. More low coordination atoms are exposed outside, either along the edges or at the corners of small-sized AuNCs. In catalysis reactions, the coordination number is considered to be significant for the catalyst [12–14]. Häkkinen et al. explained how the quantum size effect can influence the catalytic efficiency using the CO oxidation reaction as an example [10]. Several ligand-protected AuNCs with diameter ranging from 1.2 to 2.4 nm were chosen. According to the Langmuir–Hinshelwood mechanism, after the partial removal of protecting ligands, effective absorption of O_2 producing O_2^* species is accomplished through the transfer of electron from AuNCs. The authors found that the HOMO–LUMO (highest occupied and lowest unoccupied molecular orbitals) energy gap, largely determined by the quantum size effect, has a crucial effect. According to the results, the HOMO–LUMO energy gap correlates with binding energy of O_2^*. It is also claimed that the oxidation reaction of CO proceeds effectively only on the smallest AuNC which binds to O_2^* appreciably. The size-dependent catalytic performance demonstrates the importance of clarifying the structure–property relationships and also proves the necessity to synthesize AuNCs with precise size.

A full control over the synthesis of AuNCs with precise size is rather important because it is highly helpful for clarifying the properties of AuNCs. Unlike only polydisperse AuNC samples obtained earlier, more than 70 monodisperse AuNCs have been successfully synthesized now [15]. Several groups contributed significantly to this field. Especially, Jin et al. prepared substantial ligand-protected AuNCs with precise size [16]. They first proposed a synthesis protocol called "one pot for one size" [17]. Through utilizing rigid ligands, Jin et al. adopted a kinetically controlled method and successfully obtained $Au_{25}(SC_2H_4Ph)_{18}$ nanoclusters for the first time [18]. Notably, the product was rightly obtained in one pot without any purification. Wu also synthesized this Au_{25} nanocluster with other thiols through using the one-phase method [19]. Later, Jin et al. successfully prepared $Au_{144}(SR)_{60}$ [20], $Au_{38}(SR)_{24}$ [21], $Au_{64}(SR)_{32}$ [22], $Au_{99}(SR)_{42}$ [23], $Au_{333}(SR)_{79}$ [24], $Au_{36}(SR)_{24}$ [25], $Au_{28}(TBBT)_{20}$ [26], $Au_{133}(TBBT)_{52}$ [27], $[Au_{23}(SC_6H_{11})_{16}]^-$ [28], $Au_{40}(SR)_{24}$ [29], and so on through combining multiple methods [17]. Currently, there exist several universal methods to synthesize AuNCs with precise size experimentally [17]. Besides, other groups also contributed a lot independently [30–42].

With these beautiful and highly symmetric structures, extensive studies have focused on their applications. Zhu et al. performed substantial experiments on exploring their catalytic properties [43–46]. Besides their utilization in catalysis, AuNCs can also be used for other purposes due to other properties. For example, AuNCs have no toxicity and are chemically inert [47]. Also, monolayer cover of the AuNCs can be tuned to display hydrophilic or hydrophobic behavior with different charges [48, 49]. These properties broaden the application background to the biological field for AuNCs as nanocarriers for drugs in therapeutics [50].

To better realize the potential of AuNCs, either as catalysts or as drug carriers or for other utilizations, one prerequisite is to have a thorough understanding of their structures. This is important because it is valuable for explaining the experimental results as well as for predicting new AuNCs. Initially, little was known about the real structures of AuNCs. Studies on the structures of AuNCs benefited a lot from the knowledge of self-assembled monolayers (SAMs) on which organosulfur compounds thiol and disulfide adsorb [51]. Considering the similar chemical environment, a rather plausible guess of the structure of AuNCs would also be ligands cap plus Au core [52]. Later, according to low-temperature scanning tunneling microscopy (STM) experiments on Au surfaces, a new model of RSAuSR was considered to be present [53]. Almost simultaneously, Häkkinen proposed a novel thiol bonding pattern, which first claimed –SR motif as $-(AuSR)_x$ [54]. This work signifies that the chemistry of shell Au atoms may differ with the core Au atoms. In 2007, a pioneering work reported the first ever single-crystal X-ray structure of Au_{102} cluster, which brought the studies on the structure of AuNCs into an almost totally new era [55]. Sooner after, Au_{25} cluster was also prepared experimentally. Based on these first two crystal structures of AuNCs, several key theories on understanding and predicting stable AuNCs have been proposed since then. These models laid the most important foundation for this field of today. Thus, for the purpose of gaining the understanding of these models, this chapter will focus on the theoretical aspect of how to interpret the stability of core-shell AuNC structures. First, these key models will be briefly introduced. Then, the structures of the fcc packing AuNCs will be revisited. Given the immense attentions paid to this field, we would like to emphasize that this chapter mainly focuses on the theoretical models used to interpret the stability of AuNCs. An unbiased scenario of this field can be seen in several well-organized review articles [7, 15, 17, 56–70].

5.2 Theoretical Models of Structures of AuNCs

5.2.1 "Divide and Protect Model" Concept

In 1999, the first theoretical model of the structure of AuNC was suggested [52]. In this model, $Au_{38}(SCH_3)_{24}$ was chosen as an example. Its metal core was found to be fcc-type through optimization using density functional theory (DFT). The remaining thiol groups were in the $-SCH_3$ form as monolayer covering the Au_{38} core. At that time, this was a rather sound guess of the geometric architecture of AuNCs. Until 2006, a low-temperature STM experiment performed on Au surface suggested that RSAuSR complex could also be the absorbate, which implied that this complex might also cover the Au core as a shell in AuNC [53]. Independently and almost simultaneously, Häkkinen performed another theoretical study proposing a novel motif covering the Au core [54]. $Au_{38}(SCH_3)_{24}$ was again used as the sample cluster. $(AuSCH_3)_4$ was suggested as the shell cap and the rest of the Au_{14} formed the core of this AuNC. This work represents a significant improvement for understanding AuNCs because it suggests that different bonding statuses of Au atoms may coexist in the outer ligand and the inner core. The "divide and protect" concept just originates from this work. This work also suggests that the overall structure of AuNC may be present in a highly symmetric form, which helps to maximize the number of Au–Au interactions, rather than in a disordered form proposed earlier [71, 72].

5.2.2 Inherent Structure Rule

The proposal of the "divide and protect model" concept was a breakthrough in this field at that time because no experimental evidence of how atoms arrange in the cluster was available then. In 2007, Jadzinsky et al. successfully obtained the crystal structure of $Au_{102}(p\text{-MBA})_{44}$ [55]. This was the first structural determination study that proved the value of the divide and protect model concept. Later, Jin and Murray independently prepared $[Au_{25}(SR)_{18}]^-$, which was another major advance in understanding the structure of AuNC [18, 73]. These three studies confirmed the structural pattern of core plus protecting-ligand and refreshed the views of what is the shape of Au core and what is the protecting ligand. According to the crystal structures, the motifs may include –RS–Au–RS– or –RS–Au–RS–Au–RS–.

Although major advances have been made, the question of whether there exists a general and underlying formula for AuNCs arouses the interest of researchers. Through analyzing the available structures of AuNCs, Pei et al. proposed an inherent structure rule in 2008 [74]. They claimed that the structures of future possible $Au_m(SR)_n$ AuNCs could be divided into several parts by following a general formula of

$$Au_m(SR)_n = Au_a[Au(SR)_2]_b[Au_2(SR)_3]_c[Au_3(SR)_4]_d\ldots,$$

where a represents the number of Au atoms in the core, b is the number of $Au(SR)_2$ motifs, and c and d are the number of $Au_2(SR)_3$ and $Au_3(SR)_4$, respectively. This rule is consistent with the previously proposed divide and protect rule. According to this inherent rule, the outer motif can be monomeric, dimeric, trimeric, and so on based on the number of Au atoms in this motif. And m is equal to the sum of a + b + 2*c + 3*d . . .. This rule fits well with the available Au_{102} and Au_{25} clusters and has been used in the prediction of new AuNCs [75, 76].

5.2.3 Superatom Complex (SAC) Model

This model adapts from the initial Jellium model. Jellium model was initially used in understanding alkali metal clusters [56] in which the valence electrons were considered to be distributed unbiasedly within the cluster, instead of over confined to particular atoms. According to Jellium model, the orbitals of these free electrons spread over the whole metal atoms rather than an individual atom [77, 78]. Analogy to the atomic theory, this produces an electron configuration of $1S^2|1P^6|1D^{10}|2S^2 1F^{14}|2P^6 1G^{18}|2D^{10} 3S^{21} H^{22}|$..., which means that a stable metal cluster should have the full occupation of the outermost orbital. Orbitals of each shell configuration that distribute over the entire gold core and shape like the atomic orbitals have been found in a lot of AuNCs [79–81]. A notation called the "magic number" is generally used to denote the number of these free electrons. Typical magic numbers include 2, 8, 18, 20, 40, and 58.

In 2008, Häkkinen further extended the Jellium model to explain the high stability of spherical gold clusters, including $Au_{102}(SR)_{44}$, $Au_{39}(PR_3)_{14}Cl_6^-$, $Au_{11}(PR_3)_7Cl_3$, and $Au_{13}(PR_3)_{10}Cl_2^{3+}$ [82]. In these ligand-protected AuNCs, the number of free electrons n^* correlates to the number of Au atoms or the number of Au($6s^1$) electrons N, the number of ligands M, and the charge of the AuNC q. This can be described using a simple formula

$$n^* = N - M - q$$

Note that the protecting ligands can be either withdrawing (e.g., thiol groups or halogen) or coordination groups like PH_3. For the latter, the binding of PH_3 to the Au core is via weak coordination and it is not included in the calculation of the number of free electrons.

5.2.4 Superatom Network (SAN) Model

Although the SAC model holds for most of AuNCs, some of the AuNCs with four free electrons, such as $[Au_{18}(SR)_{14}]$ (4e), $Au_{20}(SR)_{16}$ (4e), and $Au_{24}(SR)_{20}$ (4e), do not seem to follow this model [30, 83, 84]. To understand the stability of these AuNCs, the SAN model was proposed by Yang and Cheng et al [85]. SAN can handle only 4e clusters if they are shell-closed electronic structures. Based on the SAN model, the 4e clusters can be depicted as a network of two-electron superatom gold cores. The authors analyzed the Au–Au distance within and between the two-electron superatom gold cores, the delocalized multicentered bonding, and the nucleus-independent chemical shift to support this model.

5.2.5 Grand Unified Model (GUM)

Understanding the stability of AuNCs has always been an interesting topic since it is one of the most fundamental problems in the research.

Each of the models for interpreting AuNCs' stabilities proposed so far is applicable only to a certain subset of AuNCs. Thus, it is demanding to understand their stabilities through a unified model.

Through analyzing crystal structures of the available 71 AuNCs, Gao et al. proposed grand unified model [86]. In this model, inspired by the quark model, gold atoms can be assigned to be three flavors to represent the possible valence states 1.0e, 0.5e, and 0e. Then, these three flavors constitute two composite particles of this model. These two composite particles include triangular elementary block Au_3(2e) and tetrahedral elementary block Au_4(2e). Both of them satisfy the duet rule, which is similar to the full occupation of the valence shell. In conceiving this model, the authors also tried different possible valence states and different potential isoelectronic species to make the fundamental concept of this model solid. The results of formation energies show that Au_3(2e) and Au_4(2e) seem to be the only two possible elementary blocks because they have highly negative formation energies. To satisfy the duet rule and the confined number of gold atoms of these two elementary blocks, different numbers of flavors could be combined. This produces 10 variants for triangular elementary blocks and 15 variants for tetrahedral elementary blocks. Later, to conveniently explain the stability of those AuNCs with the icosahedral Au_{13} motif, a secondary block Au_{13}(8e) was introduced into this model [87]. This elementary block also owns highly negative formation energy. Considering this block, another eight variants should be included.

GUM aims at understanding the stability of AuNCs from the perspective of Au core. Having built the underlying architecture of GUM, the next step is to apply it in understanding a specific AuNC. To achieve this, the protecting ligands should be detached first. It is clear that the bonding between the ligands and the Au core can be either chemical bond or coordination based on the chemical nature of the bonding atom. Here, thiolate ligands, phosphine ligands, and halogen ligands were considered as examples. For these three ligands, the Au core would transfer 0.5e valence electron, 1e valence electron, and 0e valence electron when detaching them. Xu et al. provided several prototypical liganded clusters for interpreting how to apply GUM stepwise. It will also be introduced briefly in the next section. This is rather helpful to gain the knowledge of the beautiful structural pattern of AuNCs. Through GUM, it would be also much easier to understand the stability of AuNCs. In addition, GUM can also be used for other purposes. Xu et al. used GUM in understanding the structural evolution of AuNCs and in predicting new AuNCs as well [88]. $Au_{60}(SR)_{36}$, $Au_{68}(SR)_{40}$, and $Au_{76}(SR)_{44}$ have been successfully predicted based on GUM [88]. GUM can also be used in understanding the structural isomerism of AuNCs [89]. Besides, GUM can also rationally design new ligand-protected gold clusters [90, 91].

5.3 Rethinking the Structure of Gold Nanoclusters with fcc-Based Kernel through GUM

AuNCs can show different packing types such as fcc and hcp. Since the discovery of $Au_{36}(SR)_{24}$, dozens of fcc type nanoclusters have been crystallized [81]. Zeng et al. synthesized two fcc-based Au_{40}(*o*-MBT)$_{24}$ and $Au_{52}(TBBT)_{32}$ in 2015 and found that their Au kernels can be segregated into two helixes consisting of several fused tetrahedral Au_4 which is exactly one of the elementary blocks of GUM [92]. Later in 2016, the researchers in the same group produced another four fcc-based $Au_{28}(TBBT)_{20}$, $Au_{36}(TBBT)_{24}$, $Au_{44}(TBBT)_{28}$, and $Au_{52}(TBBT)_{32}$ nanoclusters; helix-like units were also found in these four AuNCs [93]. This is also the case for $Au_{34}(S\text{–}c\text{–}C_6H_{11})_{22}$ and $Au_{42}(S\text{–}c\text{–}C_6H_{11})_{26}$ [94]. All these studies share one point of view that helix-like Au kernels could be obtained based on the inhomogeneous distribution of Au–Au bond lengths. However, for some other fcc-based AuNCs, the decompositions of the Au kernels are not so obvious. For example, the Au cores of both $Au_{30}S(S\text{-}t\text{-}Bu)_{18}$ and $Au_{30}(S\text{-}t\text{-}Bt)_{18}$ were recognized as interpenetrating bicuboctahedral geometry [95, 96]. In the structure determination study of $[Au_{23}(SC_6H_{11})_{16}]^-$ and $Au_{24}(SAdm)_{16}$, the authors saw the Au_{13} kernel as cuboctohedral [28, 97]. In addition, the Au_{10} core of Au_{21}(S–Adm)$_{15}$ was treated as two octahedrons sharing one edge and the average Au–Au bond length for whole AuNC was calculated [42]. The average bond length was found to be roughly equal to that of bulk gold in this work [42]. Although structure evolution for AuNCs containing fcc-based kernels was reported very recently for further understanding of structural evolvement [81], the above studies suggest that the current knowledge about the structural information of Au kernel still needs to be enriched to check if the inhomogeneous distribution of Au–Au bond lengths also exists in non-helix-like AuNC cores. This point motivates this work partly. Through investigating the inhomogeneous distribution of Au–Au distances, examinations can also be done to see if the inhomogeneous distribution is in accordance with the distribution of the elementary blocks of GUM. To achieve these goals, the Au–Au distances within the Au cores and some other calculations were counted and analyzed.

Thiolate-protected gold clusters involved in this work were optimized using DFT. TPSS exchange-correlation functional was used in the optimization. This functional has been proven to be capable of obtaining the reasonable structures [98–100]. Initial geometry inputs were obtained from the experimental crystal structures. During the calculations, the ligands of AuNC were replaced by –SH to reduce the computational cost. Effective core basis set LANL2DZ was used for Au atoms and a larger SDD basis set was used as well for the purpose of comparisons and validations. Other atoms, including S, C, and H, were described by using 6-31G* basis set. Self-consistent calculations were considered as convergence when a criterion of 10^{-6} Hartree was reached. MP2 has been already proven to be able to reproduce aurophilic interaction [2, 101, 102]. Thus, MP2 calculations were also done in order to describe electron correlation accurately. All the calculations were done using Gaussian 09 package [103].

5.3.1 Segregation of Sample AuNCs Based on GUM

According to GUM, each Au atom and SR group possess +1e and –1e valence electrons, respectively. In the first step, all the thiol ligands were detached from the Au kernel accompanied with the electron transfer of 0.5e from each Au kernel atom to the corresponding S atom, that is, *step a* of Fig. 5.1. This step also presents that $[Au_{23}SR_{16}]^-$ kernel is composed of three units, that is, Au_3, Au_7, and Au_3. Distances between two Au atoms from the same or two different units were measured to check if there was any underlying principle. *Step b* of Fig. 5.1 also shows that these three units can further be decomposed into two Au_4 blocks and two Au_3 blocks, where the 1e valence electron of the Au atom in magenta was separated into two 0.5e of each Au_4.

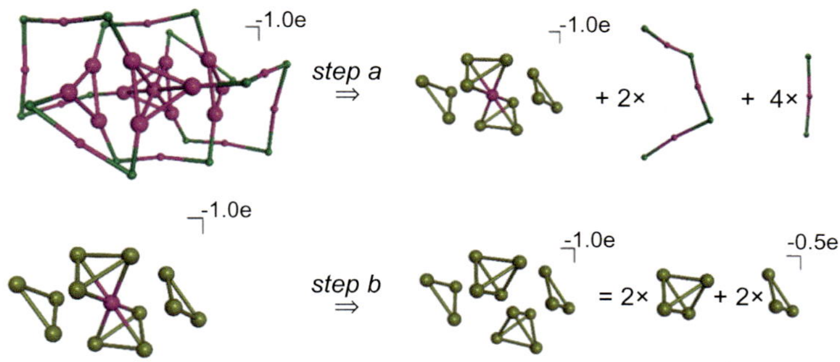

Figure 5.1 The structural segregation of $[Au_{23}SR_{16}]^-$. All the –R groups are omitted for clarity. Color: Au – magenta and dark yellow, S – dark green. Based on GUM, Au atoms are shown in different colors to represent different numbers of valence electrons in each Au atom, that is, magenta Au: 1e valence electron and dark yellow Au: 0.5e valence electron.

5.3.2 Validation of Calculations

We first show the results of validation of calculations before presenting further results. Since structural information is focused particularly in this work, the average bond length differences between the corresponding experimental and computational results were obtained for the purpose of comparison. Two types of representative bond lengths, that is, intra-unit Au–Au bond length and inter-unit Au–Au bond length, were counted. Inter-unit Au–Au bond length refers to the distance between two Au atoms in two different units, while intra-unit Au–Au bond length represents the Au–Au distance within a unit. Only the reasonably short inter-unit Au–Au bond was considered here. Eight relatively small AuNCs were selected as samples to get intra-unit and inter-unit Au–Au bond lengths. The results are shown in Fig. 5.2.

Two pseudopotential basis sets, that is, LANL2DZ and SDD, were tested. Figure 5.2 illustrates the differences between calculated average bond length and the corresponding experimental results. We can find that the bond length differences calculated using LANL2DZ were closer to those obtained from crystal structures in the majority cases. Thus, LANL2DZ was used exclusively in the calculations discussed in the following section.

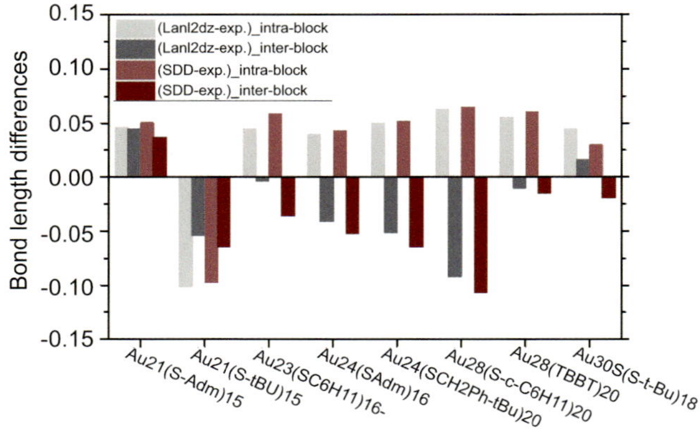

Figure 5.2 Differences of average intra-unit and inter-unit Au–Au bond lengths using LANL2DZ basis set and SDD basis set over the experimental average bond length. –SR groups were replaced by –SH during calculations to lower the computation price.

5.3.3 Bond Length and Bond Order

As stated above, three units of products of *step a* in Fig. 5.1 can be obtained according to GUM [86]. In this section, we would show if the inter-unit or intra-unit Au–Au distances fit with the decomposition results of GUM.

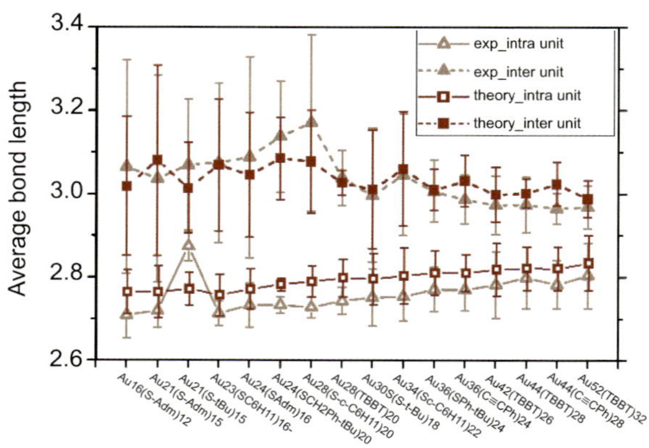

Figure 5.3 Experimental and calculated bond lengths of two Au atoms within a unit (intra-unit) and between units (inter-unit). –SR groups were replaced by –SH during calculations to lower the computational price.

Bulk gold and gold nanoparticles with diameters above 2 nm typically possess fcc-based structure. It has also been stated that the fcc-based Au core structure would collapse when the size goes down to nanocluster whose diameter is below 2 nm [16, 57]. During this process, structural distortions may take place. One important piece of information to reflect the structural variations of AuNCs is their structural parameter. Figure 5.3 shows the results of averaged experimental and theoretical intra-unit and inter-unit Au–Au distances. An obvious gap can be observed between intra-unit and inter-unit Au–Au lengths for both experimental and theoretical results. The presence of the bond length gap demonstrates that the fcc-based AuNC kernel is less uniformed over the bulk gold. This gap also indicates that units defined in this work can actually be separated structurally. The results of separating the Au kernel will just be identical when applying GUM or counting the bond length difference between intra-unit and inter-unit Au–Au distances. We can find in Fig. 5.3 that the intra-unit Au–Au distance generally presents an increasing trend with the enlargement of the AuNC size, while the inter-unit Au–Au distance has the opposite trend. These two contrary tendencies imply that the bigger the cluster, the more likely the bulk. In addition, we also note that although inter-unit or intra-unit Au–Au distance may increase or decrease along with the cluster size, the calculated average Au–Au bond length within the Au core does not change too much from 2.90 Å, as the short dashed line shows in Fig. 5.3. It is only this overall average value that is close to the distance in bulk gold, that is, 2.88 Å [42].

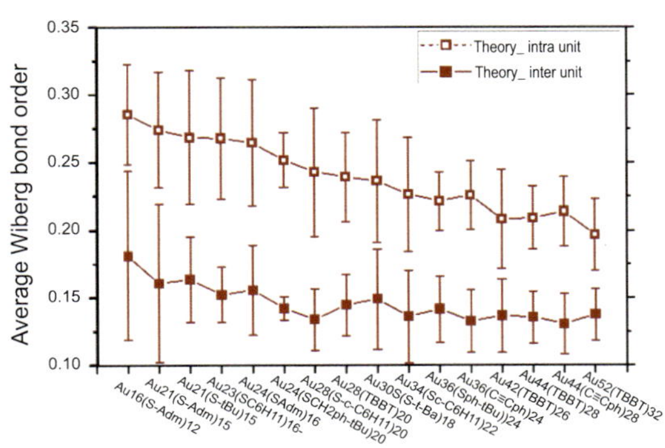

Figure 5.4 Wiberg bond order of two Au atoms within a unit (intra-unit) and between units (inter-unit). –SR groups were replaced by –SH during calculations to lower the computation price.

Besides the bond length, bond order (BO) is also a significant parameter to reflect the diversity of chemical bonds. In this work, Wiberg BOs of intra-unit and inter-unit Au–Au bonds were calculated (Fig. 5.4).

As with the average intra/inter-unit Au–Au bond length, there also exists inhomogeneous distribution for the corresponding Wiberg BO. The intra-unit Au–Au Wiberg BO generally follows the tendency that the longer the averaged bond length, the lower the corresponding averaged Wiberg BO. Following this reverse trend between bond length and bond order, the positive correlation between the inter-unit Au–Au distance and the corresponding BO seems odd at the first glance. Prior to clarifying this point, we need to address that all the given bond lengths of Fig. 5.3 and the Wiberg BO of Fig. 5.4 are average values by counting the respective types of Au–Au bond. When a cluster has fewer Au atoms, it also has limited-sized kernel. Under this condition, elementary block Au_3^+ or Au_4^{2+} of GUM or their fused small-sized unit possibly exists as an independent part, which implies that their orientations are susceptible to steric environments. The Au kernels of optimized nanoclusters $Au_{21}SH_{15}$ and $Au_{52}SH_{32}$ are in Fig. 5.5a,b as an example. To reflect the relative orientations, representative inter-unit Au–Au bond lengths were also listed. The first Au kernel of Fig. 5.5 shows that two Au_3^+ blocks and one Au_4^{2+} block stack face to face. We can find that the distances of Au_1–Au_2 and Au_1–Au_3 differ by 0.605 Å. And these two bond lengths should be roughly equal at first thought for a highly symmetric Au core. This distortion should come from the influence of outer protecting ligands. In addition, the shortest and longest inter-unit Au–Au distances are 2.775 Å and 3.474 Å, respectively. Compared with the kernel of Fig. 5.5a, the Au–Au distance of $Au_{52}SH_{32}$ shown in Fig. 5.5b are more uniform, with the shortest and longest inter-unit Au-Au distances of 2.929 Å and 3.085 Å, respectively. Parts of the inter-unit Au–Au bond lengths were shortened by the steric environment which in turn increase the corresponding average bond order. This is why the inter-unit Au–Au bonds do not seem to follow reverse tendency between bond length and bond order. To show directly the deviations of inter-unit Au–Au bond lengths of Fig. 5.5a,b, their standard deviations (SDs) were calculated and were 0.229 Å and 0.044 Å, respectively. This result reflects that kernels of large nanoclusters are generally more symmetric than that of small-sized nanoclusters. Unlike the inter-unit bond length, the intra-unit Au–Au bond lengths are just the Au–Au distances within stable elementary blocks, for example, Au_3^+ block and Au_4^{2+} block. These elementary blocks are supposed to be relatively stable [15, 86, 102]. To confirm this, SDs of intra-unit bond lengths for Fig. 5.5(a,b) were also calculated and were 0.063 Å and 0.066 Å, respectively.

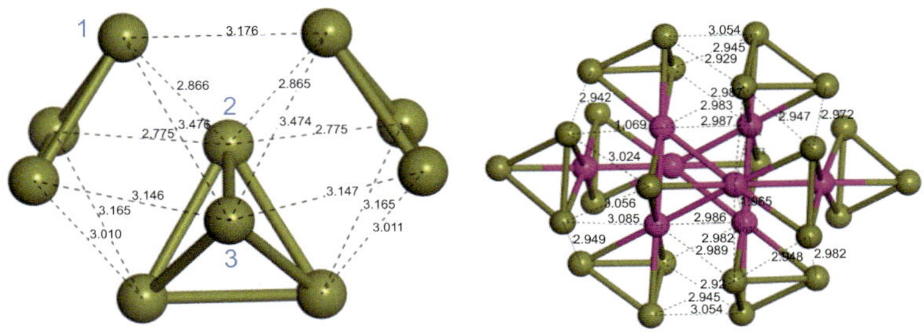

(a) (b)

Figure 5.5 Au kernels extracted from optimized (a) $Au_{21}SH_{15}$ and (b) $Au_{52}SH_{32}$ at TPSS/LANL2DZ level of theory. Only some representative bond lengths are given in this figure for clarity.

5.4 Conclusion

Gold is an amazing element that exhibits several distinct properties compared with other elements in the same group. This is largely caused by the much stronger relativistic effects of gold. One significant change brought by relativity is the contraction of 6s orbital and the expansion of 5d orbitals. It is rather necessary to include relativity for AuNCs because when the size of AuNCs narrows to as low as 2 nm, they become molecule-like. And the interactions between frontier molecular orbitals dominate among the interactions with other clusters or molecules at this scale. Changing the size of AuNCs can also impact the ratio between surface and volume, which is believed to be important for catalysis reactions. This also demonstrates the necessity to prepare AuNCs with precise size. Besides the experimental contributions, it is equally important to understand the stability of synthesized AuNCs theoretically because it is valuable for the understanding of the structure–property relationship. Several theoretical models have been proposed for this purpose. Among them, the "divide and protect" concept proposes that different bonding statuses of Au atoms may coexist in the outer ligand and the inner core. A general inherent structure rule proposed later suggests that ligand-protected AuNCs can be further divided into several parts. This rule has been confirmed by the crystal structure prepared afterward. SAC model proposed by Häkkinen suggests that the free electrons of AuNCs should be consistent with a series of magic numbers. While SAC model is sound, some AuNCs do not exhibit one of the magic numbers. For those AuNCs with 4e, their stabilities can be explained by the SAN model. GUM model was developed in the hope of providing a unified model to interpret the stability of AuNCs. This model suggests that the Au core of AuNCs can be seen as proper combinations of three elementary blocks Au_3(2e), Au_4(2e), and Au_{13}(8e). At last, through applying GUM, experimental crystal structures of fcc-based AuNCs were analyzed. These structures show inhomogeneous distribution of Au–Au bond lengths present between the intra-unit and the inter-unit Au–Au distance, which provides the straightforward evidence of the validity of GUM. Calculations of Wiberg bond order confirm this point as well. On the basis of these results, further segregation of Au kernels by considering their valence electron numbers can be done according to GUM. To understand the positive correlation between inter-unit Au–Au bond lengths and their bond orders, two representative Au cores extracted from $Au_{21}SH_{15}$ and $Au_{52}SH_{32}$ were analyzed. The results show that small-sized Au kernels are inclined to tune their orientations when bonding with the outer protecting ligands. And the correlation between inter-unit Au–Au bond length and its bond order reflects the nature of Au–Au bond in AuNCs.

Although much information about the synthesis, electronic structure, geometry, and applications of AuNCs is known now, more aspects of AuNCs await to be discovered yet. To our knowledge, the "static" properties of AuNCs have been discussed appreciably from both experimental and theoretical views. However, the detailed mechanisms of several "dynamic" processes are rarely addressed, for example, the nucleation and the growth mechanism, the transformation mechanism between two clusters of different size, the isomerization mechanism between two clusters of equal size, and detailed mechanism of ligand exchange processes. These processes are important because they help a lot toward the controllable synthesis of these unique materials.

Acknowledgments

We would like to thank the funding support from National Science Foundation of China (21773287) and helpful discussion with Gangli Wang.

References

1. Pyykkö, P. (2008). Theoretical chemistry of gold. III. *Chem. Soc. Rev.*, **37**, pp. 1967–1997.
2. Pyykkö, P. (2004). Theoretical chemistry of gold. *Angew. Chem. Int. Ed.*, **43**, pp. 4412–4456.
3. Gorin, D. J. and Toste, F. D. (2007). Relativistic effects in homogeneous gold catalysis. *Nature*, **446**, pp. 395.
4. Figgen, D., Rauhut, G., Dolg, M. and Stoll, H. (2005). Energy-consistent pseudopotentials for group 11 and 12 atoms: Adjustment to multi-configuration Dirac–Hartree–Fock data. *Chem. Phys.*, **311**, pp. 227–244.
5. Schwerdtfeger, P., Hermann, H. L. and Schmidbaur, H. (2003). Stability of the gold(I)–phosphine bond. A comparison with other group 11 elements. *Inorg. Chem.*, **42**, pp. 1334–1342.
6. Pyykkö, P. (2012). Relativistic effects in chemistry: More common than you thought. *Annu. Rev. Phys. Chem.*, **63**, pp. 45–64.
7. Tang, Q., Hu, G., Fung, V. and Jiang, D.-e. (2018). Insights into interfaces, stability, electronic properties, and catalytic activities of atomically precise metal nanoclusters from first principles. *Acc. Chem. Res.*, **51**, pp. 2793–2802.
8. Pyykkö, P. and Bond G. C. (2016). Introduction to the physical and chemical properties of gold, in *Gold Nanoparticles for Physics, Chemistry and Biology*, World Scientific (Europe), pp. 29–49.
9. Ni, B. and Wang, X. (2016). Chemistry and properties at a sub-nanometer scale. *Chem. Sci.*, **7**, pp. 3978–3991.
10. Lopez-Acevedo, O., Kacprzak, K. A., Akola, J. and Häkkinen, H. (2010). Quantum size effects in ambient CO oxidation catalysed by ligand-protected gold clusters. *Nat. Chem.*, **2**, p. 329.
11. Beletskiy, E., Kung, M. C. and Kung, H. H. (2016). Catalytic properties of gold nanoparticles, in *Gold Nanoparticles for Physics, Chemistry and Biology*, World Scientific (Europe), pp. 285–318.

12. Zhu, B., Xu, Z., Wang, C. and Gao, Y. (2016). Shape evolution of metal nanoparticles in water vapor environment. *Nano Lett.,* **16**, pp. 2628–2632.

13. Duan, M., Yu, J., Meng, J., Zhu, B., Wang, Y. and Gao, Y. (2018). Reconstruction of supported metal nanoparticles in reaction conditions. *Angew. Chem. Int. Ed.* **130**, pp. 6574–6579.

14. Li, X.-Y., Zhu, B., Qi, R. and Gao, Y. (2019). Real-time simulation of nonequilibrium nanocrystal transformations. *Adv. Theory Simul.,* **2**, p. 1800127.

15. Xu, W. W., Zeng, X. C. and Gao, Y. (2018). Application of electronic counting rules for ligand-protected gold nanoclusters. *Acc. Chem. Res.,* **51**, pp. 2739–2747.

16. Li, G. and Jin, R. (2013). Atomically precise gold nanoclusters as new model catalysts. *Acc. Chem. Res.,* **46**, pp. 1749–1758.

17. Jin, R., Zeng, C., Zhou, M. and Chen, Y. (2016). Atomically precise colloidal metal nanoclusters and nanoparticles: Fundamentals and opportunities. *Chem. Rev.,* **116**, pp. 10346–10413.

18. Zhu, M., Lanni, E., Garg, N., Bier, M. E. and Jin, R. (2008). Kinetically controlled, high-yield synthesis of Au_{25} clusters. *J. Am. Chem. Soc.,* **130**, pp. 1138–1139.

19. Wu, Z., Suhan, J. and Jin, R. (2009). One-pot synthesis of atomically monodisperse, thiol-functionalized Au_{25} nanoclusters. *J. Mater. Chem.,* **19**, pp. 622–626.

20. Qian, H. and Jin, R. (2009). Controlling nanoparticles with atomic precision: The case of $Au_{144}(SCH_2CH_2Ph)_{60}$. *Nano Lett.,* **9**, pp. 4083–4087.

21. Qian, H., Zhu, Y. and Jin, R. (2009). Size-focusing synthesis, optical and electrochemical properties of monodisperse $Au_{38}(SC_2H_4Ph)_{24}$ nanoclusters. *ACS Nano,* **3**, pp. 3795–3803.

22. Zeng, C., Chen, Y. , Li, G. and Jin, R. (2014). Magic size Au_{64}(S-c-$C_6H_{11})_{32}$ nanocluster protected by cyclohexanethiolate. *Chem. Mater.,* **26**, pp. 2635–2641.

23. Li, G., Zeng, C. and Jin, R. (2014). Thermally robust $Au_{99}(SPh)_{42}$ nanoclusters for chemoselective hydrogenation of nitrobenzaldehyde derivatives in water. *J. Am. Chem. Soc.,* **136**, pp. 3673–3679.

24. Qian, H., Zhu, Y. and Jin, R. (2012). Atomically precise gold nanocrystal molecules with surface plasmon resonance. *Proc. Natl. Acad. Sci. USA,* **109**, p. 696.

25. Zeng, C., Qian, H., Li, T., Li, G., Rosi, N. L., Yoon, B., Barnett, R. N., Whetten, R. L., Landman, U. and Jin, R. (2012). Total structure and electronic properties of the gold nanocrystal $Au_{36}(SR)_{24}$. *Angew. Chem. Int. Ed.*, **51**, pp. 13114–13118.

26. Zeng, C., Li, T., Das, A., Rosi, N. L. and Jin, R. (2013). Chiral structure of thiolate-protected 28-gold-atom nanocluster determined by X-ray crystallography. *J. Am. Chem. Soc.*, **135**, pp. 10011–10013.

27. Zeng, C., Chen, Y., Kirschbaum, K., Appavoo, K., Sfeir, M. Y. and Jin, R. (2015). Structural patterns at all scales in a nonmetallic chiral $Au_{133}(SR)_{52}$ nanoparticle. *Sci. Adv.*, **1**, p. e1500045.

28. Das, A., Li, T., Nobusada, K., Zeng, C., Rosi, N. L. and Jin, R. (2013). Nonsuperatomic $[Au_{23}(SC_6H_{11})_{16}]^-$ nanocluster featuring bipyramidal Au_{15} kernel and trimeric $Au_3(SR)_4$ motif. *J. Am. Chem. Soc.*, **135**, pp. 18264–18267.

29. Qian, H., Zhu, Y. and Jin, R. (2010). Isolation of ubiquitous $Au_{40}(SR)_{24}$ clusters from the 8 kDa gold clusters. *J. Am. Chem. Soc.*, **132**, pp. 4583–4585.

30. Zhu, M., Qian, H. and Jin, R. (2009). Thiolate-protected Au_{20} clusters with a large energy gap of 2.1 eV. *J. Am. Chem. Soc.*, **131**, pp. 7220–7221.

31. Yao, Q., Yu, Y., Yuan, X., Yu, Y., Xie, J. and Lee, J. Y. (2013). Two-phase synthesis of small thiolate-protected Au_{15} and Au_{18} nanoclusters. *Small*, **9**, pp. 2696–2701.

32. Yu, Y., Chen, X., Yao, Q., Yu, Y., Yan, N. and Xie, J. (2013). Scalable and precise synthesis of thiolated Au_{10-12}, Au_{15}, Au_{18}, and Au_{25} nanoclusters via pH controlled CO reduction. *Chem. Mater.*, **25**, pp. 946–952.

33. Song, Y., Wang, S., Zhang, J., Kang, X., Chen, S., Li, P., Sheng, H. and Zhu, M. (2014). Crystal structure of selenolate-protected $Au_{24}(SeR)_{20}$ nanocluster. *J. Am. Chem. Soc.*, **136**, pp. 2963–2965.

34. Wan, X.-K., Yuan, S.-F., Lin, Z.-W. and Wang, Q.-M. (2014). A chiral gold nanocluster Au_{20} protected by tetradentate phosphine ligands. *Angew. Chem. Int. Ed.*, **53**, pp. 2923–2926.

35. Yu, Y., Luo, Z., Chevrier, D. M., Leong, D. T., Zhang, P., Jiang, D.-e. and Xie, J. (2014). Identification of a highly luminescent $Au_{22}(SG)_{18}$ nanocluster. *J. Am. Chem. Soc.*, **136**, pp. 1246–1249.

36. Yuan, X., Zhang, B., Luo, Z., Yao, Q., Leong, D. T., Yan, N. and Xie, J. (2014). Balancing the rate of cluster growth and etching for gram-scale synthesis of thiolate-protected Au_{25} nanoclusters with atomic precision. *Angew. Chem. Int. Ed.*, **53**, pp. 4623–4627.

37. Chen, S., Wang, S., Zhong, J., Song, Y., Zhang, J., Sheng, H., Pei, Y. and Zhu, M. (2015). The structure and optical properties of the $[Au_{18}(SR)_{14}]$ nanocluster. *Angew. Chem. Int. Ed.,* **54**, pp. 3145–3149.

38. Song, Y., Fu, F., Zhang, J., Chai, J., Kang, X., Li, P., Li, S., Zhou, H. and Zhu, M. (2015). The magic Au_{60} nanocluster: A new cluster-assembled material with five Au_{13} building blocks. *Angew. Chem. Int. Ed.,* **54**, pp. 8430–8434.

39. Wan, X.-K., Yuan, S.-F., Tang, Q., Jiang, D.-e. and Wang, Q.-M. (2015). Alkynyl-protected Au_{23} nanocluster: A 12-electron system. *Angew. Chem. Int. Ed.,* **54**, pp. 5977–5980.

40. Yang, S., Chai, J., Song, Y., Kang, X., Sheng, H., Chong, H. and Zhu, M. (2015). A new crystal structure of Au_{36} with a Au_{14} kernel cocapped by thiolate and chloride. *J. Am. Chem. Soc.,* **137**, pp. 10033–10035.

41. Yu, Y., Li, J., Chen, T., Tan, Y. N. and Xie, J. (2015). Decoupling the CO-reduction protocol to generate luminescent $Au_{22}(SR)_{18}$ nanocluster. *J. Phys. Chem. C,* **119**, pp. 10910–10918.

42. Chen, S., Xiong, L., Wang, S., Ma, Z., Jin, S., Sheng, H., Pei, Y. and Zhu, M. (2016). Total structure determination of $Au_{21}(S\text{-}Adm)_{15}$ and geometrical/electronic structure evolution of thiolated gold nanoclusters. *J. Am. Chem. Soc.* , **138**, pp. 10754–10757.

43. Huang, P., Chen, G., Jiang, Z., Jin, R., Zhu, Y. and Sun, Y. (2013). Atomically precise Au_{25} superatoms immobilized on CeO_2 nanorods for styrene oxidation. *Nanoscale,* **5**, pp. 3668–3672.

44. Liu, C., Zheng, D., Hu, W., Zhu, Q., Tian, Z., Zhao, J., Zhu, Y. and Ma, J. (2017). Tuning the collective switching behavior of azobenzene/Au hybrid materials: Flexible versus rigid azobenzene backbones and Au(111) surfaces versus curved Au nanoparticles. *Nanoscale,* **9**, pp. 16700–16710.

45. Wang, L., Chai, X., Cheng, X. and Zhu, Y. (2018). Structure-specific catalytic oxidation with O_2 by isomers in $Au_{28}(SR)_{20}$ nanoclusters. *ChemistrySelect,* **3**, pp. 6165–6169.

46. Cai, X., Saranya, G., Shen, K., Chen, M., Si, R., Ding, W. and Zhu, Y. (2019). Reversible switching of catalytic activity by shuttling an atom into and out of gold nanoclusters. *Angew. Chem. Int Ed. Engl,* **58**, pp. 9964–9968.

47. Connor, E. E., Mwamuka, J., Gole, A., Murphy, C. J. and Wyatt, M. D. (2005). Gold nanoparticles are taken up by human cells but do not cause acute cytotoxicity. *Small,* **1**, pp. 325–327.

48. McIntosh, C. M., Esposito, E. A., Boal, A. K., Simard, J. M., Martin, C. T. and Rotello, V. M. (2001). Inhibition of DNA transcription using cationic

mixed monolayer protected gold clusters. *J. Am. Chem. Soc.,* **123**, pp. 7626–7629.

49. Han, G., Chari, N. S., Verma, A., Hong, R., Martin, C. T. and Rotello, V. M. (2005). Controlled recovery of the transcription of nanoparticle-bound DNA by intracellular concentrations of glutathione. *Bioconjug. Chem.,* **16**, pp. 1356–1359.
50. Paciotti, G. F., Kingston, D. G. I. and Tamarkin, L. (2006). Colloidal gold nanoparticles: A novel nanoparticle platform for developing multifunctional tumor-targeted drug delivery vectors. *Drug Dev. Res.,* **67**, pp. 47–54.
51. Castner, D. G., Hinds, K. and Grainger, D. W. (1996). X-ray photoelectron spectroscopy sulfur 2p study of organic thiol and disulfide binding interactions with gold surfaces. *Langmuir,* **12**, pp. 5083–5086.
52. Häkkinen, H., Barnett, R. N. and Landman, U. (1999). Electronic structure of passivated $Au_{38}(SCH_3)_{24}$ nanocrystal. *Phys. Rev. Lett.,* **82**, p. 3264.
53. Maksymovych, P., Sorescu, D. C. and Yates Jr, J. T. (2006). Gold-adatom-mediated bonding in self-assembled short-chain alkanethiolate species on the Au(111) surface. *Phys. Rev. Lett.,* **97**, p. 146103.
54. Häkkinen, H., Walter, M. and Grönbeck, H. (2006). Divide and protect: Capping gold nanoclusters with molecular gold–thiolate rings. *J. Phys. Chem. B,* **110**, pp. 9927–9931.
55. Jadzinsky, P. D., Calero, G., Ackerson, C. J., Bushnell, D. A. and Kornberg, R. D. (2007). Structure of a thiol monolayer-protected gold nanoparticle at 1.1 Å resolution. *Science,* **318**, p. 430.
56. Pei, Y. and Zeng, X. C. (2012). Investigating the structural evolution of thiolate protected gold clusters from first-principles. *Nanoscale,* **4**, pp. 4054–4072.
57. Qian, H., Zhu, M., Wu, Z. and Jin, R. (2012). Quantum sized gold nanoclusters with atomic precision. *Acc. Chem. Res. ,* **45**, pp. 1470–1479.
58. Jiang, D.-e. (2013). The expanding universe of thiolated gold nanoclusters and beyond. *Nanoscale,* **5**, pp. 7149–7160.
59. Fernando, A., Weerawardene, K. L., Karimova, N. V. and Aikens, C. M. (2015). Quantum mechanical studies of large metal, metal oxide, and metal chalcogenide nanoparticles and clusters. *Chem. Rev.,* **115**, pp. 6112–6216.
60. Chakraborty, I. and Pradeep, T. (2017). Atomically precise clusters of noble metals: Emerging link between atoms and nanoparticles. *Chem. Rev.,* **117**, pp. 8208–8271.

61. Ma, Z., Wang, P., Xiong, L. and Pei, Y. (2017). Thiolate-protected gold nanoclusters: Structural prediction and the understandings of electronic stability from first principles simulations. *WIREs Comput. Mol. Sci.,* **7**, p. e1315.
62. Aikens, C. M. (2018). Electronic and geometric structure, optical properties, and excited state behavior in atomically precise thiolate-stabilized noble metal nanoclusters. *Acc. Chem. Res.,* **51**, pp. 3065–3073.
63. Gan, Z., Xia, N. and Wu, Z. (2018). Discovery, mechanism, and application of antigalvanic reaction. *Acc. Chem. Res.,* **51**, pp. 2774–2783.
64. Higaki, T., Li, Q., Zhou, M., Zhao, S., Li, Y., Li, S. and Jin, R. (2018). Toward the tailoring chemistry of metal nanoclusters for enhancing functionalities. *Acc. Chem. Res.,* **51**, pp. 2764–2773.
65. Kang, X., Chong, H. and Zhu, M. (2018). $Au_{25}(SR)_{18}$: The captain of the great nanocluster ship. *Nanoscale,* **10**, pp. 10758–10834.
66. Lei, Z., Wan, X.-K., Yuan, S.-F., Guan, Z.-J. and Wang, Q.-M. (2018). Alkynyl approach toward the protection of metal nanoclusters. *Acc. Chem. Res.,* **51**, pp. 2465–2474.
67. Wang, S., Li, Q., Kang, X. and Zhu, M. (2018). Customizing the structure, composition, and properties of alloy nanoclusters by metal exchange. *Acc. Chem. Res.,* **51**, pp. 2784–2792.
68. Weerawardene, K. L. D. M., Häkkinen, H. and Aikens, C. M. (2018). Connections between theory and experiment for gold and silver nanoclusters. *Annu. Rev. Phys. Chem.,* **69**, pp. 205–229.
69. Yan, J., Teo, B. K. and Zheng, N. (2018). Surface chemistry of atomically precise coinage–metal nanoclusters: From structural control to surface reactivity and catalysis. *Acc. Chem. Res.,* **51**, pp. 3084–3093.
70. Pei, Y., Wang, P., Ma, Z. and Xiong, L. (2019). Growth-rule-guided structural exploration of thiolate-protected gold nanoclusters. *Acc. Chem. Res.,* **52**, pp. 23–33.
71. Garzón, I., Rovira, C., Michaelian, K., Beltrán, M., Ordejón, P., Junquera, J., Sánchez-Portal, D., Artacho, E. and Soler, J. M (2000). Do thiols merely passivate gold nanoclusters? *Phys. Rev. Lett.* **85**, p. 5250.
72. Häkkinen, H. (2016). Theoretical studies of gold nanoclusters in various chemical environments: When the size matters, in *Gold Nanoparticles for Physics, Chemistry and Biology,* World Scientific (Europe), pp. 437–481.
73. Heaven, M. W., Dass, A., White, P. S., Holt, K. M. and Murray, R. W. (2008). Crystal structure of the gold nanoparticle $[N(C_8H_{17})_4][Au_{25}(SCH_2CH_2Ph)_{18}]$. *J. Am. Chem. Soc.,* **130**, pp. 3754–3755.

74. Pei, Y., Gao, Y. and Zeng, X. C. (2008). Structural prediction of thiolate-protected Au_{38}: A face-fused bi-icosahedral Au core. *J. Am. Chem. Soc.,* **130**, pp. 7830–7832.

75. Pei, Y., Lin, S., Su, J. and Liu, C. (2013). Structure prediction of $Au_{44}(SR)_{28}$: A chiral superatom cluster. *J. Am. Chem. Soc.,* **135**, pp. 19060–19063.

76. Xu, W. W., Gao, Y. and Zeng, X. C. (2015). Unraveling structures of protection ligands on gold nanoparticle $Au_{68}(SH)_{32}$. *Sci. Adv.,* **1**, p. e1400211.

77. Mingos, D. M. P. (1985). Theoretical analyses and electron counting rules for high nuclearity clusters. *J. Chem. Soc. Chem. Commun.,* pp. 1352–1354.

78. Wales, D. J. and Mingos, D. M. P. (1989). Splitting of cluster orbitals. *Inorg. Chem.,* **28**, pp. 2748–2754.

79. Ivanov, S. A., Arachchige, I. and Aikens, C. M. (2011). Density functional analysis of geometries and electronic structures of gold-phosphine clusters. The case of $Au_4(PR_3)_4^{2+}$ and $Au_4(\mu_2\text{-}I)_2(PR_3)_4$. *J. Phys. Chem. A,* **115**, pp. 8017–8031.

80. Guidez, E. B., Mäkinen, V., Häkkinen, H. and Aikens, C. M. (2012). Effects of silver doping on the geometric and electronic structure and optical absorption spectra of the $Au_{25-n}Ag_n(SH)_{18}^-$ (n = 1, 2, 4, 6, 8, 10, 12) bimetallic nanoclusters. *J. Phys. Chem. C,* **116**, pp. 20617–20624.

81. Xiong, L., Yang, S., Sun, X., Chai, J., Rao, B., Yi, L., Zhu, M. and Pei, Y. (2018). Structure and electronic structure evolution of thiolate-protected gold nanoclusters containing quasi face-centered-cubic kernels. *J. Phys. Chem. C,* **122**, pp. 14898–14907.

82. Walter, M., Akola, J., Lopez-Acevedo, O., Jadzinsky, P. D., Calero, G., Ackerson, C. J., Whetten, R. L., Grönbeck, H. and Häkkinen, H. (2008). A unified view of ligand-protected gold clusters as superatom complexes. *Proc. Natl. Acad. Sci. USA,* **105**, p. 9157.

83. Reilly, S. M., Krick, T. and Dass, A. (2010). Surfactant-free synthesis of ultrasmall gold nanoclusters. *J. Phys. Chem. C,* **114**, pp. 741–745.

84. Zhu, M., Qian, H. and Jin, R. (2010). Thiolate-protected $Au_{24}(SC_2H_4Ph)_{20}$ nanoclusters: Superatoms or not? *J. Phys. Chem. Lett.,* **1**, pp. 1003–1007.

85. Cheng, L., Yuan, Y., Zhang, X. and Yang, J. (2013). Superatom networks in thiolate-protected gold nanoparticles. *Angew. Chem. Int. Ed. Engl.,* **52**, pp. 9035–9039.

86. Xu, W. W., Zhu, B., Zeng, X. C. and Gao, Y. (2016). A grand unified model for liganded gold clusters. *Nat. Commun.,* **7**, p. 13574.

87. Xu, W. W., Zeng, X. C. and Gao, Y. (2017). Au_{13}(8*e*): A secondary block for describing a special group of liganded gold clusters containing icosahedral Au_{13} motifs. *Chem. Phys. Lett.,* **675**, pp. 35–39.

88. Xu, W. W., Li, Y., Gao, Y. and Zeng, X. C. (2016). Unraveling a generic growth pattern in structure evolution of thiolate-protected gold nanoclusters. *Nanoscale,* **8**, pp. 7396–7401.

89. Xu, W. W., Zeng, X. C. and Gao, Y. (2018). The structural isomerism in gold nanoclusters. *Nanoscale,* **10**, pp. 9476–9483.

90. Xu, W. W., Zeng, X. C. and Gao, Y. (2017). $Au_3(\mu_3\text{-}S)(0e)$ elementary block: new insights into ligated gold clusters with μ_3-sulfido motifs. *Nanoscale,* **9**, pp. 8990–8996.

91. Wenwu X. and Yi G. (2018). Thiolate-protected hollow gold nanospheres. *Acta Phys. Chim. Sin.*, **34**, pp. 770–775.

92. Zeng, C., Chen, Y., Liu, C., Nobusada, K., Rosi, N. L. and Jin, R. (2015). Gold tetrahedra coil up: Kekulé-like and double helical superstructures. *Sci. Adv.,* **1**, p. e1500425.

93. Zeng, C., Chen, Y., Iida, K., Nobusada, K., Kirschbaum, K., Lambright, K. J. and Jin, R. (2016). Gold quantum boxes: On the periodicities and the quantum confinement in the Au_{28}, Au_{36}, Au_{44}, and Au_{52} magic series. *J. Am. Chem. Soc.,* **138**, pp. 3950–3953.

94. Dong, H., Liao, L., Zhuang, S., Yao, C., Chen, J., Tian, S., Zhu, M., Liu, X., Li, L. and Wu, Z. (2017). A novel double-helical-kernel evolution pattern of gold nanoclusters: Alternate single-stranded growth at both ends. *Nanoscale,* **9**, pp. 3742–3746.

95. Crasto, D., Malola, S., Brosofsky, G., Dass, A. and Häkkinen, H. (2014). Single crystal XRD structure and theoretical analysis of the chiral $Au_{30}S(S\text{-}t\text{-}Bu)_{18}$ cluster. *J. Am. Chem. Soc.,* **136**, pp. 5000–5005.

96. Dass, A., Jones, T., Rambukwella, M., Crasto, D., Gagnon, K. J., Sementa, L., De Vetta, M., Baseggio, O., Aprà, E., Stener, M. and Fortunelli, A. (2016). Crystal structure and theoretical analysis of green gold $Au30(S\text{-}t\text{Bu})_{18}$ nanomolecules and their relation to $Au_{30}S(S\text{-}t\text{Bu})_{18}$. *J. Phys. Chem. C,* **120**, pp. 6256–6261.

97. Crasto, D., Barcaro, G., Stener, M., Sementa, L., Fortunelli, A. and Dass, A. (2014). $Au_{24}(SAdm)_{16}$ nanomolecules: X-ray crystal structure, theoretical analysis, adaptability of adamantane ligands to form $Au_{23}(SAdm)_{16}$ and $Au_{25}(SAdm)_{16}$, and its relation to $Au_{25}(SR)_{18}$. *J. Am. Chem. Soc.,* **136**, pp. 14933–14940.

98. Jiang, D. E., Overbury, S. H. and Dai S. (2013). Structure of $Au_{15}(SR)_{13}$ and its implication for the origin of the nucleus in thiolated gold nanoclusters. *J. Am. Chem. Soc.,* **135**, pp. 8786–8789.

99. Tao, J. M., Perdew, J. P., Staroverov, V. N. and Scuseria, G. E. (2003). Climbing the density functional ladder: Nonempirical meta-generalized gradient approximation designed for molecules and solids. *Phys. Rev. Lett.,* **91**, p. 146401.

100. Pei, Y., Shao, N., Li H,. Jiang, D. E. and Zeng X. C. (2011). Hollow polyhedral structures in small gold-sulfide clusters. *ACS Nano,* **5**, pp. 1441–1449.

101. Pyykkö, P. (2002). Relativity, gold, closed-shell interactions, and $CsAu{\cdot}NH_3$. *Angew. Chem. Int. Ed.,* **41**, pp. 3573–3578.

102. Takagi, N., Ishimura, K., Matsui, M., Fukuda, R., Matsui, T., Nakajima, T., Ehara, M. and Sakaki, S. (2015). How can we understand Au_8 cores and entangled ligands of selenolate- and thiolate-protected gold nanoclusters $Au_{24}(ER)_{20}$ and $Au_{20}(ER)_{16}$ (E = Se, S; R = Ph, Me)? A theoretical study. *J. Am. Chem. Soc.*, **137**, pp. 8593–8602.

103. Frisch, M., Trucks, G., Schlegel, H., Scuseria, G., Robb, M., Cheeseman, J., Scalmani, G., Barone, V., Mennucci, B., Petersson, G. A. et al. (2009). Gaussian 09, rev. D. 01. Gaussian, Inc., Wallingford CT.

Chapter 6

Optical Properties of Atomically Precise Gold Nanoclusters: Transition from Excitons to Plasmons

Tatsuya Higaki and Rongchao Jin
Department of Chemistry, Carnegie Mellon University,
4400 Fifth Avenue, Pittsburgh, PA 15213, USA
rongchao@andrew.cmu.edu

The optical properties constitute a major topic of metal nanoparticle research. Gold nanoparticles (NPs) are particularly attractive owing to their elegant colors that are tunable with particle size and shape. While the size-dependent optical properties of regular gold NPs (e.g., 5–100 nm) have been extensively studied, ultrasmall nanoparticles remain largely under-explored due to major difficulties in the synthesis of high-quality particles. Significantly, recent progress toward atomically precise nanochemistry has opened exciting avenues for exploring new optical properties of ultrasmall particles at the unprecedented atomic level (i.e., how adding or removing one atom changes the optical properties). This chapter illustrates the optical absorption of some typical sizes of atomically precise

Atomically Precise Nanoclusters
Edited by Yan Zhu and Rongchao Jin

ISBN 978-981-4877-08-4 (Hardcover), 978-1-003-11990-6 (eBook)
www.jennystanford.com

nanoparticles (often called nanoclusters in order to differentiate them from the conventional nanoparticles without atomic precision). The grand exciton-to-plasmon transition has been discovered to occur between Au_{246} and Au_{279}.

6.1 Introduction

Since the advent of nanotechnology in 2000, gold NPs have captured significant attention due to their diverse applications. This surge in new applications has led to a major effort to understand the fundamental underlying properties, in particular the optical properties. Conventional gold NPs are well known to be metallic, that is, thousands to millions of free electrons (depending on the particle size) roaming freely inside the particle. When light is used to excite such NPs, a strong resonant absorption band is observed, with λ_{peak} being at 520–570 nm depending on the size of spherical NPs (Fig. 6.1) [1]. This excitation mode is called surface plasmon resonance (SPR) and it is unique to metallic nanoparticles. The nature of SPR is due to a collective excitation of free electrons in the particle upon absorbing light.

With decreasing size, the SPR band blue shifts and also becomes weaker (Fig. 6.1). Indeed, below ~3 nm, SPR fades out and new optical absorption features emerge, for example, multiple bands (as opposed to single SPR). This is due to the quantum size effect [2]. Naturally, many fundamental questions arise, such as (1) What are the origins of those multibands? (2) Does the evolution from metallic to nonmetallic states occur smoothly or abruptly? (3) If abruptly, at what precise size (i.e., the number of gold atoms in the particle) does the transition occur? Kubo indeed predicted in the 1960s a smooth progression of the electronic structure of ultrasmall metal particles with size and the HOMO–LUMO (highest occupied and lowest unoccupied molecular orbitals) gap should follow the scaling law $E_g \sim 1/n$, where n is the number of gold atoms [3]. However, experimental synthesis of ultrasmall particles has long remained a major challenge due to difficulties in controlling the particle monodispersity. For a size-dispersed sample, it was apparently impossible to investigate the precise scaling relationship. Thanks to the recently developed atomically precise nanochemistry [1], it has now become possible to tackle the above-mentioned fundamental questions, albeit half a century has passed since Kubo's theoretical prediction.

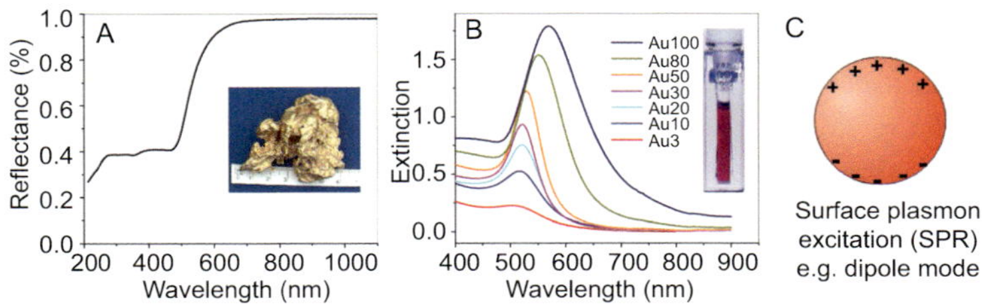

Figure 6.1 (A) Reflectance spectrum of bulk gold, (B) extinction spectra of gold colloids (3–100 nm diameters), and (C) surface plasmon excitation (e.g., dipole mode) in metal nanoparticles [1].

6.2 Optical Properties of Small-Sized Gold Nanoclusters

6.2.1 $Au_{25}(SR)_{18}$ Nanoclusters

Among the reported sizes of atomically precise nanoclusters, $Au_{25}(SR)_{18}$ (SR = thiolate) is perhaps the most studied one, due in part to its early discovery. The success in size-focusing synthesis of $Au_{25}(SR)_{18}$ in high yield and molecular purity [4] finally led to crystallization and structure determination [5].

The $Au_{25}(SR)_{18}$ structure (Fig. 6.2A) is based on a centered icosahedral Au_{13} core, which is capped by an exterior shell composed of the remaining 12 Au atoms, and the entire cluster is encapsulated by 18 thiolate ligands. The 12 surface gold atoms and 18 ligands are assembled into six dimeric staples (-SR-Au-SR-Au-SR-), which protect the Au_{13} core in D_{2h} symmetry.

The availability of the crystal structure of $Au_{25}(SR)_{18}$ permitted a structure–property correlation by performing DFT calculations; thus, the origins of multibands were revealed (Fig. 6.2B–D). Due to quantum confinement caused by the ultrasmall size (1 nm) of $Au_{25}(SR)_{18}$, the electronic structure of $Au_{25}(SR)_{18}$ is significantly quantized (Fig. 6.2C). This is in striking contrast to the continuous band structure of plasmonic Au nanoparticles or bulk gold. The first optical transition (at 680 nm, Fig. 6.2B) corresponds to a

LUMO←HOMO transition, which is essentially an intraband (*sp*←*sp*) transition (note that the term *sp* band is inherited from solid state theory for the convenience of discussing the nonmetallic to metallic state evolution; one should bear in mind that the *sp* band in clusters is significantly quantized). Interestingly, the HOMO set (nearly threefold degenerate, above the dense d-band, see Fig. 6.2B) has essentially *s* character; thus, transitions arising out of the other occupied HOMO-n orbitals (belonging to d-band) are interband transitions (*sp*←*d*). The HOMO and LUMO orbitals are comprised almost exclusively of atomic orbital contributions from 13 Au atoms in the icosahedral Au_{13} core, rather than 12 exterior Au atoms. Thus, the first 680 nm peak in the absorption spectrum can be viewed as a transition that is due entirely to the electronic and geometric structure of the Au_{13} core.

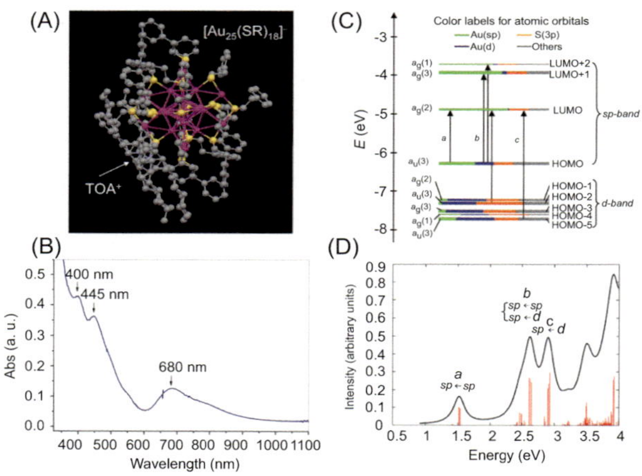

Figure 6.2 (A) X-ray structure of $Au_{25}(SR)_{18}^-$(counterion: tetraoctylammonium, TOA+). Color labels: magenta = Au, yellow = S, and gray = C; all hydrogen atoms are omitted for clarity. (B) Experimental optical absorption spectrum. (C) Kohn–Sham orbital level diagram for $Au_{25}(SR)_{18}$. (D) Theoretically simulated optical absorption spectrum [5].

6.2.2 Single-Atom Effect on Optical Properties

We chose the $Au_{24}(TBBM)_{20}$ case (TBBM = SCH_2Ph-*p*-^{t}Bu) to illustrate the single-atom effect [6]. Although its size is only one atom less than $Au_{25}(PET)_{18}$ (PET = SCH_2CH_2Ph), drastic effects were observed in terms of structure and optical properties. The Au_{24} structure exhibits a prolate shape (Fig. 6.3), as opposed to the quasi-spherical shape of Au_{25}. The core of Au_{24} is a Au_8 bi-tetrahedron, in which the two tetrahedra are anti-prismatically joined together through two triangular faces of the tetrahedra, giving rise to a prolate structure [6]. The Au_8 kernel is protected by four tetrameric staples [i.e., $Au_4(SR)_5$].

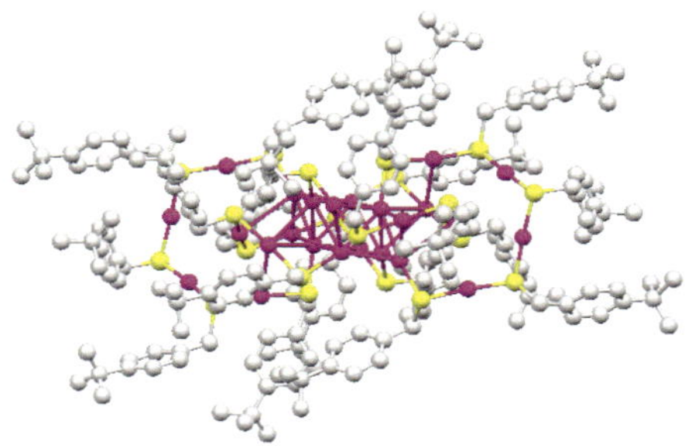

Figure 6.3 Total structure of $[Au_{24}(SCH_2Ph\text{-}^tBu)_{20}]^0$. (Color labels: magenta = Au, yellow = S, and gray = C; all H atoms are omitted for clarity) [6].

The optical absorption spectrum of Au_{24} exhibits a distinct peak at ~500 nm and less prominent ones at shorter wavelengths (Fig. 6.4A) [6]. The number of peaks is less than that of Au_{25} due to lower symmetry of Au_{24}. DFT calculations reproduced the optical spectrum and thus explained the origins of the electronic transitions involved (Fig. 6.4B) [6]. The experimental peak at ~500 nm is mainly contributed by the HOMO–LUMO electronic transition, but also in less part by the HOMO-2 to LUMO transition. Figure 6.4C,D illustrates the schematic diagrams of HOMO and LUMO orbitals, which are quite different from the superatomic orbitals ($1P^6$ and $1D^0$) in $Au_{25}(SR)_{18}^-$. The HOMO–LUMO transition (at 500 nm) of $Au_{24}(SR)_{20}$ is also significantly blue-shifted compared to that of $Au_{25}(SR)_{18}$ (at 680 nm). This is due to the smaller core (Au_8 vs Au_{13}). Generally, for small sizes of nanoclusters, the HOMO–LUMO gap energies do not scale with size in a monotonic trend, but rather a zigzag behavior was observed, which is largely caused by the diverse structures of small nanoclusters [7].

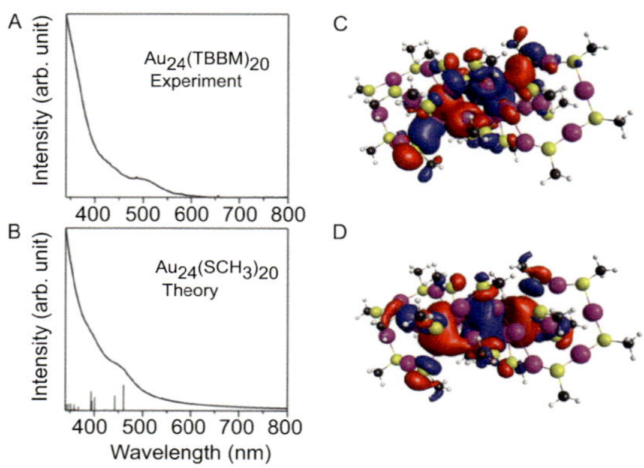

Figure 6.4 (A) Experimental UV–Vis absorption spectrum of $Au_{24}(SR)_{20}$, (B) theoretical absorption spectrum, (C) schematic diagrams of HOMO, and (D) LUMO of the cluster [6].

6.3 Optical Properties of Large-Sized Gold Nanoclusters

The optical absorption peaks in small-sized nanoclusters all involve single-electron transitions, and hence excitons. These excitonic peaks are in striking contrast to the plasmon resonance peaks; the latter involve collective excitation of all the free electrons (e.g., thousands to millions) in the metallic particle. Mapping out the precise evolution from the metallic (or plasmonic) state to the semiconducting (or excitonic) state is of major importance because it will not only reveal the origin of SPR and the formation of metallic bonding, but also provide new opportunities for the applications of such nanomaterials. In the past decades, however, the precise transition and its effects on the properties remained unclear due to major difficulties in the synthesis of atomically precise nanoparticles. Significantly, the recent work by Higaki et al. has revealed a sharp transition from nonmetallic to metallic states between Au_{246} and Au_{279}, which came as a surprise as it is against the five-decade-long theoretical prediction of a smooth evolution ($1/n$) [8–11]. Below we discuss the optical properties of nanoclusters over the transition size regime.

As discussed in Section 6.1, significant changes in optical properties are expected to occur when the transition from nonmetallic to metallic states occurs (Fig. 6.5). Plasmonic nanoparticles show a strong peak at ~520 nm (i.e., 2.38 eV in Fig. 6.5B), whereas nonmetallic (excitonic) nanoclusters show multiple absorption peaks like in organic molecules (Fig. 6.5A) [1]. The photoexcitation of plasmonic nanoparticles induces collective excitation of conductive electrons, in contrast to the single-electron excitation for excitonic nanoclusters.

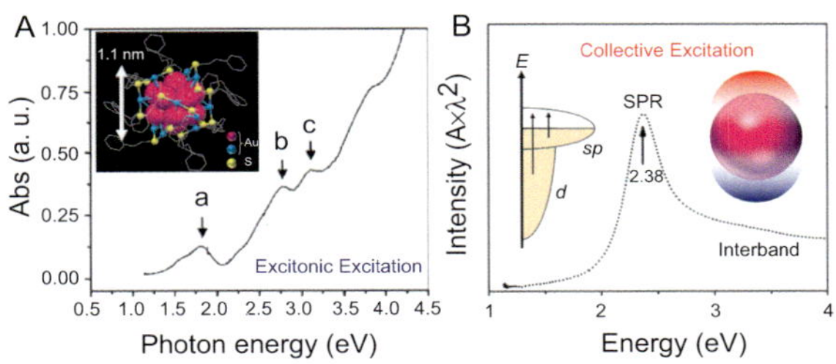

Figure 6.5 Typical steady-state absorption spectra for (A) excitonic Au nanoclusters and (B) plasmonic Au nanoparticles [1].

Further probing on the exciton-to-plasmon evolution can be pursued by femtosecond laser spectroscopy. As reported in the previous studies [12, 13], relatively monodisperse plasmonic Au nanoparticles in the size range of 5–100 nm show electron dynamics unique to the continuous electronic band structure. When these plasmonic nanoparticles are excited with a pump pulse, the conduction electrons are thermalized within the timescale of 10–100 fs via electron–electron scattering. Then, the absorbed energy of light is released via two subsequent processes with different timescales. For the faster process, the excited state electrons rapidly transfer the energy to the lattice vibrations (i.e., phonons) via interaction between negative electron and positive nuclei within a typical timescale of 1–10 ps, which is called electron–phonon coupling. The slower process occurs as soon as the phonons are excited within the particle in order to transfer the thermal energy of the excited phonons to the environment at the timescale of >100 ps, which is called phonon–phonon coupling. For the metallic-state nanoparticles, the increase in pump pulse energy will lead to longer relaxation times, that is, power-dependent dynamics, which can serve as a distinct feature of the metallic state. The fundamental coupling times for electron–phonon and phonon–phonon interactions decrease with particle size shrinking due to more efficient energy dissipation by higher surface-to-volume ratios in smaller nanoparticles. In terms of spectral features, the transient absorption spectra of plasmonic (spherical) nanoparticles show a single ground-state bleaching (GSB) signal at the SPR position, which is the same as the peak position in the steady-state absorption spectrum.

On the other hand, nonmetallic (excitonic) nanoclusters show multiple GSB bands after excitation by pump pulse [14]. The excitonic nanoclusters possess discrete energy levels with non-zero bandgap, and the photoexcited electron experiences relaxation from excited state to ground state. The lifetime of the process depends on the size and structure of the nanocluster. Unlike the power-dependent dynamics in plasmonic nanoparticles, the excitonic nanoclusters exhibit the lifetimes independent of pump fluence [8].

Therefore, several features can be used as the criteria to identify the exciton or plasmon state of Au nanoclusters with increasing size: (1) the number of absorption peaks in the steady-state absorption spectrum, (2) the number of GSB bands in the transient absorption spectrum, and (3) laser power dependence of lifetimes.

Using such criteria, for example, rule (1), we find that the steady-state absorption spectrum of Au_{246} shows multiple peaks at 400, 460, 600, and 800 nm (Fig. 6.6A), similarly to the cases of small Au nanoclusters in nonmetallic state [11]. In contrast, Au_{279} shows a single peak at 506 nm (Fig. 6.6A), similarly to the case of plasmonic nanoparticles [10]. Thus, a sharp exciton-to-plasmon transition is identified to occur between Au_{246} and Au_{279}.

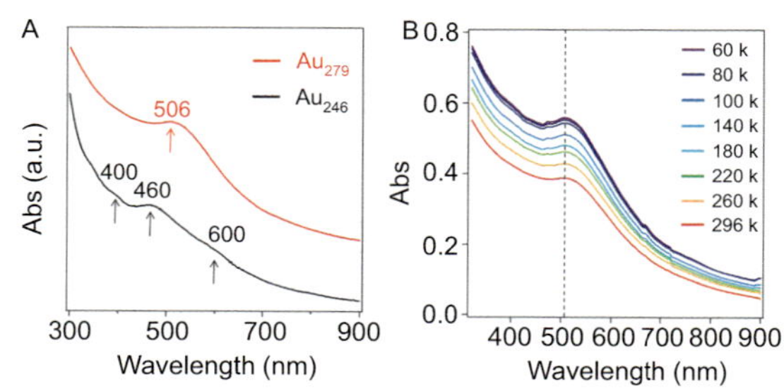

Figure 6.6 (A) Steady-state absorption spectra for Au_{279} (in red) and Au_{246} (in black). (B) Temperature-dependent optical absorption of Au_{279} [10].

The Kubo criterion of the metallic state is $E_g = k_B T$ (where k_B is Boltzmann's constant and T the temperature in Kelvin), which would imply a temperature dependence [3]. Thus, we tested the temperature effect of Au_{279} [10]. The absorption spectrum of Au_{279} at variable temperatures exhibits no peak shifting, narrowing, or splitting between 296 K and 60 K (Fig. 6.6B). The peak shifting, narrowing, or splitting at cryogenic temperature would indicate excitonic transitions. Therefore, all the observations in steady-state spectroscopic analyses indicate the excitonic nature of Au_{246} (i.e., nonmetallic) but plasmonic nature of Au_{279} (i.e., metallic state) [10].

In the following section, further analysis by time-resolved spectroscopy on Au_{246} and Au_{279} is made. Details on the sharp transition from nonmetallic Au_{246} to metallic Au_{279} are revealed by their significantly different electron dynamics.

6.3.1 Case of $Au_{246}(SR)_{80}$

The excited state dynamics of Au_{246} was probed in the range between 480 nm and 800 nm by femtosecond transient absorption (TA) after photoexcitation at 470 nm (Fig. 6.7) [11]. The TA spectra of Au_{246} show net GSB signals at around 530 and 610 nm in the first 10 ps (Fig. 6.7A,B). A broad excited state absorption (ESA) is also observed between 650 and 800 nm (Fig. 6.7B). The GSB around 530 and 610 nm correspond to the absorption peaks in the steady-state spectroscopy (Fig. 6.7A). The TA data map shows that the signal does not completely decay to zero even at 50 ps (Fig. 6.7C), indicating the presence of an additional decay component. The kinetic traces reveal that 90% of the TA signal decays within 15 ps, while the remaining 10% decays within the following 200 ps.

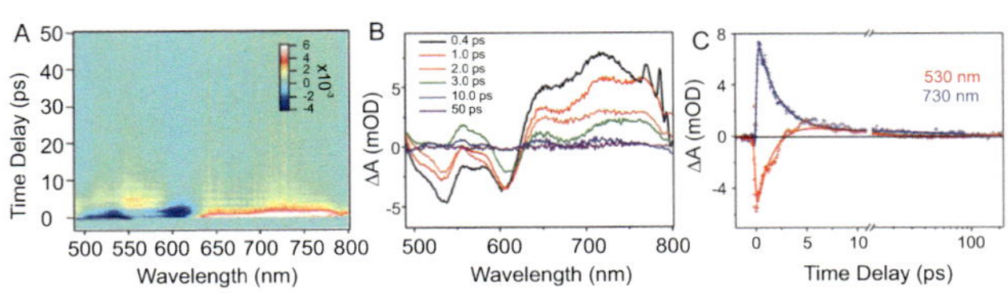

Figure 6.7 (A) TA data map of Au_{246} with pump at 470 nm. (B) TA spectra as a function of time delay. (C) Kinetics traces at 530 and 740 nm [11].

Global fitting of the obtained TA data reveals three decaying components with different lifetimes (i.e., 1.9, 4.6, and 73 ps) (Fig. 6.8A) [11]. Given that 470 nm excitation is significantly larger than the bandgap, the faster process of 1.9 ps is assigned to the relaxation from higher excited state to the lower excited state (i.e., internal conversion). The 4.6 ps and 73 ps processes are assigned to the relaxation from the lowest excited state to the ground state and the hot ground-state relaxation, respectively, considering the disappearance of GSB signals after 15 ps. To further probe the nonmetallic state of Au_{246}, TA spectra are recorded with different pump pulse energies (Fig. 6.8B). The relaxation dynamics of Au_{246} is totally independent of laser fluence where the extracted decay lifetimes show no change with laser fluence increasing from 60 to 300 nJ/pulse, which is in contrast to the case of power-dependent dynamics of Au_{279} (Fig. 6.8B) [10, 11]. Taken together, the observed dynamics demonstrates the nonmetallic state of Au_{246}.

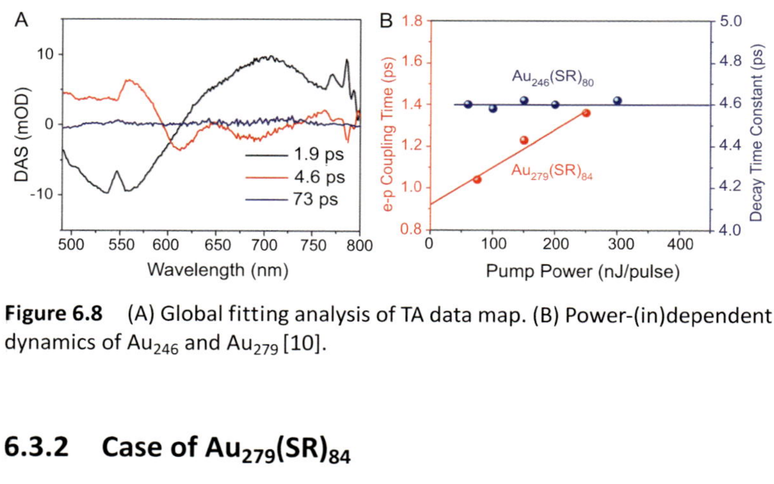

Figure 6.8 (A) Global fitting analysis of TA data map. (B) Power-(in)dependent dynamics of Au_{246} and Au_{279} [10].

6.3.2 Case of $Au_{279}(SR)_{84}$

The analysis of Au_{279} was also performed by TA spectroscopy in terms of its plasmonic electron dynamics (Fig. 6.9) [10]. Upon photoexcitation at 360 nm, TA spectra of Au_{279} show a net GSB band centered at ~530 nm (single band) and broad ESA below 490 nm and above 600 nm (Fig. 6.9A,B). The sharpening of the GSB peak is observed with time evolution due to the phonon induction after photoexcitation, which is also reported for the case of larger plasmonic nanoparticles [12]. The TA data map (Fig. 6.9A)

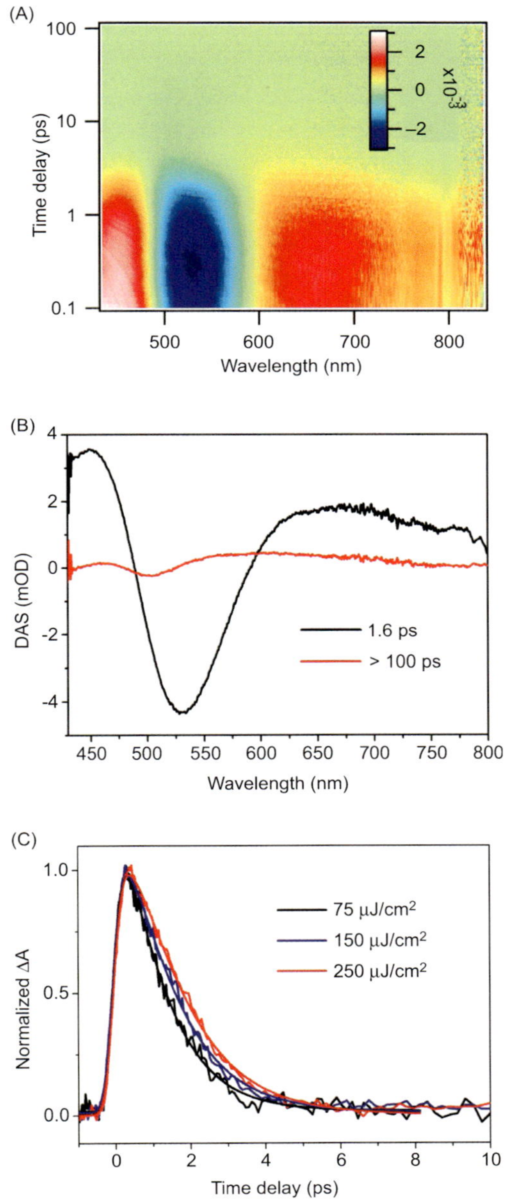

Figure 6.9 (A) TA data map of Au_{279}, (B) global-fitting analysis and (C) power-dependent dynamics [10].

shows the disappearance of most signals within 5 ps. Global fitting analysis has revealed two lifetimes for the relaxation processes (Fig. 6.9B). The fast process is assigned to electron–phonon coupling, and the decay time constant (τ = 1.6 ps) is consistent with previous reports [12, 13]. The long-lived process (τ > 100 ps) is attributed to phonon–phonon coupling. Laser power dependence is also studied for the electron dynamics of Au_{279} (Fig. 6.9C). Upon photoexcitation with different pump pulse energies, the time constants for the relaxation increase, indicating the plasmonic nature of the Au_{279}.

TA spectra of Au_{279} is also probed in near-infrared (NIR) region for further studies (Fig. 6.10) [10]. Upon photoexcitation at 360 nm, a strong ESA peak at 930 nm was observed with its rapid decay process within the timescale of 5 ps (Fig. 6.10A), which is similar to the dynamics probed in the UV–Vis region. However, the dynamics probed at both 770 and 930 nm do not completely decay to zero until 400 ps (Fig. 6.10B). The observed lifetime is relatively long considering its size (~2.2 nm) because smaller nanoparticles should show faster decay process due to their larger surface-to-volume ratios. For example, 4 nm AuNPs in aqueous media were reported to show a time constant of 10 ps, whereas the observed lifetime of ~2.2 nm Au_{279} is ~300 ps [10]. The long-lived phonon–phonon process in Au_{279} indicates its unusual electron dynamics due to its size near the transition regime. Further studies are expected to investigate such a "nascent" state of the plasmonic Au_{279}.

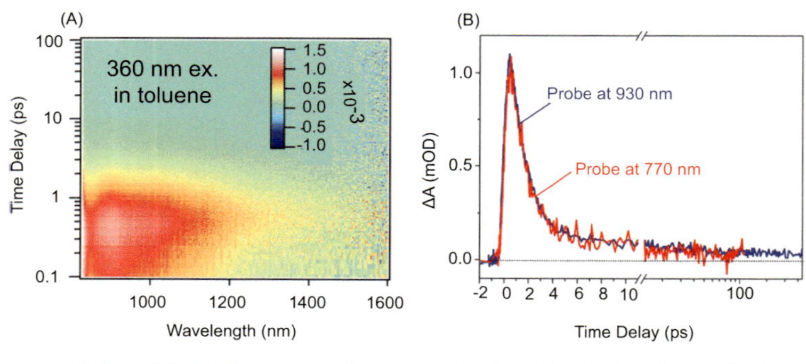

Figure 6.10 (A) TA data map of Au_{279} probed in the NIR region. (B) Kinetic traces at 770 and 930 nm for Au_{279} [10].

6.4 Conclusions and Future Perspectives

The success in mapping out the evolution of nonmetallic to metallic states between Au_{246} and Au_{279} (both 2.2 nm in diameter) constitutes a milestone in metal nanoparticle research. First, it reveals a sharp transition as opposed to the theoretically predicted smooth progression ($1/n$) and thus came as a surprise. Second, the Kubo criterion for metallic state indicates a temperature dependence ($E_g = k_B T$), but experimentally no such effect was found. We believe that the discrepancies should be caused by the neglected electron–electron correlation in the Kubo treatment; apparently, with increasing density of electronic states (i.e., smaller gaps), the electron correlation (e.g., screening) will become strong, which leads to collapse of discrete quantum states, giving rise to the collective state of electrons. The new discoveries call for a revisit to the electronic structure modeling of large-sized nanoclusters.

For future work, first the shape effect on the transition from nonmetallic to metallic is worthy to be pursued. Compared to the spherical case (both Au_{246} and Au_{279}), nanoclusters of non-spherical shapes (e.g., 1D rod or 2D oblate shapes) are rare; thus, the synthesis will be more challenging and need new strategies.

Second, the potential ligand effect on the transition also deserves a future effort. Other than the thiolate system, to what extent ligands such as phosphine and alkynyls would affect the transition? The synthesis of large-sized nanoclusters with ligands other than thiolates, in combination with spectroscopic analysis, will provide insights into this potentially interesting issue.

Third, it remains unclear to what extent the detailed atomic arrangement of gold nanoclusters plays a role in the transition. As discussed previously, Au_{246} has a decahedral structure, whereas Au_{279} is face-centered cubic. Recent theoretical work indicates the potential effect of symmetry of the gold core on the transition [15]; this remains to be probed in future experimental work.

Last but not least, the phonon dynamics in large-sized nanoclusters is still not well understood. The scaling relationship of phonon frequency with the number of atoms in the cluster remains to be mapped out, especially how this evolves to the well-known "phonon frequency $\sim 1/R$" law in plasmonic NPs [12].

Overall, the probing of metal-to-nonmetal transition and origin of plasmons and metallic bonding will impact the understanding of new physicochemical properties of nanoclusters, which will in turn promote their applications in energy transfer and utilization as well as catalysis.

Acknowledgments

This work is financially supported by the National Science Foundation (DMR-1808675).

References

1. Jin, R., Zeng, C., Zhou, M. and Chen, Y. (2016). Atomically precise colloidal metal nanoclusters and nanoparticles: Fundamentals and opportunities, *Chem. Rev.*, **116**, pp. 10346–10413.
2. Jin, R. (2010). Quantum sized, thiolate-protected gold nanoclusters, *Nanoscale*, **2**, pp. 343–362.
3. Kubo, R. (1962). Electronic properties of metallic fine particles. I, *J. Phys. Soc. Jpn.*, **17**, pp. 975–986.
4. Zhu, M., Lanni, E., Garg, N., Bier, M. E. and Jin, R. (2008). Kinetically controlled, high-yield synthesis of Au_{25} clusters, *J. Am. Chem. Soc.*, **130**, pp. 1138–1139.
5. Zhu, M., Aikens, C. M., Hollander, F. J., Schatz, G. C. and Jin, R. (2008). Correlating the crystal structure of a thiol-protected Au_{25} cluster and optical properties, *J. Am. Chem. Soc.*, **130**, pp. 5883–5885.
6. Das, A., Li, T., Li, G., Nobusada, K., Zeng, C., Rosi, N. L. and Jin, R. (2014). Crystal structure and electronic properties of a thiolate-protected Au_{24} nanocluster, *Nanoscale*, **6**, pp. 6458–6462.
7. Jin, R. (2015). Atomically precise metal nanoclusters: Stable sizes and optical properties, *Nanoscale*, **7**, pp. 1549–1565.
8. Zhou, M., Zeng, C., Chen, Y., Zhao, S., Sfeir, M. Y., Zhu, M. and Jin, R. (2016). Evolution from the plasmon to exciton state in ligand-protected atomically precise gold nanoparticles, *Nat. Commun.*, **7**, p. 13240.
9. Zeng, C., Chen, Y., Kirschbaum, K., Lambright, K. J. and Jin, R. (2016). Emergence of hierarchical structural complexities in nanoparticles and their assembly, *Science*, **354**, pp. 1580–1584.
10. Higaki, T., Zhou, M., Lambright, K. J., Kirschbaum, K., Sfeir, M. Y. and Jin, R. (2018). Sharp transition from nonmetallic Au_{246} to metallic Au_{279}

with nascent surface plasmon resonance, *J. Am. Chem. Soc.,* **140**, pp. 5691–5695.

11. Zhou, M., Zeng, C., Song, Y., Padelford, J. W., Wang, G., Sfeir, M. Y., Higaki, T. and Jin, R. (2017). On the non-metallicity of 2.2 nm $Au_{246}(SR)_{80}$ nanoclusters, *Angew. Chem. Int. Ed.*, **56**, pp. 16257–16261.
12. Hartland, G. V. (2011). Optical studies of dynamics in noble metal nanostructures, *Chem. Rev.*, **111**, pp. 3858–3887.
13. Link, S. and El-Sayed, M. A. (2003). Optical properties and ultrafast dynamics of metallic nanocrystals, *Annu. Rev. Phys. Chem.*, **54**, pp. 331–366.
14. Yau, S. H., Varnavski, O. and Goodson, T. (2013). An ultrafast look at Au nanoclusters, *Acc. Chem. Res.*, **46**, pp. 1506–1516.
15. Malola, S., Kaappa, S. and Häkkinen, H. (2019). The role of nanocrystal symmetry in the crossover region from molecular to metallic gold nanoparticles, *J. Phys. Chem. C*, **123**, pp. 20655–20663.

Chapter 7

Gold Nanoclusters with Atomic Precision: Optical Properties

Lin Xiong and Yong Pei
Department of Chemistry, Key Laboratory of Environmentally Friendly Chemistry and Applications of Ministry of Education, Xiangtan University, Hunan Province, Xiangtan 411105, P. R. China
ypnku78@gmail.com

As a bridge connecting molecules and macroscopic materials, clusters hold the key to the door from microscopic to macroscopic world. Due to the quantum confinement effect, clusters exhibit unique physical and chemical properties, especially optical properties. Advances in the atomic precision thiolate-protected gold nanoclusters over the last decade have provided a good opportunity to study the optical properties of noble metal nanoclusters. Despite extensive researches on the optical properties of clusters in recent years, there is still a lack of systematic description of the relevant aspects. In this chapter, we focus on the thiolate-protected gold clusters and present an overview of their optical absorption, optical emission, and nonlinear optical properties. Among them, the light absorption and emission of clusters are the most widely studied

Atomically Precise Nanoclusters
Edited by Yan Zhu and Rongchao Jin

ISBN 978-981-4877-08-4 (Hardcover), 978-1-003-11990-6 (eBook)
www.jennystanford.com

optical properties, but there are still some unresolved problems that have plagued researchers. For instance, what is the origin of surface plasmon resonance (SPR)? What is the mechanism of fluorescence emission? Moreover, gold nanoclusters have become promising candidates in the field of high-resolution multiphoton imaging and optical confinement applications due to their nonlinear optical properties (NLO). However, studying on the NLO properties of clusters is still rare, and so it is worth further exploration. In the last part, we briefly discuss the optical stability, chiral origin, and optical activity of gold nanoclusters. Overall, this chapter provides deep insights into the fundamental questions of optical properties of gold nanoclusters.

7.1 Introduction

As a precious metal, gold has a long and profound history. Bulk solid gold exists almost as a single substance in nature due to its chemical inertia and was used as currency and for making utensils thousands of years ago. Gold nanoparticles exhibit the quantum confinement effect [1] when their size is reduced to the Fermi wavelength of electrons (viz., the de Broglie wavelength of electrons at the Fermi energy), which results in distinctive physicochemical properties that are not observed in their bulk counterparts and molecular compounds [2, 3]. As a result, electrons have dispersed energy levels and molecular properties, including a HOMO–LUMO (highest occupied and lowest unoccupied molecular orbitals) gap [1, 4], molecular magnetism [5, 6], molecular optical chirality [7], strong photoluminescence (PL) [8], and nonlinear optical properties [9, 10]. Among the numerous properties of gold nanoparticles, optical properties have become an important subject due to scientific interest and practical applications. In fact, the optical properties of gold nanoparticles have been exploited for a long time and can be traced back to the ancient Roman era in the 4th century AD. Ancient Roman glassmakers developed an exquisite craft for making cups that could display different colors under different lighting conditions. The shape of such a cup is a carved pattern separated from the cup body; it is commonly called a "cage cup" because the carved pattern is similar to a cage covering the outside of the cup.

The most typical example of this type of cup is the Lycurgus Cup [11] (Fig. 7.1). The unique features of the Lycurgus Cup are not only its decorative quality, but also its special optical properties; it displays different colors under different lighting conditions. For example, when light illuminates from outside of the cup, the cup appears to be green (Fig. 7.1a); when light illuminates from inside of the cup, the cup appears to be wine red (Fig. 7.1b).

Figure 7.1 The Lycurgus Cup, an exquisite handicraft made in the ancient Roman era. The cup displays different colors under different lighting conditions. For example, it appears to be (a) green under reflected light and (b) wine red under transmitted light (Image courtesy and copyright: The British Museum).

Such a fascinating phenomenon demonstrates that when art and science are combined, they often produce extraordinary results; however, it also perplexed people for centuries, and it was only in modern times that an explanation for the phenomenon was finally clarified. Researchers examined the cup by transmission electron microscopy (TEM) and found that the glass contained trace gold and silver alloy particles with diameters of only 50–100 nm, about one-thousandth the diameter of human hair. It is these nanoparticles that produce the magical discoloration effect of the Lycurgus Cup [12]. In addition, the surfaces of ceramics made in Mesopotamia in the 9th century AD exhibit different colors (a rainbow phenomenon) from different perspectives; this is also due to the presence of copper nanoparticles in the enamel of the surface. In 1857, gold nanoparticles synthesized by Michael Faraday turned the gold colloidal solution red [13]. This was the first scientific point of view that demonstrated that the special properties of nanoparticles originate from their tiny size and was also the pioneering work on nanoparticles in modern science. Since the 1940s, the popularity of electron microscopy has enabled scientists to better characterize nanoparticles and to study the relationship between their size, shape, and properties. Since 1990, with the development of synthetic methods, it has been possible to synthesize nanoparticles with specific components, specifications, and properties in large quantities. As expected, an outburst of research occurred on the synthesis and characterization of gold nanoparticles over the next decade [10]. Nanogold, which includes gold nanorods, gold nanowires, and gold nanoclusters with atomic precision, has been synthesized and characterized in succession. It is against this background that an excellent platform for studying the optical properties of gold nanoparticles is provided. With the rapid development of nanotechnology, from the first synthesis and characterization of gold nanorods in 1994 [14] to the present, the progress of the nanoparticle synthesis technology has allowed for the precise control of the chemical composition, size, and shape of nanoparticles. On this basis, nanoparticles are widely used in human production activities for the benefit of mankind. In this chapter, the progress of the findings on the optical properties of gold nanoclusters is reviewed.

7.2 Optical Properties

7.2.1 Absorption Properties

The study on nanoparticles has a fairly long history that can be traced back to the time of Faraday. Since Faraday first prepared gold sol [13] (also called "potable gold") in 1857, researchers have systematically studied the properties of colloids of various elements and their compounds. Among these colloids, gold colloids have been favored by researchers because of their various colors. It is now known that the colorful colloidal solution of gold nanoparticles is due to the strong absorption band of Au nanoparticles in visible to near-infrared bands, which is not observed in their bulk materials. The root cause for such a phenomenon is attributed to the optical properties of nanoparticles, particularly SPR. It is well known that when light illuminates the surface of metal particles, if the frequency of the incident light matches the frequency of the electronic vibration of particles, the particles will absorb the incident light intensely, and SPR will then occur. The SPR phenomenon of traditional gold nanoparticles can explain their absorption, scattering, surface-enhanced Raman scattering, and many other properties. Hence, SPR is an important optical property of gold nanoparticles and has attracted significant attention from researchers for many years. Wood first discovered SPR in 1902 [15]. In 1908, German physicist Gustav Mie explained the optical properties of spherical particles by solving the Maxwell equation in polar coordinates, which is called the Mie theory [16]. The SPR properties of spherical nanoparticles are isotropic due to their highly symmetrical structure, which results in a single absorption peak in their absorption spectrum [17, 18]. In addition, the SPR properties of nanoparticles are highly dependent on their size. For example, the SPR peak will gradually shift to red with the increase in size. As the diameter of spherical nanoparticles increases from 10 to 100 nm, their SPR peak will correspondingly shift from 520 to 570 nm. When the size of nanoparticles is reduced to a small enough range (<3 nm), the quantum confinement effect begins to emerge and results in these nanoparticles having discrete electronic levels, thus exhibiting physical and chemical properties similar to those of molecules. In this case, the metal-like state of nanoparticles will gradually weaken and eventually disappear, and will exhibit a multi-peak property similar to the molecular-like state in the optical absorption spectrum. At such a small size level, even if an atom is different, the properties of nanoparticles become extremely sensitive.

Therefore, the precise determination of the structures of nanoparticles at the atomic level is very important in the study of the mechanism of their optical properties. Understanding the structure–properties relationship of nanoparticles at the atomic level has become the main goal of nanochemists. To determine this relationship, it is necessary to explore the precise atomic structures of nanoparticles.

Metal nanoparticles (MNPs) with atomic accuracy are usually called metal clusters to distinguish them from traditional MNPs. Of the different types of metal clusters, gold nanoclusters (AuNCs) that are composed of a specific number of gold atoms and thiolate ligands [denoted as $Au_m(SR)_n$] have received considerable attention and found broad applications in biomedicine [19–23], sensing

[24–27], optoelectronics [28], and catalysis [29–31] due to their distinctive physicochemical properties. Their size is usually in the range of 1–3 nm. As a mesoscopic system, unlike microscopic particles with uncertain properties (e.g., electrons), clusters have precise atomic compositions that can be measured by various measurement techniques. Furthermore, the structures of AuNCs can be precisely tailored at the atomic level to obtain some functional materials with special properties. For these reasons, it is very convenient for researchers to systematically study the structure–properties relationship of clusters. Recently, the amount of atomic structures of extra-small AuNCs has been determined benefit from the rapid development and improvement of various characterization techniques and synthesis methods. Atomically accurate, ligand-protected gold nanoclusters have undergone widespread research in the nanoscience field over the past two decades due to their special properties. The crystal structures of a large number of clusters have been synthesized from the initial rise of citrate, glutathione, and organophosphorus ligand-protected gold nanoparticles to gold nanoclusters protected by thiol ligands that have shown excellent stability in various environments, and alkyne-protected gold clusters have been characterized in recent years [10, 32]. Compared with other ligand-protected clusters, thiol ligand-protected gold clusters have excellent stability and have therefore become template clusters for studying the quantum mechanism of nano-size clusters.

Since Jadzinsky et al. first synthesized $Au_{102}(p\text{-MBA})_{44}$ clusters in 2007 [33], great progresses have been made in the synthesis, characterization, and structure determination of thiolate-protected gold clusters [10, 34–45]. A series of theoretical work has also been carried out to examine the synthesis of clusters and to determine their structure [3, 46–51]. On this basis, the exploration of the properties of nanoclusters has been performed, which has contributed to a deep understanding of the relationship between the structure and properties of gold nanoclusters. Much of the work on the spectral study of clusters was carried out by Aikens et al. [48, 52–57], who first performed a time-dependent density functional theory (TDDFT) study on Au_{25} using a simplified model and reproduced the experimental spectrum [57]. Unlike the SPR absorption exhibited by spherical gold nanocrystals at around 2.4 eV, the TDDFT results revealed that Au_{25} unexpectedly exhibits multiple absorption peaks similar to those of molecules, with at least three identifiable peaks located at 1.8, 2.75, and 3.1 eV (Fig. 7.2a) [57]. The absorption peak *a* corresponds to the absorption at 670 nm on the experimental spectrum, which is attributed to the transition from HOMO to LUMO, and is essentially an in-band transition from sp to sp. Molecular orbital composition analysis obviously demonstrates that the energy level of Au_{25} is quantized. In addition, the Kohn–Sham molecular orbital maps of Au_{25} show that HOMO and LUMO exhibit characteristics similar to the p and d atomic orbitals, respectively, and therefore, in a certain sense, Au_{25} is considered to be a superatom with an 8e structure. Furthermore, because the HOMO and LUMO orbital maps of Au_{25} are almost localized in the core (the distribution of electrons does not change significantly before and after excitation), the absorption peak of Au_{25} at 670 nm in the experimental spectrum can be attributed to the local excitation of the Au_{13} core of the Au_{25} cluster.

Based on the crystal structure of $Au_{28}(TBBT)_{20}$ [58] reported by Jin et al., Häkkinen et al. discussed its optical properties and electronic structure by the DFT (density functional theory) method [59]. Similar to Au_{25}, Au_{28} also has a dispersed electronic energy level and its spectrum also exhibits multi-peak characteristics similar to molecules. In addition, the analysis of the core and the simplified structure of Au_{28} has revealed that although the Au_{28} core is a non-spherical structure, both the core and the ligand-protected Au_{28} structures are all superatomic.

Interestingly, another cluster $[Au_{23}(SC_6H_{11})_{16}]^-$, which also has an 8e structure, is not a superatom [60]. Nobusada et al. performed DFT calculations on simplified $Au_{23}(SCH_3)_{16}^-$, and the results reasonably reproduced the experimental spectral profile (Fig. 7.2c). However, as shown in the insets in Fig. 7.2c, the molecular orbitals are not only localized above the core, but are also distributed above the Au atom of the ligand. Jin et al. claimed that the HOMO of Au_{23} is not a d-like orbital, on the contrary, it is a hybrid orbital composed of the atomic orbitals of Au. With the continuous development and improvement of synthetic methods, increasingly more cluster structures have been determined, and theoretical studies on the optical properties of these clusters have also been implemented.

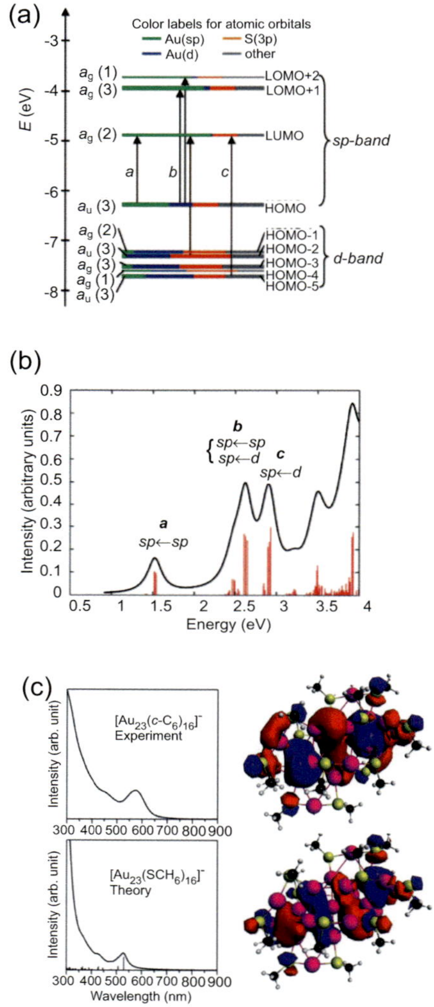

Figure 7.2 (a) Kohn–Sham orbital energy level diagram for $Au_{25}(SH)_{18}^-$. The different colors of each orbital represent the contributions of different types of atomic orbits, such as Au (6sp) in green, Au (5d) in blue, S (3p) in yellow, and others in gray (orbits with contributions of <1% are not shown here). (b) The theoretical absorption spectrum of $Au_{25}(SH)_{18}^-$. Copyright 2008 American Chemical Society. (c) Experimental and simulated UV–Vis spectra of Au_{23} and its HOMO and LUMO molecular orbital schematics. Figures (a) and (b) reprinted from Ref. [57], Copyright 2008, and Fig. (c) from Ref. [60], Copyright 2013, by permission of the American Chemical Society.

Despite the tremendous progress in experimentation and theory, endless problems have emerged. In particular, many fundamental questions about the evolution of geometry and electronic structures have arisen.

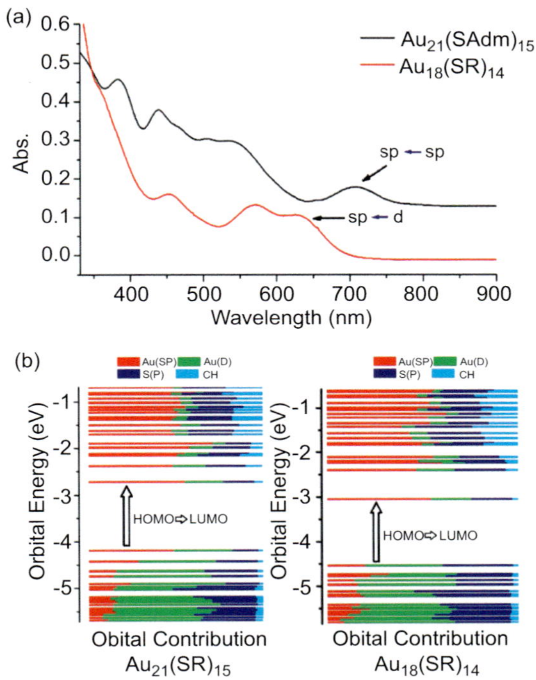

Figure 7.3 (a) UV–Vis spectra of $Au_{21}(SR)_{15}$ and $Au_{18}(SR)_{14}$ nanoclusters. (b) Kohn–Sham orbitals of $Au_{21}(SCH_3)_{15}$ (left) and $Au_{18}(SCH_3)_{14}$ (right) nanoclusters. The orbital is drawn to indicate the relative contributions (line length with color labels) of the atomic orbitals of Au(6sp) (red), Au(5d) (green), S(3p) (blue), and other orbital contributions from C and H atoms (cyan).

First, the fluorescence of gold clusters mostly originates from the sp $\leftarrow$ sp intraband and sp $\leftarrow$ d interband transitions of gold atoms. Some studies have shown that HOMO to LUMO transitions of small-sized clusters and Au complexes are typical interband electronic transitions from the d band to the sp band [61, 62]. On the contrary, the transition from HOMO to LUMO of larger clusters (e.g., 8e or more free valence electrons) mainly involves the transition from the sp band to the sp band [57, 63]. Thus, it is unclear whether the transition of the electronic excitation mode from sp ← d to sp ← sp is caused by the increase in cluster size/Au-kernels or other factors, such as the increased number of free metal electrons. Our research group used the DFT method to explore the electronic structure of several small-sized clusters [64]. It was found that the HOMO to LUMO transitions of clusters with 4e or fewer free valence electrons (including $Au_{15}(SR)_{13}$ [65], $Au_{18}(SR)_{14}$ [61, 62], $Au_{20}(SR)_{16}$ [66], and $Au_{22}(SR)_{18}$ [67]) are interband transitions from d to sp (Fig. 7.3a). Interestingly, the HOMO to LUMO transition of $Au_{21}(SR)_{15}$ with a valence electron number of six mainly involves sp to sp transition (Fig. 7.3b). The results of this study indicate that the change in the frontier orbital transition mode is due to the increase in the number of valence electrons. The $Au_{21}(SR)_{15}$ between 4e and 8e is an edge cluster whose transition mode is changed [64].

Second, the "cluster sequence" is an interesting phenomenon in cluster structures in which the clusters appear to follow a certain growth mechanism and grow in a particular pathway, and the study of the optical properties of such clusters is very interesting. A typical example is the $Au_{28}(TBBT)_{20}$ [58], $Au_{36}(TBBT)_{24}$ [68], $Au_{44}(TBBT)_{28}$ [69], $Au_{52}(TBBT)_{32}$ [70] cluster sequence. In a previous study, Tsukuda et al. synthesized Au_{76}(4-MEBA)$_{44}$ [4-MEBA: 4-(2-mercaptoethyl)benzoic acid] [71], which showed a strong (3×10^5 M^{-1} cm^{-1}), near-infrared absorption band at 1340 nm. Note that although Au_{76} is different from the ligands of this cluster series, they all follow the same structural formula, namely, $Au_{20+8N}(SR)_{16+4N}$. Based on the atomic structures of Au_{28}, Au_{36}, Au_{44}, and Au_{52}, Ma et al. predicted a novel class of thiol-protected gold clusters with a one-dimensional linear growth mode containing of a face-centered cubic (fcc) core structure [72, 73]. Obviously, when $N = 7$, the cluster formula is exactly $Au_{76}(SR)_{44}$. The simulated optical absorption spectra of $Au_{44}(SR)_{28}$ and $Au_{76}(SR)_{44}$ predicted by Pei and Ma et al. reproduced the experimental spectra well. TDDFT calculations show that the aspect ratio of such clusters significantly affects their optical absorption properties. As shown in Fig. 7.4, as the value of N increases, the center position of the first characteristic peak in the optical absorption spectrum gradually red-shifts, and the corresponding oscillator strength also increases.

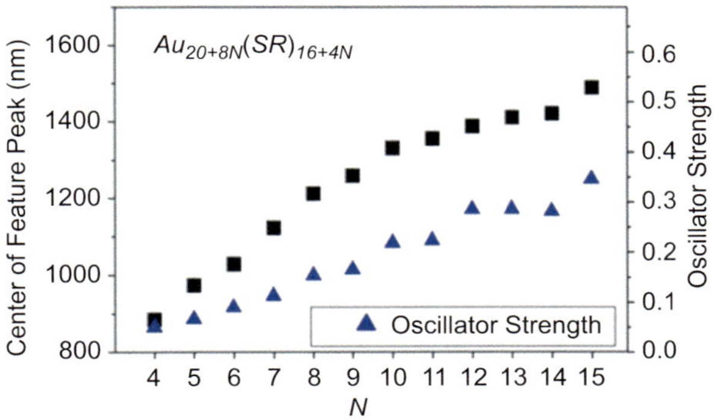

Figure 7.4 The center position and oscillator strength of the first feature absorption peak in the optical absorption spectra of various $Au_{20+8N}(SR)_{16+4N}$ clusters with N = 4–15.

In addition, Wang et al. predicted the structure of a series of large-sized, thiol-protected gold clusters $Au_{145-3N}(SR)_{60-2N}$ (N = 1–8) [74], and their work also conveyed the optical absorption properties of the lowest energy isomers of the cluster series. Future research may focus on the study of the optical properties of such cluster sequences.

Third, in terms of the electronic structure of gold clusters, even small differences in size cause changes in the electron energy level, resulting in changes in optical absorption properties. With the determination of increasingly more atomically precise ligand-protected gold clusters and more theoretical studies being carried out, the question naturally arises: Where is the boundary between clusters with discrete energy levels and nanoparticles with SPR? In other words, what is the critical grain size for the transition from the molecular state to the metallic state? As the initial form of condensed matter, clusters play a key role in the transformation of matter from atoms and molecules to macroscopic materials [45]. The evolution of MNPs from the metal state to the molecular state is a key issue in nanoscience research, as it is of great significance to reveal the origin of metal bonds and the generation of SPR. Because of their definite size and structure, thiol ligand-stabilized gold clusters are ideal research objects to study the transition from the non-metal to the metal state and provide an excellent opportunity to explore the evolutionary mechanism from the quantum state to the bulk-like state, as well as the influencing factors of this evolution.

As mentioned previously, small-sized nanoclusters exhibit discrete electronic transitions in the absorption spectrum, while plasma gold nanocrystals have an optical plasma absorption band at around 520 nm [10]. Therefore, the absorption spectrum can be utilized to distinguish the state of gold nanoparticles.

Previous studies have pointed out the size ranges of nanoparticles that begin to present plasma characteristics. Häkkinen et al. studied the experimental clusters by an *ab initio* method based on a theoretical model and explained the excitation behavior of electrons when the core of the clusters increased from 1.5 to 2.0 nm [75]. They found that clusters with a metal core size of 2.0 nm exhibit strong plasma absorption peaks at 540 nm, while in a smaller cluster with a 1.5 nm gold core, the plasmon-like resonance at 540 nm is confined in the metal core by the molecular layer. This study revealed that the threshold for the generation of localized surface plasmon resonance (LSPR) in a thiol ligand-stabilized gold cluster system is approximately 200–400 Au atoms. Subsequently, Häkkinen et al. synthetically analyzed a series of gold clusters protected by SC_{12} using temperature-dependent optical spectroscopy [76]. It was found that there was an obvious transition from molecule-like to bulk-like between $Au_{144}(SC_{12})_{60}$ and $Au_{187}(SC_{12})_{68}$ (Fig. 7.5). Interestingly, combined studies using powder XRD (X-ray diffraction) and DFT calculations revealed that the transition from fcc to non-fcc (icosahedral-based or M-D_h-based) structures occurs in the same size range.

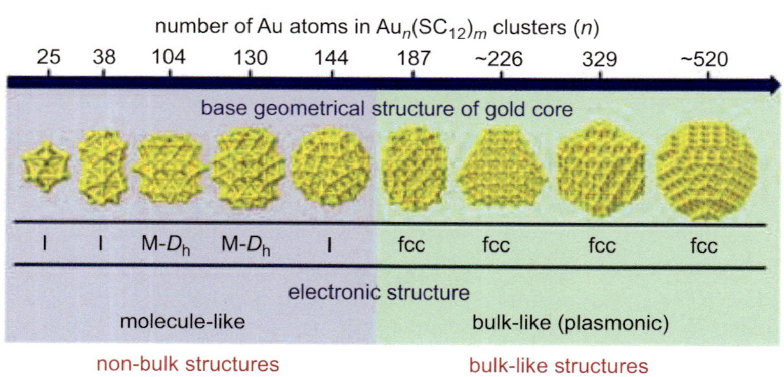

Figure 7.5 Transition from molecule-like to bulk-like electronic properties as a function of the number of gold atoms. Reprinted with permission from Ref. [76], Copyright 2015, American Chemical Society.

In addition, Xu et al. have reported the threshold number of Au atoms (*N*) at which the transition from molecule-like to metal-like properties occurs to be ~263 by linear fitting of the HOMO–LUMO gaps of various thiolate-protected gold nanomolecules (expressed in the original text as RS–AuNPs) against $N^{-1/3}$ (where *N* is the number of Au atoms in RS–AuNPs) [77].

In recent years, the crystal structures of huge clusters such as $Au_{103}S_2$(S-Nap)$_{41}$ (S-Nap: 2-naphthalenethiolates) [78], $Au_{108}S_{24}(PPh_3)_{16}$ [79], Au_{130}(*p*-MBT)$_{50}$ (*p*-MBT: 4-methylbenzenethiolate) [80], Au_{133}(*p*-TBBT)$_{52}$ (*p*-TBBT: 4-*tert*-butylbenzenethiolate) [81, 82], $Au_{144}(SCH_2Ph)_{60}$ [83], and Au_{146}(*p*-MBA)$_{57}$ (*p*-MBA: *para*-mercaptobenzoic acid) [84] have been obtained. Although the optical properties of these large-sized clusters with atomic precision have been analyzed, the results indicate that these clusters fall into the nonmetallic state, and so a fundamental problem in the nanometer field—how many gold atoms will ultimately lead to a metallic state—remains controversial. Thus, to understand the transition from the nonmetallic state to the metallic state, it is necessary to obtain larger sizes of the gold nanocrystalline structure. Despite the great experimental efforts to measure large-sized gold clusters, it is difficult to measure the overall structure because ionizing it under ESI conditions is challenging, and a part of the surface ligand can be easily lost under laser energy; this has hindered the understanding of the origin of SPR.

Fortunately, in 2016, Zeng et al. reported a giant Au_{246}(*p*-MBT)$_{80}$ gold nanocluster (about 2.2 nm in diameter), hereinafter referred to as Au_{246} [85]. The Au_{246} cluster was synthesized by a two-step method and contains 100 more gold atoms than the largest Au_{146} previously reported [84]. Optical absorbance spectra of Au_{246} revealed a prominent peak at 470 nm and several weak humps at 400 and 600 nm, indicating the non-plasmonic nature of the giant AuNC. Subsequently, Jin et al. carried out further nanoscience research on the properties of Au_{246} to determine the transition behavior from the molecular to the plasma state [86]. However, based on the analysis of the optical absorption spectra, femtosecond transient absorption (TA) spectra, and electrochemical measurements of Au_{246}, they found that even such a large cluster still has molecular properties. It is noteworthy that X-ray diffraction crystal structure analysis has shown that Au_{246} is a nearly spherical nanoparticle, and so the anisotropic shape does not lead to peak splitting. Additionally, in previous studies, $Au_{\sim253}$, which is only a few atoms different from Au_{246}, has been found to exhibit a plasma state [76].

The study of the transition of nanoparticles from nonmetallic to metallic states has experienced difficulties. Fortunately, in 2017, Dass et al. obtained the largest thiol ligand-protected gold cluster to date by using a three-step synthesis method [87] [Step 1: crude synthesis, Step 2: thermochemical treatment (etching), and Step 3: isolation of molecularly pure Au_{279} nanocrystals and crystallization]. The heavy atom molecular formula, $Au_{279}S_{84}$, was established by a low-temperature X-ray diffraction study. This composition was independently verified by ESI (electrospray ionization) mass spectrometry (vide infra), and its molecular formula was established as $Au_{279}(SPh\text{-}t\text{Bu})_{84}$ (denoted as Au_{279}). Dass et al. conducted an in-depth analysis of the structure of the giant cluster, including the structure of the metal core and the ligand layer [88]. Spectroscopic analysis indicated that Au_{279} exhibited a SPR peak at 510 nm. The following year, Jin et al. also reported the crystal structure of Au_{279} and discussed its optical properties in depth [89]. As an example, femtosecond transient absorption spectroscopic analysis was performed by pumping at 360 nm and probing in the 430–820 nm range to probe the excited-state dynamics of Au_{279}. As shown in Fig. 7.6a,b, they found that the sharpening of the ground-state bleaching (GSB) peak with time evolution corresponds to the phonon induction after photoexcitation, similar to larger sized plasmonic nanoparticles [90, 91]. To obtain a deeper understanding of the nature of Au_{279}, TA measurements were performed under different pump fluencies (Fig. 7.6c). In contrast to laser power-independent dynamics in Au_{246} and smaller Au nanoclusters, all the relaxation dynamics in Au_{279} are dependent on the laser fluence (Fig. 7.6d). This result further supports that Au_{279} has a plasma-like electronic structure. However, unlike typical plasma nanoparticles, when Au_{279} is excited at 360 nm, a broad ESA up to 1600 nm can be observed. In addition, the long-lived τ_{ph-ph} (~300 ps) in Au_{279} also implies that the heavy gold cluster is in a primary plasmonic state [89].

The sharp transition from the nonmetallic state of Au_{246} to the metallic state of Au_{279} was accomplished by only 33 gold atoms, which was unexpected by researchers. As mentioned in the article by Jin. et al., future theoretical work should explore how the tiny increment from Au_{246} to Au_{279} (only 33 gold atoms) triggers the transition from the nonmetallic to the metallic state [89].

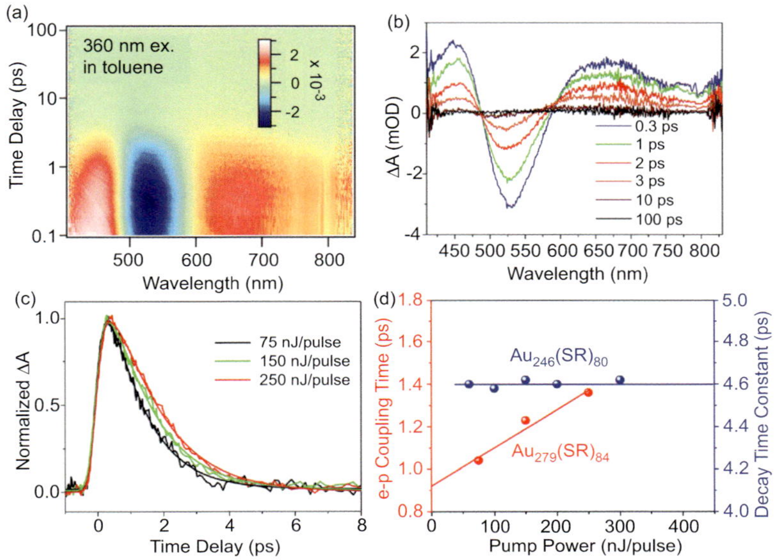

Figure 7.6 (a) TA data map of Au_{279} pumped at 360 nm in toluene. (b) TA spectra as a function of time delay between 0.3 and 100 ps. (c) Normalized decay kinetics around 530 nm as a function of laser fluence. (d) Extracted τ_{e-ph} of Au_{246} (in blue, 470 nm pump) and Au_{279} (in red, 360 nm pump) as a function of pump fluence. Reprinted with permission from Ref. [89], Copyright 2018, American Chemical Society.

It is worth noting that the single-crystal XRD results show that Au_{279} nanoparticles have a core–shell structure, and the core is an fcc octahedron (Fig. 7.7). Although Au_{246} is also a shell structure, its atomic arrangement is not the fcc configuration (Fig. 7.7). In combination with the previous study of a series of dodecanethiolate-protected gold clusters by Häkkinen et al., in which it was found that gold nanoclusters began to exhibit metallic states starting from Au_{187} in the fcc configuration [76], the question arises: in the field of gold nanoclusters, is the number of gold atoms at which nanoparticles begin to exhibit plasma properties configuration dependent? To answer this question, it is necessary to obtain single-crystal structures of nanoclusters with the same atomic compositions and different configurations. Unfortunately, no such clusters have yet been reported.

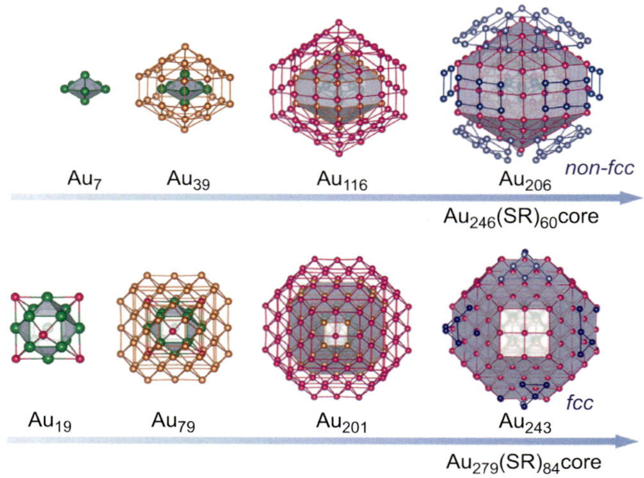

Figure 7.7 Schematic diagrams of the shell-like core structures of $Au_{246}(SR)_{60}$ and $Au_{279}(SR)_{84}$.

7.2.2 Photoluminescence

Photoluminescence (PL) is one of the most fascinating properties of MNPs. Among the many MNPs, the optical properties of gold nanoparticles/nanoclusters are the most widely studied. Because bare gold nanoparticles are unstable in solution, the most common research object at present is gold nanoparticles capped by ligands. Benefiting from the high affinity of thiols for the gold surface, thiol ligands stand out in various protected ligands, making thiolate-protected gold clusters [denoted as $Au_n(SR)_m$] ideal targets for studying the luminescence properties of gold nanoparticles [92]. Luminescent $Au_n(SR)_m$ has the advantages of strong luminosity, low toxicity, ultrafine size, and good biocompatibility and is widely used in the fields of cell labeling, biosensing, and phototherapy [10]. Therefore, the preparation of $Au_n(SR)_m$ with good luminescence properties and high quantum yield has important practical significance. At present, there are two main preparation methods: bottom-up and top-down. Gold clusters created by the former method are usually prepared by Au(I)–thiolate complexes and strong reductants, a typical representative of which is the Brust method

[14]. However, the quantum yield of gold clusters prepared by this method rarely exceeds 0.1%. The latter method is based on the decomposition of large-sized gold clusters or large Au(I)–thiolate complexes. Although gold clusters with strong luminescence can be produced by this method, it is difficult to apply this method because of its slow reaction time and difficult post-processing process (separating the target products from the reaction system). Thus, it is imperative to find new strategies to enhance the luminescence of gold clusters based on the bottom-up preparation method.

The strategies currently used to enhance the luminous efficiency of gold nanoclusters are as follows: (1) capping the metal core surface with different types of ligands, (2) changing the core size or doping the metal core with other metal atoms, and (3) utilizing aggregation-induced emission (AIE). The three strategies are briefly summarized below.

7.2.2.1 Capping the gold core with different ligands

Although the theoretical calculations of the absorption spectra of clusters tend to save computational resources by simplifying ligands, the ligands on the surface of clusters, unlike internal atoms, have completely different chemical environments, which may lead to different properties and functions of clusters. Recent studies have shown that electron transfer may occur between the ligands and the gold core; thus, the type, structure, and arrangement of the ligands on the surface of the cluster have important effects on its optical properties [93, 94]. Based on this, the regulation of the optical properties of the cluster can be achieved by regulating the types of ligands on the surface of the cluster.

Wang et al. studied two types of reactions [95]: (1) monothiol tiopronin AuMPCs with the dithiol molecule 2,3-dimercaptopropanesulfonate (DMPS) and (2) DMPS Au dithiol clusters (AuDTCs) with tiopronin monothiol ligands. It was found that when excessive amounts of DMPS molecules were added to the tiopronin AuMPCs solution, the tiopronin molecules were effectively released from the original AuMPCs. This process was accompanied by a decrease in the near-infrared luminescence of the tiopronin AuMPCs and an increase in the absorption band at 282 nm in the absorption spectrum. In contrast, in the second type of reaction, the tiopronin molecules did not undergo ligand replacement with AuDTCs, but were added directly to the DMPS monolayer, which enhanced the near-infrared luminescence (Fig. 7.8a). Similarly, in a subsequent study [96], Wang et al. designed two types of monolayer reactions to study the effect of gold–thiolate bonding on the near-infrared luminescence of gold nanoclusters (Fig. 7.8b). The two reactions were: (1) the reaction of 1,4-dithiol durene (durene-DT) with Au_{25} protected by phenylethanethiolate (PhC_2S) ligands and (2) the reaction of PhC_2S monothiols with durene-DT stabilized Au nanoclusters (DTCs). In the first reaction, the addition of durene-DT enhanced the near-infrared luminescence of Au_{25} and weakened the clear absorption bands in the absorption spectra. This change in optical properties was found to be related to ligand exchange via proton NMR (nuclear magnetic resonance). Conversely, the addition of PhC_2S to AuDTCs resulted in a decrease in the near-infrared luminescence of the cluster. Mass spectrometry and NMR analysis revealed that the two reactions had similar final products. These results indicate that the near-infrared luminescence of gold clusters can be significantly enhanced by the bonding interaction of 1,4-dithiol with Au atoms. In addition, Wang et al. proposed a heat treatment method to improve the quantum efficiency of the near-infrared emission of gold nanoclusters (Fig. 7.8c) [97]. Pradeep et al. used two Au_{25} templates ($Au_{25}SBB_{18}$ [98] and $Au_{25}PET_{18}$ [57, 99], where SBB is 4-(*tert*-butyl)benzyl-mercaptan) to enhance the charge density on the surface of clusters by generating an additional thiol shell on the surface, thus enhancing the visible light [100]. In the case of $Au_{25}SBB_{18}$, by adding the [S(AuSBB)$_n$] unit, $(Au_{29}SBB_{22}S)^{2-}$ was obtained after the reaction, which led to enhanced illumination.

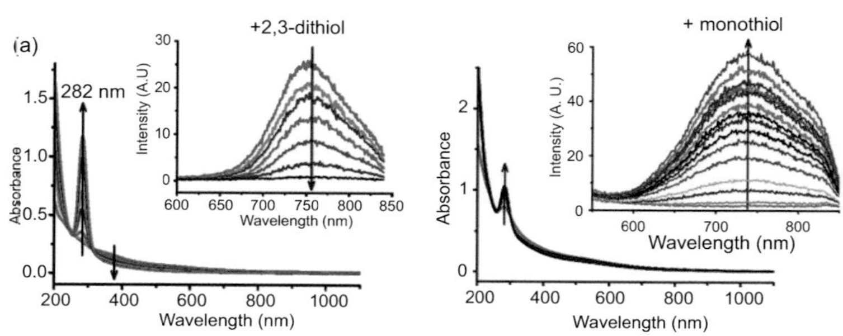

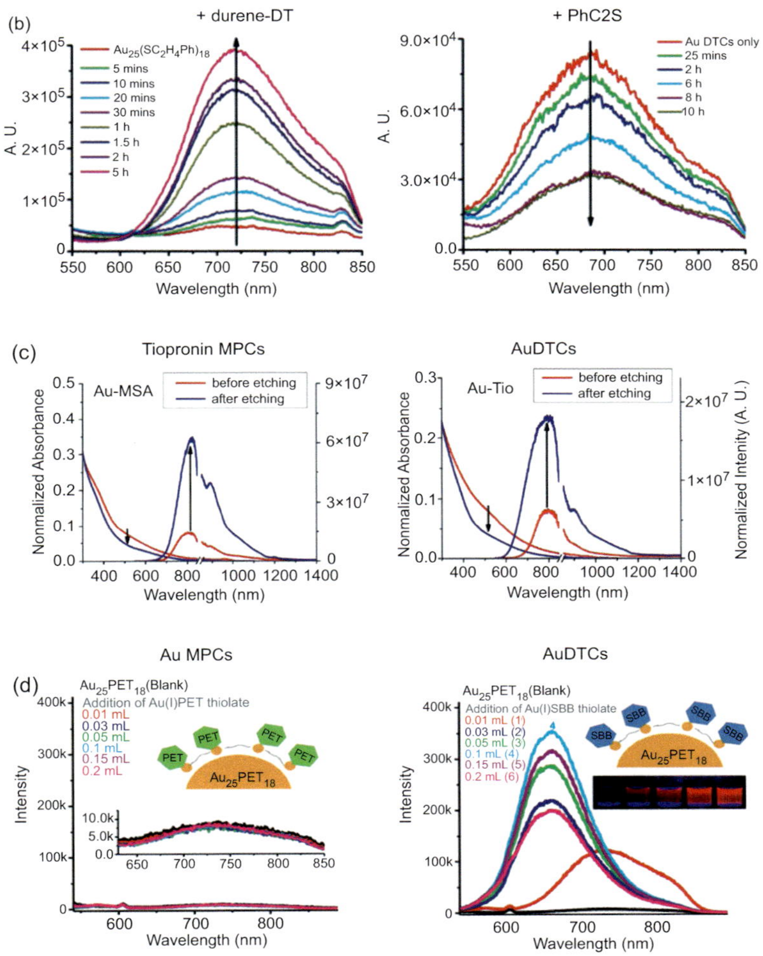

Figure 7.8 Changes in optical properties caused by ligand strategies. (a,b) Ligand-induced changes in the optical properties of clusters. (c) Changes in optical properties caused by thermal treatment methods. (d) Emission spectra of $Au_{25}PET_{18}$ with the addition of different amounts of Au(I)PET (A) and Au(I) SBB (B) thiolates (λ_{ex} = 514 nm). Illustrations show thiol complexes covering the surface of the $Au_{25}PET_{18}$ cluster. Photographs of $Au_{25}PET_{18}$ solution after adding different amounts of Au(I)SBB thiolate are also shown. Figure (a) reprinted by permission of the American Chemical Society from Ref. [95], Copyright 2011. Figures (b), (c), and (d) reprinted by permission of the Royal Society of Chemistry from Refs. [96], [97], and [100], Copyright 2012, 2014, and 2015, respectively.

For $Au_{25}PET_{18}$, thermal treatment was carried out by adding Au(I)PET and Au(I)SBB thiolates to $Au_{25}PET_{18}$ solution, and it was found that the two reaction products had distinct emission spectra. In Fig. 7.8d, compared to the case in which Au(I)PET was added, the $Au_{25}PET_{18}$ solution to which Au(I)SBB was added was observed to have a 35-fold increase in luminescence after thermal treatment. The stronger electron-donating properties of Au(I)SBB as compared to Au(I)PET may have been responsible for the enhanced luminescence [100].

7.2.2.2 Tailoring core size and doping

Doping is an effective way to improve the luminescence of gold clusters [101, 102]. Negishi et al. synthesized $Au_{25-n}Ag_n(SC_{12}H_{25})_{18}$ using different Au–Ag mixing ratios and analyzed the optical properties of this series of doped clusters. It was found that Ag doping can significantly affect the electronic structure and luminescence properties of Au_{25} [101]. Wang et al. reported that the doped rod-like Au_{25} has a very high quantum yield (40.1%) [103]. Ag-doped Ag_xAu_{25-x} clusters were synthesized by two different methods (**I** and **II**). Among these clusters, Ag_xAu_{25-x} ($x \leq 12$) obtained by method **I** had only weak fluorescence [quantum yield (QY) ≈ 0.2%]. In addition, Ag_xAu_{25-x} ($x \leq 13$) obtained by method **II** was highly luminescent (QY ≈ 40.1%), especially when $x = 13$. The distribution of these 13 Ag atoms was confirmed via X-ray, partial occupancy, and ^1H NMR analyses, and the atoms were found to be distributed at the two vertices, that is, the waist and center of Au_{25}. Moreover, the origin of such strong luminescence was also discussed from the perspective of electrochemical means [103].

Based on the electrochemical data, it was found that the HOMO–LUMO gap and emission wavelength of Au_{25} nanorod are similar to those of $Ag_{13}Au_{12}$, and hence the fluorescence of these two clusters may be related to the transition of HOMO–LUMO. The electronic structure analysis of Au_{25} and $Ag_{13}Au_{12}$ revealed that the HOMO and LUMO of Au_{25} are contributed by the Au 5d and Au 6s orbitals, while the HOMO and LUMO of $Ag_{13}Au_{12}$ are contributed by the Ag 5s orbital. The difference between the electronic structures of Au and Ag results in the higher electron mobility (fluorescence correlation) of $Ag_{13}Au_{12}$ than of Au_{25}, which leads to the strong fluorescence of the $Ag_{13}Au_{12}$ cluster [103].

Moreover, Bakr et al. found that the Au-doped $Ag_{29}(BDT)_{12}(TPP)_4$ cluster (BDT: 1,3-benzenedithiol; TPP: triphenylphosphine) presented enhanced stability and photoluminescence [104]. Recently, Dolg et al. studied the geometric and electronic structures, as well as the optical absorption and luminescence properties, of the Au-doped $Ag_{29}(BDT)_{12}(TPP)_4$ cluster from a computational perspective [105]. They found that the strong relativistic effect of Au atoms plays an important role in enhancing luminescence. In addition, heteronuclear Au@Ag bonds not only enhance the stability of gold clusters, but also tune their luminous intensity. Mattoussi et al. found that a small fraction of Au-doped Ag_{29} [106] has strong fluorescence emission [107]. Compared to that of a pure Ag cluster, the PL quantum yield of the bimetallic cluster increased by a factor of six (from ~3.3% to ~20%).

7.2.2.3 Aggregation-induced emission

The AIE mechanism was first proposed by Tang et al [108]. They found that a series of silole derivatives were not emitted in dilute solution. However, as their molecules aggregated in concentrated solution or were cast into solid films, they emitted strongly [109–111]. With the continued synthesis and characterization of a large amount of $Au_n(SR)_m$, the core–shell structure of gold clusters is coming into view, and the mechanism of the photoluminescence properties of gold clusters has become a popular research topic [93]. It was originally believed that the electronic behavior in the core structure of gold clusters was the main source of their luminescence properties. Later studies demonstrated that the ligand layer structure on the surface of gold clusters plays an important role in their optical properties [93, 112]. For instance, the type of ligand on the surface of Au_{25} can significantly affect its luminescence properties. The intrinsic mechanism of the luminescence properties of gold clusters may provide a glimpse into AIE.

Structurally, AIE active molecules are composed of several wheels (such as phenyl) that can freely rotate or vibrate in dilute solutions. Due to in-depth research, it is believed that the rotation or vibration of these wheels within the molecule is limited in the aggregate state, thus effectively limiting the non-radiative relaxation process of the excited electrons and resulting in the AIE of the molecule [110, 113]. The mechanism of the AIE phenomenon can be understood intuitively from Fig. 7.9 [111]: (1) Restriction of intramolecular rotation (RIR)—taking TPE as an example, the four benzene rings in the molecule can rotate freely in a poor solution, and the rotation is limited when the molecules are aggregated; (2) restriction of intramolecular vibration (RIV)—for example, THBA does not have a rotatable benzene ring like TPE, but the vibration of its symmetrical butterfly structure can result in the non-radiative relaxation of the excited electrons. Such vibrations are limited when the molecules are concentrated.

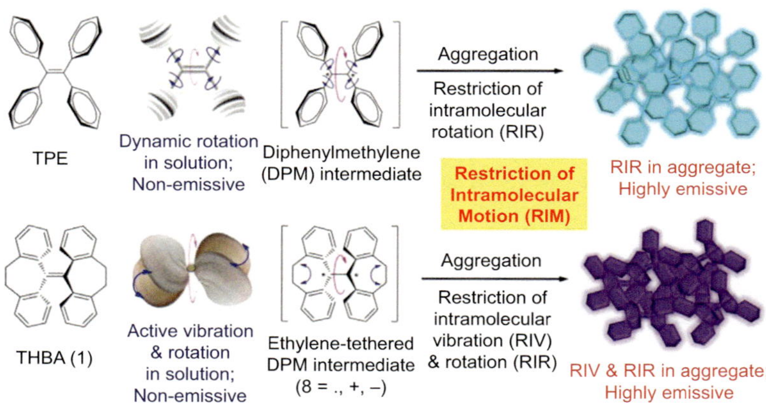

Figure 7.9 The mechanism of the AIE phenomenon [111]. The non-luminous TPE molecule in the dilute solution possesses free-rotating phenyl and emits intense light due to the RIR principle in an aggregation state. The non-radiative relaxation of electrons is caused by the vibration of the butterfly-shaped THBA wing structure in dilute solution. In an aggregation state, the non-radiative relaxation process is suppressed by the RIV principle, resulting in strong luminescence. Reprinted with permission from Ref. [111], Copyright 2015, American Chemical Society.

Based on the AIE mechanism of molecules, the research on AIE of active molecules has made great progress in recent years. For example, Tang et al. have conducted in-depth research on AIE and made considerable progress in exploring the applications of AIE [114–120]. The AIE phenomenon is not limited to organic compounds, and an important finding has been the AIE characteristics of organometallic compounds [121]. The Au(I)–SR complex is a special metal–organic complex with the AIE activity. In view of the "divide and protect" concept [122], the similarity of the structure can be easily found by comparing the structure of the Au(I)–SR complex with the thiolate-stabilized gold nanoclusters; the staple motifs on the surface of thiolate-protected gold clusters can be viewed as a special Au(I)–SR complex. This similarity prompted us to relate the AIE phenomenon of the Au(I)–SR complex to the luminescence properties of the thiolate-protected gold cluster. In addition, gold nanoclusters are usually synthesized by the reduction of Au(I)–SR complexes in solution in a bottom-up manner, and so the staple motifs on the surface of the clusters are similar to the residues of Au(I)–SR complexes. Because of this, the AIE of Au–SR complexes has been studied to better understand the luminescence properties of gold nanoclusters.

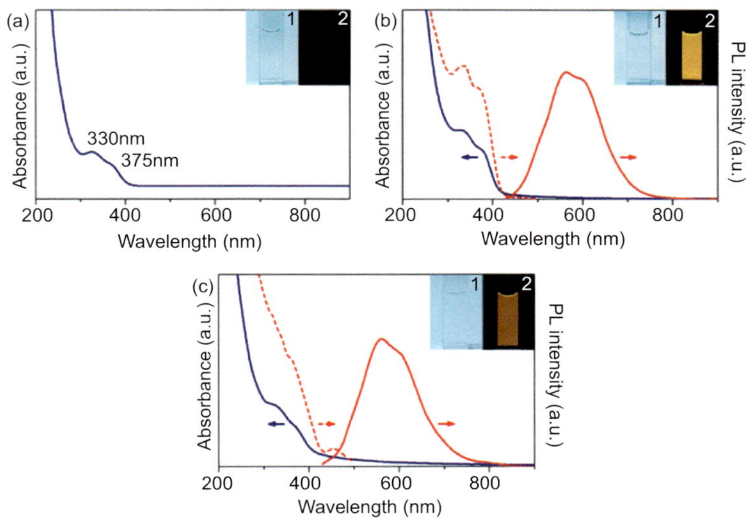

Figure 7.10 (a) UV–Vis spectrum of oligomeric Au(I)–thiolate complexes. (b) UV–Vis (solid blue lines), photoemission (solid red lines, λ_{ex} = 365 nm), and photoexcitation (dotted red lines, λ_{em} = 610 nm) spectra of Au(I)–thiolate complexes aggregated by ethanol (95% ethanol by volume) and (c) Cd^{2+} ions (with a Cd^{2+}-to-GSH ratio of 1:2). (Insets) Digital photos of aggregated complexes under (1) visible and (2) UV light. Reprinted with permission from Ref. [123], Copyright 2012, American Chemical Society.

Xie et al. discovered the AIE phenomenon of mercaptan complexes on the surface of clusters by studying Au–thiolate complexes. They found that non-luminescent oligomeric gold complexes produce strong luminescence when concentrated, and the intensity and color of the luminescence depend on the degree of aggregation [123]. In a previous study, Xie et al. introduced the photophysical properties of some organometallic compounds with AIE properties [124]. They synthesized Au–thiolate complexes using GSH as a model mercaptan ligand and found that strong luminescence was produced when these complexes aggregated. Such aggregation-induced emission can be achieved by solvent-induced aggregation and cation-induced aggregation. As shown in Fig. 7.10, the synthesized Au–thiolate complex does not illuminate under ultraviolet light (inset in Fig. 7.10a), and when these complexes aggregate, it causes strong luminescence (Fig. 7.10b,c).

The principle of solvent-induced aggregation is to introduce a poor solvent to change the polarity of the original solvent, which destroys the hydration shell of the complex. This causes the stability of the complex to be destroyed, thereby resulting in charge neutralization and the subsequent complex aggregation. According to this principle, the luminescence properties of the Au–thiolate complex can be adjusted by varying the polarity of the solvent to adjust the degree of aggregation of molecules. For example, the luminescence properties of Au–thiolate complexes can be adjusted by changing the volume fraction (f_e) of ethanol in a mixed solvent of ethanol and water (Fig. 7.11). When $f_e > 75\%$, the solution becomes cloudy and illuminates, and as f_e increases, the luminescence will be brighter. These experimental phenomena further confirm the mechanism of AIE.

It is worth mentioning that the aggregation of the Au(I)–thiolate complex can also be promoted by the aurophilicity in or between complexes [125]. On the other hand, cation-induced aggregation is the formation of complex cross-links by the electrostaticity and coordination of certain multivalent cations (e.g., Cd^{2+}) with monovalent carboxyl anions (from GSH) [126]. The as-formed cross-linkers promote the Au(I)–thiolate complexes to be closer to each other, thereby fostering a tighter aggregation of these complexes.

Based on the results of the previously discussed works, it can be foreseen that the essence of the AIE mechanism is to prevent the non-radiative relaxation process of the substance's molecules to enhance the luminescence. Therefore, it is particularly important to explore ways in which to prevent the non-radiative relaxation process of molecules.

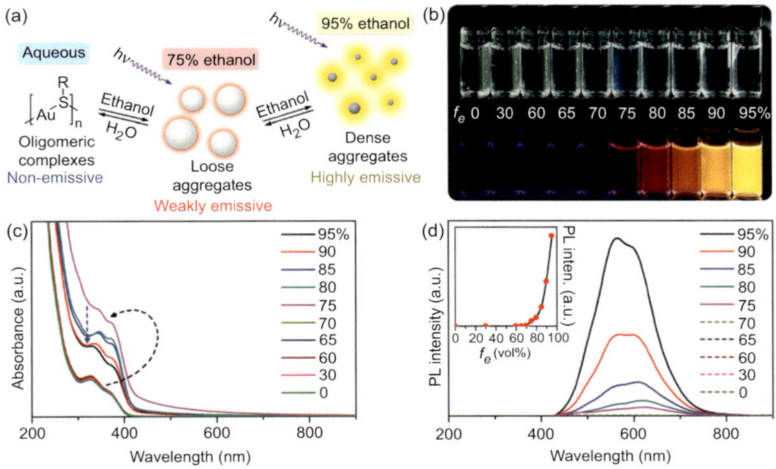

Figure 7.11 Ethanol as an example to elucidate the solvent-induced AIE properties of Au–thiolate complexes. (a) Solvent-induced AIE properties of the Au–thiolate complex. (b) Photograph of Au in a mixed solvent of ethanol and water with different f_e values under visible light (top) and ultraviolet light (bottom). (c) UV–Vis and (d) emission spectra of the Au–thiolate complex in the mixed solvent of water and ethanol with different f_e values. Reprinted with permission from Ref. [123], Copyright 2012, American Chemical Society.

Lee et al. utilized temperature-dependent luminescence measurements to study the luminescent origin of $Au_{22}(SG)_{18}$ (in which SG is glutathione) [127]. They found that the luminescence intensity increased with the decrease in the temperature, along with a shift in the luminescence maximum from 670 to 580 nm (Fig. 7.12a). The maximum luminescence and luminescence intensity were greatly enhanced when the temperature decreased to approximately 233 K (corresponding to the freezing point of the glycerol–water mixture). Moreover, Fig. 7.12b depicts the relationship between the excited-state decay of Au_{22} and the corresponding excited-state lifetime as a function of temperature. It can be seen that the lifetime of the excited state is greatly increased at 233 K. These phenomena indicate that the rigidity of the ligand on the surface of Au_{22} is important for its luminescence properties. Therefore, if a suitable method to solidify the ligand on the surface of Au_{22} can be found, it will be conducive to its luminescence at room temperature.

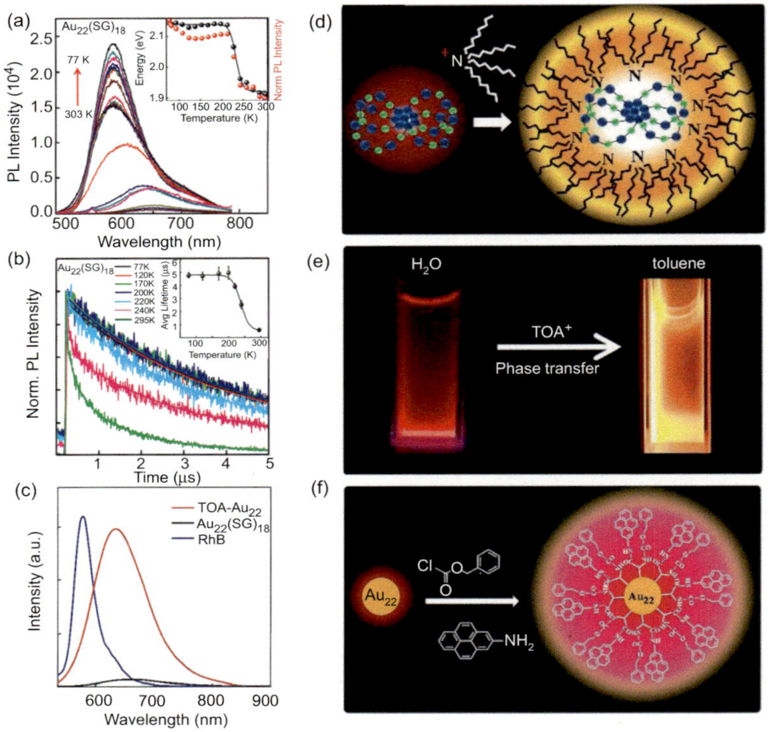

Figure 7.12 (a) Optical emission spectra of Au_{22} at different temperatures. (b) PL spectra decay of Au_{22} at different temperatures as a function of delay time. (c) Luminescence spectra of $Au_{22}(SG)_{18}$ in water, TOA-Au_{22} in toluene, and Rhodamine B. (d) Schematic diagram of TOA bonding to Au_{22}. (e) Digital photos of Au_{22} in solution (water and toluene) under UV light (365 nm). (f) Schematic representation of two molecularly functionalized Au_{22} clusters. Figures (a–d) reprinted by permission of American Chemical Society from Refs. [127] and [128], Copyright 2015. Figures (e) and (f) reprinted with permission from Ref. [129], Copyright 2016, The Royal Society of Chemistry.

To this end, Lee et al. used ion pairing with Au_{22} that can bind to the carboxyl anion of glutathione (Fig. 7.12c,d). By analyzing the influence of the alkyl chain length of the solvent and the complex ion on the luminescence, it was confirmed that the luminescence enhancement effect of Au_{22} is caused by the rigidity of the ligand layer induced by bulky counterions (Fig. 7.12e). Interestingly, similar to Au_{22}, TOA-Au_{25} exhibits only weak luminescence, but its luminescence is six times stronger than that of the original Au_{25}, and its lifetime is increased from 160 ns to 940 ns [128]. Based on this, Lee et al. believe that all of these luminescence enhancement phenomena are derived from the ligand layer. Furthermore, as shown in Fig. 7.12f, Lee et al. prepared highly luminescent water-soluble $Au_{22}(SG)_{18}$ clusters with a 30% quantum yield by functionalizing the shell of the cluster [129].

Kang and Zhu have previously outlined several influencing factors [112] that produce AIE phenomena, including solvent-induced aggregation, temperature-induced self-assembly, heteroatom-induced luminescence, polymer-induced aggregation, electron-rich induction aggregation, and crystallization-induced emission enhancement (CIEE). In addition, the AIE phenomenon exists not only in a single cluster, but also between clusters. Based on these influencing factors of AIE, the illuminating intensity, emission wavelength, and illuminating mode of cluster molecules can be precisely adjusted by changing the conditions.

In addition, it is worth mentioning that the configuration also has a certain influence on the luminescence of gold clusters [130, 131]. At present, most of the clusters with the precise atomic structure obtained via experimentation have fcc configurations [10, 45, 51]. However, in addition to the luminescence of the ligand itself, there are few reports on the luminescence of clusters with fcc configurations. Therefore, it is considered that the clusters with fcc configurations are not conducive to luminescence. Unexpectedly, Wu et al. recently reported a new non-fcc configuration of the $Au_{42}(TBBT)_{26}$ cluster prepared by a new synthetic method (ion induction) [132]. The structural configuration of the as-prepared cluster is significantly different from that which has been previously reported [133]. These two Au_{42} clusters are true structural isomers (their composition is identical). Wu et al. further investigated the emission of four pairs of clusters of the similar size and configuration, including two Au_{42} clusters, and found that the emission of fcc clusters is stronger than that of non-fcc clusters, suggesting that the arrangement of atoms affects cluster emission and that fcc clusters may be more conducive to luminescence [134]. Research on the photoluminescence properties of nanoclusters with fcc configurations should receive additional attention.

Although the AIE phenomenon has been reported by spurt type, the luminescence efficiency of AIE nanoclusters is far from sufficient as compared to that of well-developed inorganic quantum dots and luminescent dye molecules. In particular, there are a few reports of ligand-protected gold clusters of AIE [124]. For these reasons, the practical application of the AIE phenomenon of gold nanoclusters is limited. Therefore, it is of great significance to continue to explore more nanoclusters that can display AIE in future work.

7.3 Nonlinear Optical Properties

The nonlinear optical properties of gold nanoclusters have become an emerging research topic in the field of optics in recent years [135–145]. This research is mainly focused on the laws of interaction between strong light (e.g., lasers) and matter and is a branch of modern optics. The interaction between a traditional light source produced by spontaneous radiation and a substance is basically a linear category. For example, the refractive index or polarizability of a medium is independent of light intensity, and the polarization intensity $\boldsymbol{P}$ (the sum of the dipole moments of atoms per unit volume) of a medium is proportional to the electric field strength $\boldsymbol{E}$ of the light wave. The polarization intensity vector $\boldsymbol{P}$ can be defined as [146]:

$$\boldsymbol{P} = \varepsilon_0 \cdot \boldsymbol{E} + \chi^{(2)} \cdot \boldsymbol{E}^2 + \chi^{(3)} \cdot \boldsymbol{E}^3 + \ldots \tag{7.1}$$

where ε_0 and χ are the vacuum permittivity and the polarization coefficient of the medium, respectively. The first term on the right side of the formula is a linear term. However, the nonlinear terms (the second term and those following) of Eq. (7.1) cannot be ignored when the light used for the medium is strong light. $\chi^{(2)}$ and $\chi^{(3)}$ are the second-order and third-order nonlinear optical coefficients of the molecules, respectively. At this point, the polarization of the medium will be nonlinear. In 1960, Maiman manufactured the world's first ruby laser [147], which opened the door to the research of nonlinear optical properties. Since then, many nonlinear optical properties have been discovered, including optical sum frequency, optical frequency doubling, stimulated Raman scattering, two-photon absorption (TPA), saturated absorption, self-focusing, and self-defocusing.

In recent years, researchers have studied the nonlinear optical properties of gold nanoclusters, which primarily include TPA [135,

143], two-photon fluorescence (TPF) [148, 149], and second/third harmonic generation (SHG/THG) [150]. Due to these properties, gold nanoclusters have become promising candidates in the field of high-resolution multiphoton imaging and for optical confinement applications. The thiol ligand-protected gold clusters have become an important research object in the study of the nonlinear optical properties of such clusters due to their high stability and good biocompatibility.

7.3.1 Two-Photon Absorption/Emission

The relatively strong emission of gold clusters under single-photon excitation leads to the possibility of two-photon excitation emission. TPA refers to the process in which atoms in a medium simultaneously absorb two photons, transitioning from a ground state to an excited state. The quantum features of this transition have high spatial resolution and good dielectric penetration and thus have the application value in high-tech fields, such as low-power medical imaging, optical information storage, and optical limiting [4]. As early as 1961, Kaiser and Garrett discovered the phenomenon of two-photon absorption [151]. Previous reports have shown that the generation of TPA is based on the effective electron delocalization of the medium [152]. In addition, increasing the two-photon absorption cross section is the key factor to enhancing the intensity of two-photon absorption.

In 2008, Goodson et al. studied the two-photon emission and two-photon cross section of Au_{25} [135]. An effective two-photon absorption with a cross section of 2700 GM at 1290 nm and an emission peak at 830 nm was found. It was determined that as the incident laser energy increases, the fluorescence intensity of the two-photon emission is linearly related to the square of the incident light energy, indicating that the process is a typical TPA mechanism (Fig. 7.13b,c). In addition, the two-photon emission in the visible region of nanoparticles with an increase in the cluster size from 1.0 to 4.0 nm under the excitation of 800 nm is also shown in Fig. 7.13. It can be seen that the maximum emission wavelength is related to the size of nanoparticles and nanoclusters. For nanoclusters, the emission range is 500–535 nm, while nanoparticles are emitted at about 550 nm.

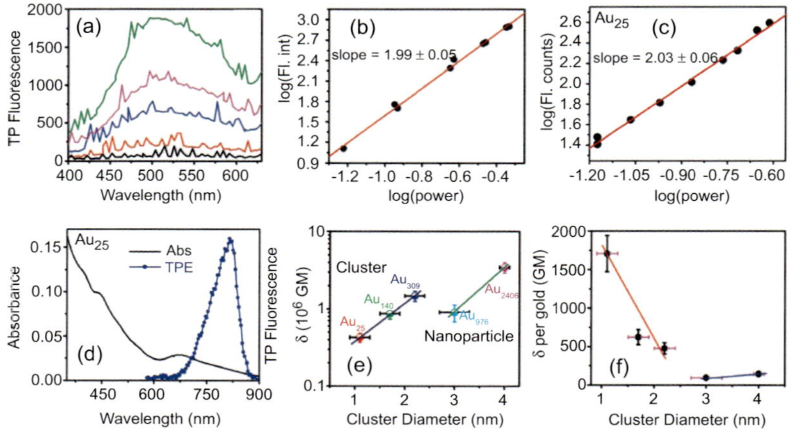

Figure 7.13 (a) Two-photon fluorescence spectra at different pump powers after excitation at 800 nm. (b,c) Power dependence of emission spectra of Au_{25} after excitation at 510 nm and 1290 nm, respectively. (d) Two-photon emission of Au_{25}. (e) TPA cross sections of gold nanoclusters/nanoparticles with different sizes. (f) The TPA cross section of each gold atom varies with the cluster size. Note that the TPA data in (c) and (d) were obtained after excitation at 800 nm. Reprinted with permission from Ref. [135], Copyright 2008, American Chemical Society.

Furthermore, the relationship between the TPA cross section and the size of clusters and nanoparticles indicates that the increase in the size of nanoclusters or nanoparticles is accompanied by the increase in the two-photon absorption cross section (Fig. 7.13e). However, as shown in Fig. 7.13f, analysis of the cross section of each atom indicates that the smaller the size of nanoclusters, the larger the cross section. In contrast with nanoclusters, nanoparticles have a smaller correlation with the size of the cross section; nevertheless, it is much larger than the value of organic macromolecules at 800 nm (~1000 GM). Such a huge two-photon absorption cross section opens up a prosperous road for the application of small-sized gold clusters in the field of nonlinear optics, such as in optical power limiting, nanolithography, and multiphoton biological imaging [135]. Therefore, the nonlinear optical properties of small-sized gold nanoclusters (<2 nm) are worth exploring. Although the NLO properties of thiol-protected gold clusters have been studied, there are few other reports on the NLO properties of Au_{25} [153].

In fact, bioimaging technology based on the two-photon absorption principle (e.g., two-photon microscopy) allows for the visualization of living tissue at depths that the conventional (single-photon) fluorescence or confocal microscopy cannot achieve. The conventional fluorescence microscopy primarily uses a single photon of visible wavelength (390–700 nm) to excite fluorescent dyes. When the fluorescent molecule is excited, the electron will relax to a stable state, during which a photon with a slightly lower energy than the excited photon will be released (the energy is lost due to vibration relaxation and other processes). Two photons with twice the wavelength are used in a two-photon microscope to excite the same or similar fluorescent dyes. However, only one photon is released when the electrons relax to a more stable orbit. This photon will have the same wavelength as in the equivalent single photon fluorescence method (i.e., slightly longer than half the excitation wavelength). The most commonly used fluorescent dye has an excitation spectrum in the range of 400–500 nm. Therefore, in the infrared spectrum, the wavelength of the same dye excited by two photons is often between 800 and 1000 nm. Imaging of the biological tissue using such a wavelength range can effectively avoid the scattering of the excitation light by the biological tissue, increasing the penetration depth, and thereby exciting the fluorescent medium. The available information indicates that infrared lasers have sufficient energy to excite fluorophores up to 1 mm in living tissue. In contrast, a single-photon confocal microscope can penetrate only about 200 μm. On the other hand, the infrared laser allows the excitation to be limited to a small focus in the sample, which can be used to scan the las. For instance, Xu et al. successfully demonstrated the application of biocompatible glutathione-capped gold clusters as one- and two-photon excitation fluorescence contrast agents for live cell imaging. Brevet et al. studied the nonlinear optical properties of Au_{15} [154] and Au_{25} [155, 156] based on TPA [137]. Two-photon emission spectra and the first hyperpolarizabilities [corresponding to $\chi^{(2)}$ in Eq. (7.1)] of the two clusters were obtained by the hyper-Rayleigh scattering technique [157]. They found that Au_{15} has a large two-photon absorption cross section of about 65,700 GM. Au_{15} may, therefore, be a good candidate for nonlinear optical microscopy (e.g., two-photon microscopy). Additionally, the first hyperpolarizability of Au_{15} and Au_{25} is two orders of magnitude larger than that of Au nanoparticles in the size range of 10–50 nm.

Another important and interesting topic is the critical size of the quantum effect of gold nanoparticles. In 2010, Goodson et al. concluded that an abrupt change in the optical properties (fluorescence, transient absorption, two-photon absorption) is observed once the diameter of the cluster is less than 2.2 nm, which is consistent with the calculated critical size of quantization [158]. This work is also of great significance for exploring the transition from nonmetallic to metallic states. In 2012, Thomas et al. explored the evolution of nonlinear optical properties from quantum gold nanoclusters to metallic plasmonic nanocrystals [136]. They found that the third-order optical nonlinearity in these ultrasmall gold clusters exhibits a significantly lower threshold for optical power limiting, which is highly beneficial for optical limiting applications.

7.3.2 Second Harmonic Generation

Second harmonic generation (SHG) is a nonlinear optical phenomenon and is the most important and widely used technique in nonlinear optical mixing. SHG occurs when the fundamental light of a certain intensity passes through the nonlinear optical crystal to generate an electromagnetic wave with twice the original frequency, and it is therefore also called the double-frequency effect. In 1961, Franken et al. first discovered the second harmonic phenomenon [150]. They used a ruby laser (0.6943 μm) to perform a second harmonic experiment on quartz crystal. It was found that the output of the 694 nm laser passed through the quartz crystal and obtained 347.1 nm of ultraviolet light. This provides a convenient method for obtaining ultraviolet light using a near-infrared laser. Since then, research on SHG has been rapidly developed and applied to achieve laser modulation and to obtain various lasers as short as UV and as long as infrared. The SHG signal is related to both the intensity of the optical electric field and the polarizability of the medium. Because the dielectric polarizability reflects the microscopic characteristics of the medium, such as its symmetry and alignment, the second harmonic signal can be used to characterize the microstructure of the medium. In recent years, second harmonic imaging has often been used in the field of biological imaging [159, 160].

Thiol ligand-protected gold clusters for use in various imaging technologies have become a research hotspot due to the high stability and biocompatibility of ligands, and they are promising candidates for imaging applications. However, the SHG of thiolate-protected gold clusters is rare. Knoppe et al. reported SHG for $Au_{25}(SCH_2CH_2Ph)_{18}$ and $Au_{38}(SCH_2CH_2Ph)_{24}$ [140]. Because SHG is directly related to the symmetry of the crystal structure, $Au_{25}(SCH_2CH_2Ph)_{18}$, which has an inversion center, does not produce a significant SHG signal. As a control, the chiral $Au_{25}(Capt)_{18}$ (Capt: captopril) has a SHG response. In contrast, the inherent chirality of $Au_{38}(SCH_2CH_2Ph)_{24}$ leads to its optical activity of the second harmonic. These experimental phenomena all indicate that SHG is sensitive to the structural symmetry of the material. In addition, two-photon fluorescence and third harmonics of Au_{25} and Au_{38} were also observed. Recently, Knoppe et al. investigated the first hyperpolarizability of a series of model clusters $[Au_m(SH)_n]^z$ (m = 18–38) through DFT [142]. Note that this work focuses on the contribution of the cluster itself to the hyperpolarizability, rather than the contribution of the ligand. The calculated results showed that there is no correlation between the first hyperpolarizability and the size, but it instead relies on the symmetry of clusters. The chiral Au_{38} cluster has a strong nonlinear optical response, whereas that of the centrosymmetric Au_{25} cluster is close to zero; this finding is consistent with previous experimental results [140]. Nevertheless, the central symmetry of the Au_{25} cluster can be effectively destroyed by ligand exchange, which causes the strong nonlinear optical response of Au_{25}, even surpassing inherent chiral clusters such as Au_{38}. These nonlinear optical responses of gold clusters render them interesting candidates for biological tissue imaging applications. It is worth mentioning that, compared to two-photon absorption, no photon absorption occurs during the second harmonic generation, and no energy level transition occurs.

Despite these findings, little research has been conducted on the nonlinear optical properties of thiolate-based, ligand-protected gold clusters. As increasingly more crystal structures are resolved, future research on the nonlinear optical properties of atomically precise metal nanoclusters will help to better clarify the structure–property relationship of these materials and will benefit their applications.

7.3.3 Ultrafast Electron Dynamics

The evolution of anything in nature is associated with the basic concept of time. The speed of development is different for different things, and so their description requires different time units. If a process is studied, some means (e.g., a camera) is needed to record the change in the process over the corresponding timescale. There is an old and interesting debate about whether a horse's four feet will leave the ground at the same time when it is running. To explore this issue, Muybridge used a technique called stop motion, which utilizes a high-speed mechanical shutter with a shutter switch on the order of milliseconds [161]. In this way, it is possible to clearly observe whether the four hooves of the horse will leave the ground at the same time when running. However, when the motion speed of an object is fast enough, there is no way to detect its properties by the conventional methods. Therefore, it is necessary to develop some advanced instruments and methods to expand the human cognition of the nature of ultrafast things. Many superfast occurrences can happen within a couple of femtoseconds, and if the flash of a camera is too slow, these fleeting shots will not be observed.

The behavior of electrons in gold nanoclusters determines their physicochemical properties [1], and so the observation and description of their electronic behavior are important for understanding their electronic structure and practical/potential application. However, at the scale of atomic molecules, the duration of the reaction of ultrafast processes, including energy transfer, electron transfer, breakage, and the formation of chemical bonds, is mostly on the order of ps–fs (picosecond–femtosecond) [162–164]. To study such short and high-speed processes, ultrafast pulses are required. The saying "workers must first sharpen their tools" is applicable, and the advent of lasers has laid the technical foundation for the generation of picosecond and femtosecond pulses. Due to the development of the laser mode-locking technology [165], the operation of picosecond and femtosecond pulsed lasers was realized in the 1970s and 1980s, respectively. Among them, the pump-probe technique is a typical representative of the ultrashort pulse technology, and it is a powerful tool to study the electronic dynamic behavior of gold nanoclusters [166]. The attenuation of the energy levels from the excited states of material molecules to the ground-state energy levels, and the transitions between the excited states, can be obtained via ultrafast time-resolved transient absorption spectroscopy [166]. This technique was mentioned in the evolution of the metallic to nonmetallic nature of gold nanoclusters in the previous section [86, 88, 89, 136, 167, 168]. Since El-Sayed et al. studied the ultrafast electron kinetics of $Au_{28}(SG)_{16}$ using femtosecond laser pulses [169], a series of ultrafast kinetic studies of gold nanoclusters have revealed their various physicochemical properties, such as metallic (e.g., plasma) or nonmetallic (e.g., molecules) properties, and the transfer of electrons and energy.

7.3.3.1 Metallic or nonmetallic state of gold nanoparticles

As mentioned previously, the study of the evolution from molecular to metallic states is of great significance for exploring the origins of metal bonds and the birth of SPR [167, 170]. Ligand-protected gold nanoclusters are an excellent choice for studying this evolution due to their precise atomic structure. However, due to the difficulty of synthesis, the ultrasmall size of the transition zone from the nonmetallic to the metallic state has only recently been explored [86–89]. Knappenberger et al. studied the electron energy relaxation of $Au_{144}(SC_6H_{13})_{60}^q$ by femtosecond time-resolved transient absorption spectroscopy and found that $Au_{144}(SC_6H_{13})_{60}^q$ exhibited the characteristics of metals [171]. However, in a subsequent study, Jin et al. explored the property transitions of a series of gold nanoclusters and found that the electron dynamics of Au_{144} do not depend on the pump power, revealing its nonmetallic nature [167]. The difference between the findings of these two research groups on the nature of Au_{144} was that the intense overlap of the probe light at 525 nm with the ESA led to an underestimation of the decay time. Interestingly, Häkkinen et al. conducted both low-temperature optical absorption spectroscopy and density functional theory calculations, which revealed that $Au_{144}(SC_{12})_{60}$ has a molecular-like electronic structure and that the critical point of the metal-to-molecule transition is between Au_{144} and Au_{187} [76]. Whetten et al. transmitted the information of the quantum state of Au_{144} by theoretical and experimental means [170]. Ramakrishna also observed no power dependence of Au_{144} when studying the exciton dynamics of gold clusters [172]. Based on these studies, Au_{144} should behave as a quantum state. It is worth noting, however, that the crystal structure of the thiol ligand-protected Au_{144} was not determined until 2018 when Jin and Wu et al. reported it [83].

In addition, other brief discussions about the metallic or nonmetallic nature of gold clusters have also been conducted. In

2015, Jin et al. used femtosecond broadband transient absorption spectroscopy to confirm that Au_{133} has similar molecular properties [81]. In 2016, Jin et al. reported the crystal structure of large-sized Au_{246} clusters [85], and subsequent kinetic analysis indicated that it still has molecular properties [86]. In 2017, along with the determination of the crystal structure of another large-sized Au_{279} cluster, Dass and Jin et al. conducted electron dynamics studies on the cluster [87, 88]. The single SPR peak in the steady-state absorption spectrum and the power-dependent lifetime in the TA spectrum revealed the metal-like properties of Au_{279} in fcc configuration. Even so, the long-life phonon–phonon coupling process and the broad excited-state absorption in the near-infrared region of Au_{279} indicate its nascent plasma properties [88, 89].

Recently, Jin et al. observed interesting electrokinetic characteristics in $Au_{333}(SR)_{79}$ nanoparticles [168]. It is worth noting that early research indicated that $Au_{333}(SR)_{79}$ (R: CH_2CH_2Ph) is a metallic nanomolecule [173]. The steady-state absorption spectrum of $Au_{333}(SR)_{79}$ protected by three different SR groups exhibited a single absorption peak at room temperature. Interestingly, the low-temperature spectrum showed a splitting of the absorption peak. Further analysis indicated that the splitting of the SPR peak was derived from the oblate core of $Au_{333}(SR)_{79}$. In addition, the femtosecond TA spectra of $Au_{333}(SR)_{79}$ presented two distinct GSB signals, which are consistent with the steady-state absorption peaks at cryogenic temperatures. Global fitting of the spectral data reveals that the electrodynamic process of $Au_{333}(SR)_{79}$ includes electron–phonon coupling (~1 ps), normal phonon–phonon coupling (>100 ps), and an unusual 4–5 ps relaxation process (a fast phonon–phonon relaxation process). These two unique features (the phonon–phonon coupling process of 4–5 ps and the two GSB signals in the TA spectrum) are independent of the protecting groups and solvents, indicating that they are derived from the core structure of $Au_{333}(SR)_{79}$.

As a result, this work considers that the splitting of plasmon and quantum effects occurs in the core of Au_{333}. Note that the atomic structure of Au_{333} has not yet been resolved. The origin mechanism of plasmon resonance and the critical point of transition from the metallic to the nonmetallic state still remain to be further explored.

7.3.3.2 Electron and energy transfer

It is well known that ligand-protected gold clusters are composed of a gold core and outer protective ligands [10, 34, 51]. The photophysical properties of gold clusters can be adjusted by their composition, size, and ligands [174]. In the cluster structure, the ligand plays a role in regulating and modifying the surface function and helps to stabilize these gold clusters [175–177]. Thus, the study of the interaction between surface ligands and metal cores assists in the profound understanding of the important role of ligands in the structure–property relationships of gold clusters. The different chemical potential between the ligand and the gold core may result in electron transfer between them [178]. Electron transfer has an important influence on the physicochemical properties of gold clusters [179, 180]. For example, by changing the structure of the cluster to affect its electron transfer process, the exciton radiation relaxation process can be promoted and luminescence enhanced.

Johansson and Gong et al. predicted the structure of the bare Au_{32} cage with high stability [181, 182] and believed that the structure of the Au_{32} cage should be very similar to that of C_{60} [183, 184]; for this reason, Au_{32} was named "golden fullerene" [181]. The source of stability of Au_{32} is attributed to its aurophilicity and aromatic nature. In a subsequent study, Wang et al. synthesized a "golden fullerene" Au_{32} cluster by using pyridinamine and triphenylphosphine as protective agents [185]. Single-crystal X-ray structure analysis revealed that the $[Au_{32}(Ph_3P)_8(dpa)_6](SbF_6)_2$ (dpa: 2,2′-dipyridylamine) cluster has the point group symmetry of S_6 and that the kernel is the Au_{32}^{8+} in the I_h configuration. Quantum chemistry studies have demonstrated that the particular stability of the structure of the cluster is derived from metal–ligand interactions. Zhou et al. found significant charge transfer and solvation kinetics in $Au_{20}(SR)_{16}$ and phosphine-protected Au_{20} nanoclusters [186, 187]. Solvent-dependent oscillations in the kinetic traces confirmed charge transfer between the surface ligand and the metal core.

7.4 Optical Stability

Due to their unique physicochemical properties, gold nanoclusters are considered to be ideal model materials for catalytic and solar energy conversion [10, 188]. However, due to the strong surface energy of the ultrasmall-sized Au cluster, it is very unstable under illumination conditions and can easily agglomerate into large-sized Au nanoparticles [175]. As a result, the photocatalytic reaction mechanism of the material is complicated and the catalytic activity is significantly reduced. Photo-agglomeration of gold clusters has been observed in photocatalytic materials based on gold clusters, and the agglomeration mechanism has been explored [189, 190]. However, determining how to effectively enhance the optical stability of gold clusters remains an unsolved key scientific problem.

In a previous study, Xu et al. discovered the photo-induced aggregation of Au nanoclusters for the first time [191], namely, the photo-induced transformation of small molecular-like Au clusters to larger metallic Au nanoparticles under different illumination conditions. Such a transformation has been demonstrated to be a photocatalytic oxidative process with a diffusion/aggregation mechanism.

7.5 Optical Rotation and Circular Dichroism (CD) of Gold Nanoclusters

Chiral phenomena are widely found in nature and are one of the basic and most fascinating properties of natural molecules. The most familiar and vivid phenomenon of chirality is human left and right hands. It is no exaggeration to say that from organic matter to inorganic matter, and from inanimate objects to life phenomena, the shadow of chirality is everywhere. For example, proteins in living organisms are almost always composed of L-amino acids. In 1948, Louis Pasteur separated the two enantiomers in the racemic crystals of tartrate and determined their chirality by measuring the optical rotation signal of the enantiomer solution [192]. Therefore, chirality is often associated with the optical activity. Chiral materials are generally considered to be optically active (optical and circular dichroism); of course, optically active substances are not all chiral. For example, Xu et al. observed optical signals in the achiral metal–organic framework (MOF), indicating that the optical activity can also be derived from achiral substances under certain conditions [193]. Because of the chiral environment in organisms, most drugs acting on organisms are chiral molecules in order to match the chirality of the target molecules in vivo.

It is worth noting that the concept of chirality does not exist only in organic matter; all substances with optical rotation have chirality, including chiral carbon nanotubes and chiral quantum dots [194]. Generally, chirality is expressed in organic molecules and originates from carbon atoms in the chiral center. However, due to the complexity of the cluster structure, the chiral origin of clusters is more elusive than that of common molecules. Therefore, the identification of a large number of clusters in recent years has been helpful for the analysis of the chiral origin of clusters. In particular, considerable progress has been made in the study of chiral thiol ligand-protected gold clusters [10, 194, 195]. In 1998, Whetten synthesized chiral $Au_{28}(SG)_{16}$ clusters by decomposing polymer Au(I)SG compounds and proposed the concept of a chiral gold cluster for the first time [196]. This section focuses on the chiral origin of gold clusters and their optical properties.

7.5.1 Origin of Chirality of Gold Clusters

Metal nanoparticles/clusters have attracted much attention due to their applications in optical devices, chiral recognition and separation, and biosensors, among others [194, 197]. Over the past decade, with the continuous exploration and efforts of chemists, the synthesis of clusters has shown an explosive momentum of development [10]. Chiral ligands have gradually been introduced into the synthesis process of clusters, and chiral gold clusters have quickly become a popular research area due to their unique physical and chemical properties [7, 10, 198]. In the early stages of research, due to the lack of a crystal structure of chiral clusters, it was speculated that the chirality of gold clusters might originate from the core structure [199, 200]. Theoretical calculations showed that the adsorption of thiolates on the surface of gold nanoparticles may cause distortion of the core structure, resulting in a decrease in the symmetry of the core and leading to chirality [201, 202]. In addition, the introduction of chiral ligands in clusters can also result in the production of chiral optical signals (e.g., circular dichroism). The chiral origin of gold clusters had been a mystery until the precise atomic structure was obtained. In 2007, a milestone work unveiled the mystery of the atomic-level precision structure of ligand-protected gold clusters [33]. Jadzinsky et al. successfully synthesized and resolved the single-crystal structure of Au_{102} using achiral *p*-mercaptobenzoic acid (*p*-MBA), which was the first identified chiral $Au_n(SR)_m$ nanocluster [33]. They found that the single-crystal structure of Au_{102} is a racemate composed of a levorotatory isomer and a dextrorotatory isomer. Although the Au_{79} kernel of the truncated decahedral in Au_{102} exhibits D_{5h} symmetry, the random arrangement of the surface –S–Au–S– staple motifs at the waist of the decahedron causes the symmetry to be destroyed, resulting in chirality. Similarly, the unit cell of $Au_{38}(2\text{-PET})_{24}$ (2-PET: 2-phenylethylthiolate) reported by Qian et al. also contains a pair of enantiomeric structures [203]. By analyzing its crystal structure, the chirality of Au_{38} was determined to be mainly due to the outer Au_{15} shell. The chiral Au_{38} cluster has both an intrinsic chiral core and a chiral induction of the outer chiral ligand. The chiral signal (CD signal) of Au_{38} under dual influence changes significantly, which further proves that the metal atomic arrangement of the inner core and the chirality of the ligand shell play crucial roles [200, 203]. In fact, gold clusters synthesized by achiral ligands are often racemates. To obtain optically active chiral clusters, these gold clusters in the form of racemates must be separated by high-performance liquid chromatography (HPLC) [204]. In addition, gold clusters with optical rotation and specific chirality can be directly synthesized by chiral ligands. As the structures of a large number of $Au_n(SR)_m$ clusters were resolved by single-crystal X-ray diffraction, many of the clusters were found to be chiral [10]. The chirality of these clusters is mainly due to the chiral arrangement of staple motifs and bridging thiolates on the surface of the kernel [52, 58, 66, 194]. It is worth mentioning that a cluster that does not have chirality itself will exhibit chiral features when a chiral ligand is introduced into the synthesis, for example, Au_{25} [205]. Zeng and Jin summarized three possible chiral sources of these gold clusters [194]: (1) chiral configuration of gold core atoms, (2) chiral arrangement of Au–S staple motifs on the gold kernel interface, and (3) intrinsic chiral or chiral arrangement of the outer R groups.

However, these three sources of chirality are based only on the speculation of the experimental results, and the true face of chiral origin is yet to be further studied. Chiral gold nanoclusters have laid the foundation for an in-depth understanding of the chiral origin of nanomaterials. In future work, the practical application of chiral gold nanoclusters, such as chiral sensing, enantioselective catalysis, chiral optics, and chiral recognition in biological systems, should be pursued.

7.5.2 Optical Properties of Chiral Gold Clusters

Chiral gold clusters exhibit a remarkable Cotton effect in the visible range, and their optical properties are mainly reflected in optical rotation and circular dichroism. The optical rotation of chiral gold clusters may result in a potential negative refractive material in the visible region, in which the refraction direction of the light wave is opposite to that of the conventional refraction. The incident and refracted waves are located on the same side of the normal direction of the interface.

Isotropic or cubic symmetry systems do not exhibit dichroism, and so anisotropy is a necessary condition for producing dichroism. Currently, in the identified gold clusters with atomic precision structures, many of the structures exhibit dichroism via a destroyed symmetry due to chirality. CD spectroscopy is widely used to study the chirality of matters and is usually associated with the optical activity. Although a large number of identified gold clusters appear as optically active species, because most of their crystal structures are in the form of racemates, the optically active structures can be obtained only by separating them. The chiral HPLC method is usually used for the separation of optical isomers. Bürgi et al. have conducted a significant amount of research work on the chirality of ligand-covered metal clusters [7, 195, 199, 204, 206–210]. The chirality and optical activity of the thiol ligand-protected gold clusters were outlined in a review published prior to those by Bürgi et al.

$Au_{38}(SCH_2CH_2Ph)_{24}$ was the first ligand-protected gold cluster for enantiomeric separation using the chiral HPLC method [204]. Bürgi et al. measured the CD spectrum of the fractions of the collected Au_{38} enantiomers. The CD curves of the two enantiomers of Au_{38} exhibit a good symmetry relationship. By comparing the CD spectra of Au_{38} protected by different ligands, it was found that the ligand had little effect on the curve of the CD spectrum. The enantioseparation of

$Au_{40}(SCH_2CH_2Ph)_{24}$ has also been carried out by chiral HPLC [210]. It is worth noting that the enantiomeric resolution of Au_{40} is not completely separated by HPLC.

As reported by Jin et al., $Au_{28}(TBBT)_{20}$ also contains a pair of enantiomers that were separated by chiral HPLC, and the CD spectra of the isomers were also measured [58]. However, the effectiveness of the commonly used HPLC method is still unsatisfactory, which limits the practical application of the optical activity of chiral clusters [208]. $Au_{102}(p\text{-MBA})_{44}$ [33] was the first identified thiol-protected gold cluster; however, the enantiomers in its crystal structure were not isolated until 2014 by Knoppe et al. [208]. Rather than chiral HPLC, a chiral phase transfer method was used to separate the enantiomers of Au_{102} [208]. In addition, Tang et al. obtained the enantiopure $Au_{20}(PP_3)_4Cl_4$ [PP_3: tris(2-(diphenylphosphino)ethyl)phosphine] using a supramolecular assembly strategy with α-cyclodextrin (α-CD) [211]. Theoretical studies indicate that the preferential self-assembly of α-CD with the organic ligand on the surface of the right-handed Au_{20} is responsible for this enantioseparation. In addition to the creation by enantioseparation, it is also possible to synthesize a homochiral cluster directly. For example, a homochiral Au_{13} nanocluster was synthesized enantioselectively by chiral ligands with stereogenic centers at the phosphorus atoms [212].

In future research, researchers may be able to focus on the following two aspects: obtaining highly optically active homochiral gold clusters and exploring the mechanism of chiral transformation. To achieve the first goal, it is necessary to find novel and efficient synthetic methods. For example, new and efficient methods must be developed for the enantioseparation of chiral clusters, and new ways to synthesize highly optically active homochiral clusters should be explored. For the latter goal, recent work by Häkkinen and Malola on the Au_{38} chiral transition may provide some instructive insights [213].

7.6 Summary and Prospects

As a bridge connecting molecules and macroscopic materials, clusters are rich in peculiar properties. The study of their structures and properties will help to explore their bright application prospects.

As a typical representative of clusters, ligand-protected gold clusters have developed rapidly in the field of nanoscience for nearly two decades. This chapter mainly outlined the optical properties of ligand-protected gold clusters, including optical absorption, photoluminescence, nonlinear optics, and circular dichroism.

Due to the quantum confinement effect, ligand-protected gold clusters have discrete energy levels, and their absorption spectra often present multiple absorption peaks. However, as the size of the cluster increases, the optical bandgap will gradually decrease. When the size reaches a certain critical point, plasma characteristics are exhibited. For example, the large-sized Au_{279} has only one single peak in the UV–Vis absorption spectrum, and transient absorption spectroscopy reveals its metallic properties. However, the critical point at which the transition from a nonmetallic state to a metallic state will occur is still unknown. The main reason for this is that there are only a handful of giant nanoclusters with more than 100 metal atoms. In future work, attempting to push the crystal size of the cluster to a greater extent may aid in the understanding of the origin of plasmon resonance.

The photoluminescence of clusters is one of their most fascinating properties. However, the quantum yield of gold clusters tends to be low compared to that of quantum dots. This chapter summarized some strategies for enhancing luminescence, such as aggregation-induced luminescence, doping, and ligand shell rigidification. Nevertheless, the origin of the luminescence of the cluster remains to be further explored.

Research reports on the nonlinear optical properties of clusters are currently rare. The research hotspots at this stage mainly focus on TPA, TPF, and SHG/THG. Due to the application prospects of clusters in high-resolution multiphoton imaging and optical confinement applications, future work may be attempted to impart more nonlinear optical properties on newly obtained clusters.

In terms of optical stability, small-sized bare gold clusters are prone to agglomeration under illumination conditions and are transformed into larger-sized gold nanoparticles. This causes the electronic structure to change, resulting in a complicated photocatalytic reaction mechanism of the material and the significant reduction of its catalytic activity. Photo-agglomeration of gold clusters has been observed in photocatalytic materials that are based on gold clusters, and the agglomeration mechanism has been explored in the existing literature. However, determining how to effectively enhance the optical stability of gold clusters remains an unsolved key scientific issue.

The circular dichroism of gold clusters benefits from the chirality of their structures. Chirality, as a structural feature of clusters, has been extensively studied. Although some possibilities for explaining the chiral origin of ligand-protected gold clusters have been proposed, the true origin of chirality remains unresolved. At present, it has been experimentally determined that chiral gold clusters are all racemic. The primary methods that are currently employed for obtaining an optically active chiral structure are the separation of enantiomers by chiral HPLC and chiral phase transfer. However, the actual effect of the two separation methods may not be satisfactory. Therefore, a future research direction may be the investigation of new methods for separating enantiomers or the exploration of a controllable method for directly synthesizing single-chiral cluster structures.

Acknowledgments

We acknowledge the financial support by National Natural Science Foundation of China (91961121, 21773201, and 21422305).

References

1. Jin, R. (2010). Quantum sized, thiolate-protected gold nanoclusters, *Nanoscale*, **2**, pp. 343–362.
2. Jin, R. (2015). Atomically precise metal nanoclusters: Stable sizes and optical properties, *Nanoscale*, **7**, pp. 1549–1565.
3. Aikens, C. M. (2010). Electronic structure of ligand-passivated gold and silver nanoclusters, *J. Phys. Chem. Lett.*, **2**, pp. 99–104.
4. Yau, S. H., Varnavski, O. and Goodson III, T. (2013). An ultrafast look at Au nanoclusters, *Acc. Chem. Res.*, **46**, pp. 1506–1516.
5. Zhu, M., Aikens, C. M., Hendrich, M. P., Gupta, R., Qian, H., Schatz, G. C. and Jin, R. (2009). Reversible switching of magnetism in thiolate-protected Au_{25} superatoms, *J. Am. Chem. Soc.*, **131**, pp. 2490–2492.

6. McCoy, R. S., Choi, S., Collins, G., Ackerson, B. J. and Ackerson, C. J. (2013). Superatom paramagnetism enables gold nanocluster heating in applied radiofrequency fields, *ACS Nano*, **7**, pp. 2610–2616.

7. Knoppe, S. and Bürgi, T. (2014). Chirality in thiolate-protected gold clusters, *Acc. Chem. Res.*, **47**, pp. 1318–1326.

8. Goswami, N., Yao, Q., Chen, T. and Xie, J. (2016). Mechanistic exploration and controlled synthesis of precise thiolate-gold nanoclusters, *Coord. Chem. Rev.*, **329**, pp. 1–15.

9. Desireddy, A., Conn, B. E., Guo, J., Yoon, B., Barnett, R. N., Monahan, B. M., Kirschbaum, K., Griffith, W. P., Whetten, R. L. and Landman, U. (2013). Ultrastable silver nanoparticles, *Nature*, **501**, pp. 399.

10. Jin, R., Zeng, C., Zhou, M. and Chen, Y. (2016). Atomically precise colloidal metal nanoclusters and nanoparticles: Fundamentals and opportunities, *Chem. Rev.*, **116**, pp. 10346–10413.

11. Freestone, I. C., Meeks, N., Sax, M. and Higgitt, C. (2007). The Lycurgus Cup—a Roman nanotechnology, *Gold Bull.*, **40**, pp. 270–277.

12. Barber, D. and Freestone, I. C. (1990). An investigation of the origin of the colour of the Lycurgus Cup by analytical transmission electron microscopy, *Archaeometry*, **32**, pp. 33–45.

13. Faraday, M. (1857). Experimental relations of gold (and other metals) to light, *Phil. Trans. R. Soc. Lond.*, **147**, pp. 145–181.

14. Brust, M., Walker, M., Bethell, D., Schiffrin, D. J. and Whyman, R. (1994). Synthesis of thiol-derivatised gold nanoparticles in a two-phase liquid–liquid system, *J. Chem. Soc. Chem. Commun.*, pp. 801–802.

15. Wood, R. W. (1902). XLII. On a remarkable case of uneven distribution of light in a diffraction grating spectrum, *Lond. Edinb. Dublin Philos. Mag. J. Sci.*, **4**, pp. 396–402.

16. Mie, G. (1908). Articles on the optical characteristics of turbid tubes, especially colloidal metal solutions, *Ann. Phys*, **25**, pp. 377–445.

17. Chris, W., Pollack, S., Cameron, D. and Kappes, M. M. (1990). Optical absorption spectroscopy of sodium clusters as measured by collinear molecular beam photodepletion, *J. Chem. Phys.*, **93**, pp. 3787–3801.

18. Creighton, J. A. and Eadon, D. G. (1991). Ultraviolet-visible absorption spectra of the colloidal metallic elements, *J. Chem. Soc. Faraday Trans.*, **87**, pp. 3881–3891.

19. Boisselier, E. and Astruc, D. (2009). Gold nanoparticles in nanomedicine: Preparations, imaging, diagnostics, therapies and toxicity, *Chem. Soc. Rev.*, **38**, pp. 1759–1782.

20. Sperling, R. A., Gil, P. R., Zhang, F., Zanella, M. and Parak, W. J. (2008). Biological applications of gold nanoparticles, *Chem. Soc. Rev.*, **37**, pp. 1896–1908.
21. Llevot, A. and Astruc, D. (2012). Applications of vectorized gold nanoparticles to the diagnosis and therapy of cancer, *Chem. Soc. Rev.*, **41**, pp. 242–257.
22. Dreaden, E. C., Alkilany, A. M., Huang, X., Murphy, C. J. and El-Sayed, M. A. (2012). The golden age: Gold nanoparticles for biomedicine, *Chem. Soc. Rev.*, **41**, pp. 2740–2779.
23. Qu, X., Li, Y., Li, L., Wang, Y., Liang, J. and Liang, J. (2015). Fluorescent gold nanoclusters: Synthesis and recent biological application, *J. Nanomater.*, **2015**, p. 4.
24. Saha, K., Agasti, S. S., Kim, C., Li, X. and Rotello, V. M. (2012). Gold nanoparticles in chemical and biological sensing, *Chem. Rev.*, **112**, pp. 2739–2779.
25. Yue, G., Su, S., Li, N., Shuai, M., Lai, X., Astruc, D. and Zhao, P. (2016). Gold nanoparticles as sensors in the colorimetric and fluorescence detection of chemical warfare agents, *Coord. Chem. Rev.*, **311**, pp. 75–84.
26. Mayer, K. M. and Hafner, J. H. (2011). Localized surface plasmon resonance sensors, *Chem. Rev.*, **111**, pp. 3828–3857.
27. Liu, M., Tang, F., Yang, Z., Xu, J. and Yang, X. (2019). Recent progress on gold-nanocluster-based fluorescent probe for environmental analysis and biological sensing, *J. Anal. Methods Chem.*, **2019**, 1095148
28. Fernando, A., Weerawardene, K. D. M., Karimova, N. V. and Aikens, C. M. (2015). Quantum mechanical studies of large metal, metal oxide, and metal chalcogenide nanoparticles and clusters, *Chem. Rev.*, **115**, pp. 6112–6216.
29. Corma, A. and Garcia, H. (2008). Supported gold nanoparticles as catalysts for organic reactions, *Chem. Soc. Rev.*, **37**, pp. 2096–2126.
30. Zhao, P., Feng, X., Huang, D., Yang, G. and Astruc, D. (2015). Basic concepts and recent advances in nitrophenol reduction by gold and other transition metal nanoparticles, *Coord. Chem. Rev.*, **287**, pp. 114–136.
31. Higaki, T., Li, Y., Zhao, S., Li, Q., Li, S., Du, X. S., Yang, S., Chai, J. and Jin, R. (2019). Atomically tailored gold nanoclusters for catalytic application, *Angew. Chem. Int. Ed.*, **131**, pp. 8377–8388.
32. Lei, Z., Wan, X.-K., Yuan, S.-F., Guan, Z.-J. and Wang, Q.-M. (2018). Alkynyl approach toward the protection of metal nanoclusters, *Acc. Chem. Res.*, **51**, pp. 2465–2474.

33. Jadzinsky, P. D., Calero, G., Ackerson, C. J., Bushnell, D. A. and Kornberg, R. D. (2007). Structure of a thiol monolayer-protected gold nanoparticle at 1.1 C resolution, *Science*, **318**, pp. 430–433.

34. Pei, Y. and Zeng, X. C. (2012). Investigating the structural evolution of thiolate protected gold clusters from first-principles, *Nanoscale*, **4**, pp. 4054–4072.

35. Gan, Z., Xia, N. and Wu, Z. (2018). Discovery, mechanism, and application of antigalvanic reaction, *Acc. Chem. Res.*, **51**, pp. 2774–2783.

36. Wang, S., Li, Q., Kang, X. and Zhu, M. (2018). Customizing the structure, composition, and properties of alloy nanoclusters by metal exchange, *Acc. Chem. Res.*, **51**, pp. 2784–2792.

37. Nieto-Ortega, B. and Bürgi, T. (2018). Vibrational properties of thiolate-protected gold nanoclusters, *Acc. Chem. Res.*, **51**, pp. 2811–2819.

38. Higaki, T., Li, Q., Zhou, M., Zhao, S., Li, Y., Li, S. and Jin, R. (2018). Toward the tailoring chemistry of metal nanoclusters for enhancing functionalities, *Acc. Chem. Res.*, **51**, pp. 2764–2773.

39. Sharma, S., Chakrahari, K. K., Saillard, J.-Y. and Liu, C. (2018). Structurally precise dichalcogenolate-protected copper and silver superatomic nanoclusters and their alloys, *Acc. Chem. Res.*, **51**, pp. 2475–2483.

40. Sakthivel, N. A. and Dass, A. (2018). Aromatic thiolate-protected series of gold nanomolecules and a contrary structural trend in size evolution, *Acc. Chem. Res.*, **51**, pp. 1774–1783.

41. Yao, Q., Chen, T., Yuan, X. and Xie, J. (2018). Toward total synthesis of thiolate-protected metal nanoclusters, *Acc. Chem. Res.*, **51**, pp. 1338–1348.

42. Chen, T., Yao, Q., Nasaruddin, R. and Xie, J. (2019). Electrospray ionization mass spectrometry: A powerful platform for noble-metal nanocluster analysis, *Angew. Chem. Int. Ed.*, **58**, pp. 11967–11977.

43. Hossain, S., Niihori, Y., Nair, L. V., Kumar, B., Kurashige, W. and Negishi, Y. (2018). Alloy clusters: Precise synthesis and mixing effects, *Acc. Chem. Res.*, **51**, pp. 3114–3124.

44. Kurashige, W., Niihori, Y., Sharma, S. and Negishi, Y. (2016). Precise synthesis, functionalization and application of thiolate-protected gold clusters, *Coord. Chem. Rev.*, **320**, pp. 238–250.

45. Chakraborty, I. and Pradeep, T. (2017). Atomically precise clusters of noble metals: Emerging link between atoms and nanoparticles, *Chem. Rev.*, **117**, pp. 8208–8271.

46. Aikens C. M. (2015). Optical properties and chirality, *Elsevier*, **9**, pp. 223–261.

47. Weerawardene, K. L. D. M., Häkkinen, H. and Aikens, C. M. (2018). Connections between theory and experiment for gold and silver nanoclusters, *Annu. Rev. Phys. Chem.*, **69**, pp. 205–229.

48. Aikens, C. M. (2018). Electronic and geometric structure, optical properties, and excited state behavior in atomically precise thiolate-stabilized noble metal nanoclusters, *Acc. Chem. Res.*, **51**, pp. 3065–3073.

49. Tang, Q., Hu, G., Fung, V. and Jiang, D.-e. (2018). Insights into interfaces, stability, electronic properties, and catalytic activities of atomically precise metal nanoclusters from first principles, *Acc. Chem. Res.*, **51**, pp. 2793–2802.

50. Xu, W. W., Zhu, B., Zeng, X. C. and Gao, Y. (2016). A grand unified model for liganded gold clusters, *Nat. Commun.*, **7**, p. 13574.

51. Xiong, L., Yang, S., Sun, X., Chai, J., Rao, B., Yi, L., Zhu, M. and Pei, Y. (2018). Structure and electronic structure evolution of thiolate-protected gold nanoclusters containing quasi face-centered-cubic kernels, *J. Phys. Chem. C*, **122**, pp. 14898–14907.

52. Weerawardene, K. D. M. and Aikens, C. M. (2016). Effect of aliphatic versus aromatic ligands on the structure and optical absorption of $Au_{20}(SR)_{16}$, *J. Phys. Chem. C*, **120**, pp. 8354–8363.

53. Nimmala, P. R., Knoppe, S., Jupally, V. R., Delcamp, J. H., Aikens, C. M. and Dass, A. (2014). $Au_{36}(SPh)_{24}$ nanomolecules: X-ray crystal structure, optical spectroscopy, electrochemistry, and theoretical analysis, *J. Phys. Chem. B*, **118**, pp. 14157–14167.

54. Guidez, E. B., Mäkinen, V., Häkkinen, H. and Aikens, C. M. (2012). Effects of silver doping on the geometric and electronic structure and optical absorption spectra of the $Au_{25-n}Ag_n(SH)_{18}^-$ (n = 1, 2, 4, 6, 8, 10, 12) bimetallic nanoclusters, *J. Phys. Chem. C*, **116**, pp. 20617–20624.

55. Aikens, C. M. (2009). Effects of core distances, solvent, ligand, and level of theory on the TDDFT optical absorption spectrum of the thiolate-protected Au_{25} nanoparticle, *J. Phys. Chem. A*, **113**, pp. 10811–10817.

56. Aikens, C. M. (2008). Origin of discrete optical absorption spectra of $M_{25}(SH)_{18}^-$ nanoparticles (M = Au, Ag), *J. Phys. Chem. C*, **112**, pp. 19797–19800.

57. Zhu, M., Aikens, C. M., Hollander, F. J., Schatz, G. C. and Jin, R. (2008). Correlating the crystal structure of a thiol-protected Au_{25} cluster and optical properties, *J. Am. Chem. Soc.*, **130**, pp. 5883–5885.

58. Zeng, C., Li, T., Das, A., Rosi, N. L. and Jin, R. (2013). Chiral structure of thiolate-protected 28-gold-atom nanocluster determined by X-ray crystallography, *J. Am. Chem. Soc.*, **135**, pp. 10011–10013.

59. Knoppe, S., Malola, S., Lehtovaara, L., Bürgi, T. and Häkkinen, H. (2013). Electronic structure and optical properties of the thiolate-protected $Au_{28}(SMe)_{20}$ cluster, *J. Phys. Chem. A*, **117**, pp. 10526–10533.

60. Das, A., Li, T., Nobusada, K., Zeng, C., Rosi, N. L. and Jin, R. (2013). Nonsuperatomic $[Au_{23}(SC_6H_{11})_{16}]^-$ nanocluster featuring bipyramidal Au_{15} kernel and trimeric $Au_3(SR)_4$ motif, *J. Am. Chem. Soc.*, **135**, pp. 18264–18267.

61. Chen, S., Wang, S., Zhong, J., Song, Y., Zhang, J., Sheng, H., Pei, Y. and Zhu, M. (2015). The structure and optical properties of the $[Au_{18}(SR)_{14}]$ nanocluster, *Angew. Chem. Int. Ed.*, **54**, pp. 3145–3149.

62. Das, A., Liu, C., Byun, H. Y., Nobusada, K., Zhao, S., Rosi, N. and Jin, R. (2015). Structure determination of $[Au_{18}(SR)_{14}]$, *Angew. Chem. Int. Ed.*, **54**, pp. 3140–3144.

63. Pei, Y., Gao, Y. and Zeng, X. C. (2008). Structural prediction of thiolate-protected Au_{38}: A face-fused bi-icosahedral Au core, *J. Am. Chem. Soc.*, **130**, pp. 7830–7832.

64. Chen, S., Xiong, L., Wang, S., Ma, Z., Jin, S., Sheng, H., Pei, Y. and Zhu, M. (2016). Total structure determination of Au_{21}(S-Adm)$_{15}$ and geometrical/electronic structure evolution of thiolated gold nanoclusters, *J. Am. Chem. Soc.*, **138**, pp. 10754–10757.

65. Jiang, D.-e., Overbury, S. H. and Dai, S. (2013). Structure of $Au_{15}(SR)_{13}$ and its implication for the origin of the nucleus in thiolated gold nanoclusters, *J. Am. Chem. Soc.*, **135**, pp. 8786–8789.

66. Zeng, C., Liu, C., Chen, Y., Rosi, N. L. and Jin, R. (2014). Gold–thiolate ring as a protecting motif in the $Au_{20}(SR)_{16}$ nanocluster and implications, *J. Am. Chem. Soc.*, **136**, pp. 11922–11925.

67. Pei, Y., Tang, J., Tang, X., Huang, Y. and Zeng, X. C. (2015). New structure model of $Au_{22}(SR)_{18}$: Bitetrahederon golden kernel enclosed by $[Au_6(SR)_6]$ Au(I) complex, *J. Phys. Chem. Lett.*, **6**, pp. 1390–1395.

68. Zeng, C., Qian, H., Li, T., Li, G., Rosi, N. L., Yoon, B., Barnett, R. N., Whetten, R. L., Landman, U. and Jin, R. (2012). Total structure and electronic properties of the gold nanocrystal $Au_{36}(SR)_{24}$, *Angew. Chem. Int. Ed.*, **51**, pp. 13114–13118.

69. Zeng, C., Chen, Y., Iida, K., Nobusada, K., Kirschbaum, K., Lambright, K. J. and Jin, R. (2016). Gold quantum boxes: On the periodicities and the quantum confinement in the Au_{28}, Au_{36}, Au_{44}, and Au_{52} magic series, *J. Am. Chem. Soc.*, **138**, pp. 3950–3953.

70. Zeng, C., Chen, Y., Liu, C., Nobusada, K., Rosi, N. L. and Jin, R. (2015). Gold tetrahedra coil up: Kekulé-like and double helical superstructures, *Sci. Adv.*, **1**, p. e1500425.

71. Takano, S., Yamazoe, S., Koyasu, K. and Tsukuda, T. (2015). Slow-reduction synthesis of a thiolate-protected one-dimensional gold cluster showing an intense near-infrared absorption, *J. Am. Chem. Soc.*, **137**, pp. 7027–7030.

72. Ma, Z., Wang, P., Zhou, G., Tang, J., Li, H. and Pei, Y. (2016). Correlating the structure and optical absorption properties of $Au_{76}(SR)_{44}$ cluster, *J. Phys. Chem. C*, **120**, pp. 13739–13748.

73. Ma, Z., Wang, P. and Pei, Y. (2016). Geometric structure, electronic structure and optical absorption properties of one-dimensional thiolate-protected gold clusters containing a quasi-face-centered-cubic (quasi-fcc) Au-core: A density-functional theoretical study, *Nanoscale*, **8**, pp. 17044–17054.

74. Wang, P., Xiong, L., Sun, X., Ma, Z. and Pei, Y. (2018). Exploring the structure evolution and core/ligand structure patterns of a series of large sized thiolate-protected gold clusters $Au_{145-3N}(SR)_{60-2N}$ (N = 1–8): A first principles study, *Nanoscale*, **10**, pp. 3918–3929.

75. Malola, S., Lehtovaara, L., Enkovaara, J. and Häkkinen, H. (2013). Birth of the localized surface plasmon resonance in monolayer-protected gold nanoclusters, *ACS Nano*, **7**, pp. 10263–10270.

76. Negishi, Y., Nakazaki, T., Malola, S., Takano, S., Niihori, Y., Kurashige, W., Yamazoe, S., Tsukuda, T. and Häkkinen, H. (2015). A critical size for emergence of nonbulk electronic and geometric structures in dodecanethiolate-protected Au clusters, *J. Am. Chem. Soc.*, **137**, pp. 1206–1212.

77. Xu, W. W. and Gao, Y. (2015). Unraveling the atomic structures of the $Au_{68}(SR)_{34}$ nanoparticles, *J. Phys. Chem. C*, **119**, pp. 14224–14229.

78. Higaki, T., Liu, C., Zhou, M., Luo, T.-Y., Rosi, N. L. and Jin, R. (2017). Tailoring the structure of 58-electron gold nanoclusters: $Au_{103}S_2$ $(S\text{-}Nap)_{41}$ and its implications, *J. Am. Chem. Soc.*, **139**, pp. 9994–10001.

79. Kenzler, S., Schrenk, C. and Schnepf, A. (2017). $Au_{108}S_{24}(PPh_3)_{16}$: A highly symmetric nanoscale gold cluster confirms the general concept of metalloid clusters, *Angew. Chem. Int. Ed.*, **56**, pp. 393–396.

80. Chen, Y., Zeng, C., Liu, C., Kirschbaum, K., Gayathri, C., Gil, R. R., Rosi, N. L. and Jin, R. (2015). Crystal structure of barrel-shaped chiral Au_{130} $(p\text{-}MBT)_{50}$ nanocluster, *J. Am. Chem. Soc.*, **137**, pp. 10076–10079.

81. Zeng, C., Chen, Y., Kirschbaum, K., Appavoo, K., Sfeir, M. Y. and Jin, R. (2015). Structural patterns at all scales in a nonmetallic chiral $Au_{133}(SR)_{52}$ nanoparticle, *Sci. Adv.*, **1**, p. e1500045.

82. Dass, A., Theivendran, S., Nimmala, P. R., Kumara, C., Jupally, V. R., Fortunelli, A., Sementa, L., Barcaro, G., Zuo, X. and Noll, B. C. (2015). $Au_{133}(SPh\text{-}tBu)_{52}$ nanomolecules: X-ray crystallography, optical, electrochemical, and theoretical analysis, *J. Am. Chem. Soc.*, **137**, pp. 4610–4613.

83. Yan, N., Xia, N., Liao, L., Zhu, M., Jin, F., Jin, R. and Wu, Z. (2018). Unraveling the long-pursued Au_{144} structure by X-ray crystallography, *Sci. Adv.*, **4**, p. eaat7259.

84. Vergara, S., Lukes, D. A., Martynowycz, M. W., Santiago, U., Plascencia-Villa, G., Weiss, S. C., De La Cruz, M. J., Black, D. M., Alvarez, M. M. and Lopez-Lozano, X. (2017). MicroED structure of $Au_{146}(p\text{-}MBA)_{57}$ at subatomic resolution reveals a twinned FCC cluster, *J. Phys. Chem. Lett.*, **8**, pp. 5523–5530.

85. Zeng, C., Chen, Y., Kirschbaum, K., Lambright, K. J. and Jin, R. (2016). Emergence of hierarchical structural complexities in nanoparticles and their assembly, *Science*, **354**, pp. 1580–1584.

86. Zhou, M., Zeng, C., Song, Y., Padelford, J. W., Wang, G., Sfeir, M. Y., Higaki, T. and Jin, R. (2017). On the non-metallicity of 2.2 nm $Au_{246}(SR)_{80}$ nanoclusters, *Angew. Chem. Int. Ed.*, **129**, pp. 16475–16479.

87. Sakthivel, N. A., Theivendran, S., Ganeshraj, V., Oliver, A. G. and Dass, A. (2017). Crystal structure of faradaurate-279: $Au_{279}(SPh\text{-}tBu)_{84}$ plasmonic nanocrystal molecules, *J. Am. Chem. Soc.*, **139**, pp. 15450–15459.

88. Sakthivel, N. A., Stener, M., Sementa, L., Fortunelli, A., Ramakrishna, G. and Dass, A. (2018). $Au_{279}(SR)_{84}$: The smallest gold thiolate nanocrystal that is metallic and the birth of plasmon, *J. Phys. Chem. Lett.*, **9**, pp. 1295–1300.

89. Higaki, T., Zhou, M., Lambright, K. J., Kirschbaum, K., Sfeir, M. Y. and Jin, R. (2018). Sharp transition from nonmetallic Au_{246} to metallic Au_{279} with nascent surface plasmon resonance, *J. Am. Chem. Soc.*, **140**, pp. 5691–5695.

90. Hartland, G. V. (2006). Coherent excitation of vibrational modes in metallic nanoparticles, *Annu. Rev. Phys. Chem.*, **57**, pp. 403–430.

91. Link, S. and El-Sayed, M. A. (2003). Optical properties and ultrafast dynamics of metallic nanocrystals, *Annu. Rev. Phys. Chem.*, **54**, pp. 331–366.

92. Häkkinen, H. (2012). The gold–sulfur interface at the nanoscale, *Nat. Chem.*, **4**, p. 443.

93. Wu, Z. and Jin, R. (2010). On the ligand's role in the fluorescence of gold nanoclusters, *Nano Lett.*, **10**, pp. 2568–2573.

94. Wang, S., Zhu, X., Cao, T. and Zhu, M. (2014). A simple model for understanding the fluorescence behavior of Au_{25} nanoclusters, *Nanoscale*, **6**, pp. 5777–5781.

95. Tang, Z., Xu, B., Wu, B., Robinson, D. A., Bokossa, N. and Wang, G. (2011). Monolayer reactions of protected Au nanoclusters with monothiol tiopronin and 2,3-dithiol dimercaptopropanesulfonate, *Langmuir*, **27**, pp. 2989–2996.

96. Tang, Z., Ahuja, T., Wang, S. and Wang, G. (2012). Near infrared luminescence of gold nanoclusters affected by the bonding of 1,4-dithiolate durene and monothiolate phenylethanethiolate, *Nanoscale*, **4**, pp. 4119–4124.

97. Conroy, C. V., Jiang, J., Zhang, C., Ahuja, T., Tang, Z., Prickett, C. A., Yang, J. J. and Wang, G. (2014). Enhancing near IR luminescence of thiolate Au nanoclusters by thermo treatments and heterogeneous subcellular distributions, *Nanoscale*, **6**, pp. 7416–7423.

98. Mathew, A., Natarajan, G., Lehtovaara, L., Hakkinen, H., Kumar, R. M., Subramanian, V., Jaleel, A. and Pradeep, T. (2013). Supramolecular functionalization and concomitant enhancement in properties of Au_{25} clusters, *ACS Nano*, **8**, pp. 139–152.

99. Heaven, M. W., Dass, A., White, P. S., Holt, K. M. and Murray, R. W. (2008). Crystal structure of the gold nanoparticle $[N(C_8H_{17})_4][Au_{25}(SCH_2CH_2Ph)_{18}]$, *J. Am. Chem. Soc.*, **130**, pp. 3754–3755.

100. Mathew, A., Varghese, E., Choudhury, S., Pal, S. K. and Pradeep, T. (2015). Efficient red luminescence from organic-soluble Au_{25} clusters by ligand structure modification, *Nanoscale*, **7**, pp. 14305–14315.

101. Negishi, Y., Iwai, T. and Ide, M. (2010). Continuous modulation of electronic structure of stable thiolate-protected Au_{25} cluster by Ag doping, *Chem. Commun.*, **46**, pp. 4713–4715.

102. Le Guével, X., Trouillet, V., Spies, C., Li, K., Laaksonen, T., Auerbach, D., Jung, G. and Schneider, M. (2012). High photostability and enhanced fluorescence of gold nanoclusters by silver doping, *Nanoscale*, **4**, pp. 7624–7631.

103. Wang, S., Meng, X., Das, A., Li, T., Song, Y., Cao, T., Zhu, X., Zhu, M. and Jin, R. (2014). A 200-fold quantum yield boost in the photoluminescence of silver-doped Ag_xAu_{25-x} nanoclusters: The 13th silver atom matters, *Angew. Chem. Int. Ed.*, **53**, pp. 2376–2380.

104. AbdulHalim, L. G., Bootharaju, M. S., Tang, Q., Del Gobbo, S., AbdulHalim, R. G., Eddaoudi, M., Jiang, D.-e. and Bakr, O. M. (2015). $Ag_{29}(BDT)_{12}(TPP)_4$: A tetravalent nanocluster, *J. Am. Chem. Soc.*, **137**, pp. 11970–11975.

105. Xie, X. Y., Xiao, P., Cao, X., Fang, W. H., Cui, G. and Dolg, M. (2018). The origin of the photoluminescence enhancement of gold-doped silver nanoclusters: The importance of relativistic effects and heteronuclear gold–silver bonds, *Angew. Chem. Int. Ed.*, **57**, pp. 9965–9969.

106. Aldeek, F., Muhammed, M. H., Palui, G., Zhan, N. and Mattoussi, H. (2013). Growth of highly fluorescent polyethylene glycol- and zwitterion-functionalized gold nanoclusters, *ACS Nano*, **7**, pp. 2509–2521.

107. Mishra, D., Lobodin, V., Zhang, C., Aldeek, F., Lochner, E. and Mattoussi, H. (2018). Gold-doped silver nanoclusters with enhanced photophysical properties, *Phys. Chem. Chem. Phys.*, **20**, pp. 12992–13007.

108. Luo, J., Xie, Z., Lam, J. W., Cheng, L., Chen, H., Qiu, C., Kwok, H. S., Zhan, X., Liu, Y. and Zhu, D. (2001). Aggregation-induced emission of 1-methyl-1,2,3,4,5-pentaphenylsilole, *Chem. Commun.*, pp. 1740–1741.

109. Mei, J., Hong, Y., Lam, J. W., Qin, A., Tang, Y. and Tang, B. Z. (2014). Aggregation-induced emission: The whole is more brilliant than the parts, *Adv. Mater.*, **26**, pp. 5429–5479.

110. Chen, Y., Lam, J. W., Kwok, R. T., Liu, B. and Tang, B. Z. (2019). Aggregation-induced emission: Fundamental understanding and future developments, *Mater. Horiz.*, **6**, pp. 428–433.

111. Mei, J., Leung, N. L., Kwok, R. T., Lam, J. W. and Tang, B. Z. (2015). Aggregation-induced emission: Together we shine, united we soar! *Chem. Rev.*, **115**, pp. 11718–11940.

112. Kang, X. and Zhu, M. (2019). Tailoring the photoluminescence of atomically precise nanoclusters, *Chem. Soc. Rev.*, **48**, pp. 2422–2457.

113. Hong, Y., Lam, J. W. and Tang, B. Z. (2009). Aggregation-induced emission: Phenomenon, mechanism and applications, *Chem. Commun.*, pp. 4332–4353.

114. He, Z., Ke, C. and Tang, B. Z. (2018). Journey of aggregation-induced emission research, *ACS Omega*, **3**, pp. 3267–3277.

115. Gao, M. and Tang, B. Z. (2017). Fluorescent sensors based on aggregation-induced emission: Recent advances and perspectives, *ACS Sensors*, **2**, pp. 1382–1399.

116. Gu, X., Kwok, R. T., Lam, J. W. and Tang, B. Z. (2017). AIEgens for biological process monitoring and disease theranostics, *Biomaterials*, **146**, pp. 115–135.

117. Gao, M. and Tang, B. Z. (2017). Aggregation-induced emission probes for cancer theranostics, *Drug Discov. Today*, **22**, pp. 1288–1294.

118. Cai, Y., Qin, A. and Tang, B. Z. (2017). Siloles in optoelectronic devices, *J. Mater. Chem. C*, **5**, pp. 7375–7389.

119. Qian, J. and Tang, B. Z. (2017). AIE luminogens for bioimaging and theranostics: From organelles to animals, *Chem*, **3**, pp. 56–91.

120. Feng, G., Kwok, R. T., Tang, B. Z. and Liu, B. (2017). Functionality and versatility of aggregation-induced emission luminogens, *Appl. Phys. Rev.*, **4**, p. 021307.

121. Hong, Y., Lam, J. W. and Tang, B. Z. (2011). Aggregation-induced emission, *Chem. Soc. Rev.*, **40**, pp. 5361–5388.

122. Häkkinen, H., Walter, M. and Grönbeck, H. (2006). Divide and protect: Capping gold nanoclusters with molecular gold–thiolate rings, *J. Phys. Chem. B*, **110**, pp. 9927–9931.

123. Luo, Z., Yuan, X., Yu, Y., Zhang, Q., Leong, D. T., Lee, J. Y. and Xie, J. (2012). From aggregation-induced emission of Au(I)–thiolate complexes to ultrabright Au(0)@Au(I)–thiolate core–shell nanoclusters, *J. Am. Chem. Soc.*, **134**, pp. 16662–16670.

124. Goswami, N., Yao, Q., Luo, Z., Li, J., Chen, T. and Xie, J. (2016). Luminescent metal nanoclusters with aggregation-induced emission, *J. Phys. Chem. Lett.*, **7**, pp. 962–975.

125. Vickery, J. C., Olmstead, M. M., Fung, E. Y. and Balch, A. L. (1997). Solvent-stimulated luminescence from the supramolecular aggregation of a trinuclear gold(I) complex that displays extensive intermolecular Au... Au interactions, *Angew. Chem. Int. Edit.*, **36**, pp. 1179–1181.

126. Raize, O., Argaman, Y. and Yannai, S. (2004). Mechanisms of biosorption of different heavy metals by brown marine macroalgae, *Biotechnol. Bioeng.*, **87**, pp. 451–458.

127. Pyo, K., Thanthirige, V. D., Kwak, K., Pandurangan, P., Ramakrishna, G. and Lee, D. (2015). Ultrabright luminescence from gold nanoclusters: Rigidifying the Au (I)–thiolate shell, *J. Am. Chem. Soc.*, **137**, pp. 8244–8250.

128. Kwak, K. and Lee, D. (2012). Electrochemical characterization of water-soluble Au_{25} nanoclusters enabled by phase-transfer reaction, *J. Phys. Chem. Lett.*, **3**, pp. 2476–2481.

129. Pyo, K., Thanthirige, V. D., Yoon, S. Y., Ramakrishna, G. and Lee, D. (2016). Enhanced luminescence of $Au_{22}(SG)_{18}$ nanoclusters via rational surface engineering, *Nanoscale*, **8**, pp. 20008–20016.

130. Lin, X., Liu C., Sun, K., Wu, R, Fu, X. and Huang, J. (2019). Structural isomer and high-yield of Pt_1Ag_{28} nanocluster via one-pot chemical wet method, *Nano Res.*, **12**, pp. 309–314.

131. Kang, X., Zhou, M., Wang, S., Jin, S., Sun, G., Zhu, M. and Jin, R. (2017). The tetrahedral structure and luminescence properties of bi-metallic $Pt_1Ag_{28}(SR)_{18}(PPh_3)_4$ nanocluster, *Chem. Sci.*, **8**, pp. 2581–2587.

132. Zhuang, S., Liao, L., Zhao, Y., Yuan, J., Yao, C., Liu, X., Li, J., Deng, H., Yang, J. and Wu, Z. (2018). Is the kernel–staples match a key–lock match? *Chem. Sci.*, **9**, pp. 2437–2442.

133. Dong, H., Liao, L. and Wu, Z. (2017). Two-way transformation between fcc- and nonfcc-structured gold nanoclusters, *J. Phys. Chem. Lett.*, **8**, pp. 5338–5343.

134. Zhuang, S., Liao, L., Yuan, J., Xia, N., Zhao, Y., Wang, C., Gan, Z., Yan, N., He, L. and Li, J. (2019). Fcc versus non-fcc structural isomerism of gold nanoparticles with kernel atom packing dependent photoluminescence, *Angew. Chem. Int. Ed*, **131**, pp. 4558–4562.

135. Ramakrishna, G., Varnavski, O., Kim, J., Lee, D. and Goodson, T. (2008). Quantum-sized gold clusters as efficient two-photon absorbers, *J. Am. Chem. Soc.*, **130**, pp. 5032–5033.

136. Philip, R., Chantharasupawong, P., Qian, H., Jin, R. and Thomas, J. (2012). Evolution of nonlinear optical properties: From gold atomic clusters to plasmonic nanocrystals, *Nano Lett.*, **12**, pp. 4661–4667.

137. Russier-Antoine, I., Bertorelle, F., Vojkovic, M., Rayane, D., Salmon, E., Jonin, C., Dugourd, P., Antoine, R. and Brevet, P.-F. (2014). Non-linear optical properties of gold quantum clusters. The smaller the better, *Nanoscale*, **6**, pp. 13572–13578.

138. Knoppe, S., Zhang, Q.-F., Wan, X.-K., Wang, Q.-M., Wang, L.-S. and Verbiest, T. (2016). Second-order nonlinear optical scattering properties of phosphine-protected Au_{20} clusters, *Ind. Eng. Chem. Res.*, **55**, pp. 10500–10506.

139. Day, P. N., Nguyen, K. A. and Pachter, R. (2010). Calculation of one-and two-photon absorption spectra of thiolated gold nanoclusters using time-dependent density functional theory, *J. Chem. Theory Comput.*, **6**, pp. 2809–2821.

140. Knoppe, S., Vanbel, M., Van Cleuvenbergen, S., Vanpraet, L., Bürgi, T. and Verbiest, T. (2015). Nonlinear optical properties of thiolate-protected gold clusters, *J. Phys. Chem. C*, **119**, pp. 6221–6226.

141. Van Steerteghem, N., Van Cleuvenbergen, S., Deckers, S., Kumara, C., Dass, A., Häkkinen, H., Clays, K., Verbiest, T. and Knoppe, S. (2016).

Symmetry breaking in ligand-protected gold clusters probed by nonlinear optics, *Nanoscale*, **8**, pp. 12123–12127.

142. Knoppe, S., Häkkinen, H. and Verbiest, T. (2015). Nonlinear optical properties of thiolate-protected gold clusters: A theoretical survey of the first hyperpolarizabilities, *J. Phys. Chem. C*, **119**, pp. 27676–27682.
143. Olesiak-Banska, J., Waszkielewicz, M., Matczyszyn, K. and Samoc, M. (2016). A closer look at two-photon absorption, absorption saturation and nonlinear refraction in gold nanoclusters, *RSC Adv.*, **6**, pp. 98748–98752.
144. Antoine, R. (2018). Atomically precise clusters of gold and silver: A new class of nonlinear optical nanomaterials, *Frontier Res. Today*, **1**, p 1001.
145. Bertorelle, F., Russier-Antoine, I., Calin, N., Comby-Zerbino, C., Bensalah-Ledoux, A., Guy, S., Dugourd, P., Brevet, P.-F., Sanader, Z. and Krstić, M. (2017). $Au_{10}(SG)_{10}$: A chiral gold catenane nanocluster with zero confined electrons. Optical properties and first-principles theoretical analysis, *J. Phys. Chem. Lett.*, **8**, pp. 1979–1985.
146. Verbiest, T., Clays, K. and Rodriguez, V. (2009). *Second-order Nonlinear Optical Characterization Techniques: An Introduction*, CRC Press, Boca Raton
147. Nelson, D. F., Collins, R. J. and Kaiser, W. (2010). Bell Labs and the ruby laser, *Phys. Today*, **63**, p. 40.
148. Bertorelle, F., Moulin, C., Soleilhac, A., Comby-Zerbino, C., Dugourd, P., Russier-Antoine, I., Brevet, P. F. and Antoine, R. (2018). Bulky counterions: Enhancing the two-photon excited fluorescence of gold nanoclusters, *Chem Phys Chem*, **19**, pp. 165–168.
149. Abeyasinghe, N., Kumar, S., Sun, K., Mansfield, J. F., Jin, R. and Goodson III, T. (2016). Enhanced emission from single isolated gold quantum dots investigated using two-photon-excited fluorescence near-field scanning optical microscopy, *J. Am. Chem. Soc.*, **138**, pp. 16299–16307.
150. Franken, P. A., Hill, A. E., Peters, C. W. and Weinreich, G. (1961). Generation of optical harmonics, *Phys. Rev. Lett.*, **7**, p. 118.
151. Kaiser, W. and Garrett, C. (1961). Two-photon excitation in CaF_2: Eu^{2+}, *Phys. Rev. Lett.*, **7**, p. 229.
152. Antoine, R. and Bonačić-Koutecký, V. (2017). *Liganded Silver and Gold Quantum Clusters. Towards a New Class of Nonlinear Optical Nanomaterials*, Springer.
153. Hu, Z. and Jensen, L. (2017). Importance of double-resonance effects in two-photon absorption properties of $Au_{25}(SR)_{18}^-$, *Chem. Sci.*, **8**, pp. 4595–4601.

154. Shibu, E. S. and Pradeep, T. (2011). Quantum clusters in cavities: Trapped Au_{15} in cyclodextrins, *Chem. Mater.*, **23**, pp. 989–999.

155. Shibu, E., Muhammed, M. H., Tsukuda, T. and Pradeep, T. (2008). Ligand exchange of $Au_{25}SG_{18}$ leading to functionalized gold clusters: Spectroscopy, kinetics, and luminescence, *J. Phys. Chem. C*, **112**, pp. 12168–12176.

156. Yu, Y., Luo, Z., Yu, Y., Lee, J. Y. and Xie, J. (2012). Observation of cluster size growth in CO-directed synthesis of $Au_{25}(SR)_{18}$ nanoclusters, *ACS Nano*, **6**, pp. 7920–7927.

157. Clays, K. and Persoons, A. (1991). Hyper-Rayleigh scattering in solution, *Phys. Rev. Lett.*, **66**, p. 2980.

158. Varnavski, O., Ramakrishna, G., Kim, J., Lee, D. and Goodson, T. (2009). Critical size for the observation of quantum confinement in optically excited gold clusters, *J. Am. Chem. Soc.*, **132**, pp. 16–17.

159. Campagnola, P. J. and Loew, L. M. (2003). Second-harmonic imaging microscopy for visualizing biomolecular arrays in cells, tissues and organisms, *Nat. Biotechnol.*, **21**, p. 1356.

160. Stockert, J. C. and Blázquez-Castro, A. (2017). *Fluorescence Microscopy in Life Sciences*, Bentham Science Publishers

161. Burns, P. (1997). The history of the discovery of cinematography, details available at http://precinemahistory.net.

162. Roberts, G. (2000). Femtosecond chemical reactions, *Philos. Trans. R. Soc. Lond. A*, **358**, pp. 345–366.

163. Baum, P., Yang, D.-S. and Zewail, A. H. (2007). 4D visualization of transitional structures in phase transformations by electron diffraction, *Science*, **318**, pp. 788–792.

164. Zewail, A. H. (2000). Femtochemistry: Atomic-scale dynamics of the chemical bond, *J. Phys. Chem. A*, **104**, pp. 5660–5694.

165. Rulliere, C. (2005). *Femtosecond Laser Pulses*, Springer

166. Kovalenko, S., Dobryakov, A., Ruthmann, J. and Ernsting, N. (1999). Femtosecond spectroscopy of condensed phases with chirped supercontinuum probing, *Phys. Rev. A*, **59**, p. 2369.

167. Zhou, M., Zeng, C., Chen, Y., Zhao, S., Sfeir, M. Y., Zhu, M. and Jin, R. (2016). Evolution from the plasmon to exciton state in ligand-protected atomically precise gold nanoparticles, *Nat. Commun.*, **7**, p. 13240.

168. Higaki, T., Zhou, M., He, G., House, S. D., Sfeir, M. Y., Yang, J. C. and Jin, R. (2019). Anomalous phonon relaxation in $Au_{333}(SR)_{79}$ nanoparticles with nascent plasmons, *Proc. Natl. Acad. Sci. U. S. A.*, **116**, pp. 13215–13220.

169. Link, S., El-Sayed, M. A., Schaaff, T. G. and Whetten, R. L. (2002). Transition from nanoparticle to molecular behavior: A femtosecond transient absorption study of a size-selected 28 atom gold cluster, *Chem. Phys. Lett.*, **356**, pp. 240–246.

170. Weissker, H.-C., Escobar, H. B., Thanthirige, V., Kwak, K., Lee, D., Ramakrishna, G., Whetten, R. and López-Lozano, X. (2014). Information on quantum states pervades the visible spectrum of the ubiquitous $Au_{144}(SR)_{60}$ gold nanocluster, *Nat. Commun.*, **5**, p. 3785.

171. Yi, C., Tofanelli, M. A., Ackerson, C. J. and Knappenberger Jr, K. L. (2013). Optical properties and electronic energy relaxation of metallic $Au_{144}(SR)_{60}$ nanoclusters, *J. Am. Chem. Soc.*, **135**, pp. 18222–18228.

172. Kwak, K., Thanthirige, V. D., Pyo, K., Lee, D. and Ramakrishna, G. (2017). Energy gap law for exciton dynamics in gold cluster molecules, *J. Phys. Chem. Lett.*, **8**, pp. 4898–4905.

173. Qian, H., Zhu, Y. and Jin, R. (2012). Atomically precise gold nanocrystal molecules with surface plasmon resonance, *Proc. Natl. Acad. Sci. U. S. A.*, **109**, pp. 696–700.

174. Li, G. and Jin, R. (2013). Atomically precise gold nanoclusters as new model catalysts, *Acc. Chem. Res.*, **46**, pp. 1749–1758.

175. Liu, S. and Xu, Y.-J. (2016). Photo-induced transformation process at gold clusters-semiconductor interface: Implications for the complexity of gold clusters-based photocatalysis, *Sci. Rep.*, **6**, p. 22742.

176. Nasaruddin, R. R., Chen, T., Yan, N. and Xie, J. (2018). Roles of thiolate ligands in the synthesis, properties and catalytic application of gold nanoclusters, *Coord. Chem. Rev.*, **368**, pp. 60–79.

177. Taylor, M. G. and Mpourmpakis, G. (2017). Thermodynamic stability of ligand-protected metal nanoclusters, *Nat. Commun.*, **8**, p. 15988.

178. Rambukwella, M., Sakthivel, N. A., Delcamp, J. H., Sementa, L., Fortunelli, A. and Dass, A. (2018). Ligand structure determines nanoparticles' atomic structure, metal-ligand interface and properties, *Front. Chem.*, **6**, p. 330.

179. Stamplecoskie, K. G. and Kamat, P. V. (2014). Size-dependent excited state behavior of glutathione-capped gold clusters and their light-harvesting capacity, *J. Am. Chem. Soc.*, **136**, pp. 11093–11099.

180. Stamplecoskie, K. G., Chen, Y.-S. and Kamat, P. V. (2014). Excited-state behavior of luminescent glutathione-protected gold clusters, *J. Phys. Chem. C*, **118**, pp. 1370–1376.

181. Johansson, M. P., Sundholm, D. and Vaara, J. (2004). Au_{32}: A 24-carat golden fullerene, *Angew. Chem. Int. Ed.*, **43**, pp. 2678–2681.

182. Gu, X., Ji, M., Wei, S. and Gong, X. (2004). AuN clusters (N = 32, 33, 34, 35): Cagelike structures of pure metal atoms, *Phys. Rev. B*, **70**, p. 205401.

183. Kroto, H. W., Allaf, A. and Balm, S. (1991). C_{60}: Buckminsterfullerene, *Chem. Rev.*, **91**, pp. 1213–1235.

184. Kroto, H. W., Heath, J. R., O'Brien, S. C., Curl, R. F. and Smalley, R. E. (1985). C_{60}: Buckminsterfullerene, *Nature*, **318**, pp. 162–163.

185. Yuan, S. F., Xu, C. Q., Li, J. and Wang, Q. M. (2019). A ligand-protected golden fullerene: The dipyridylamido Au_{32}^{8+} nanocluster, *Angew. Chem. Ind. Ed.*, **131**, pp. 5967–5970.

186. Zhou, M., Vdović, S., Long, S., Zhu, M., Yan, L., Wang, Y., Niu, Y., Wang, X., Guo, Q. and Jin, R. (2013). Intramolecular charge transfer and solvation dynamics of thiolate-protected $Au_{20}(SR)_{16}$ clusters studied by ultrafast measurement, *J. Phys. Chem. A*, **117**, pp. 10294–10303.

187. Zhou, M., Long, S., Wan, X., Li, Y., Niu, Y., Guo, Q., Wang, Q.-M. and Xia, A. (2014). Ultrafast relaxation dynamics of phosphine-protected, rod-shaped Au_{20} clusters: Interplay between solvation and surface trapping, *Phys. Chem. Chem. Phys.*, **16**, pp. 18288–18293.

188. Abbas, M. A., Kim, T.-Y., Lee, S. U., Kang, Y. S. and Bang, J. H. (2015). Exploring interfacial events in gold-nanocluster-sensitized solar cells: Insights into the effects of the cluster size and electrolyte on solar cell performance, *J. Am. Chem. Soc.*, **138**, pp. 390–401.

189. Xiao, F.-X., Zeng, Z., Hsu, S.-H., Hung, S.-F., Chen, H. M. and Liu, B. (2015). Light-induced in situ transformation of metal clusters to metal nanocrystals for photocatalysis, *ACS Appl. Mater. Interfaces*, **7**, pp. 28105–28109.

190. Ramasamy, P., Guha, S., Shibu, E. S., Sreeprasad, T. S., Bag, S., Banerjee, A. and Pradeep, T. (2009). Size tuning of Au nanoparticles formed by electron beam irradiation of Au_{25} quantum clusters anchored within and outside of dipeptide nanotubes, *J. Mater. Chem.*, **19**, pp. 8456–8462.

191. Weng, B., Lu, K.-Q., Tang, Z., Chen, H. M. and Xu, Y.-J. (2018). Stabilizing ultrasmall Au clusters for enhanced photoredox catalysis, *Nat. Commun.*, **9**, p. 1543.

192. Pasteur, L. (1848). Memoires sur la relation qui peut exister entre la forme crystalline et al composition chimique, et sur la cause de la polarization rotatoire, *Compt. Rend.*, **26**, pp. 535–538.

193. Xu, L.-L., Zhang, H.-F., Li, M., Ng, S. W., Feng, J.-H., Mao, J.-G. and Li, D. (2018). Chiroptical activity from an achiral biological metal–organic framework, *J. Am. Chem. Soc.*, **140**, pp. 11569–11572.

194. Zeng, C. and Jin, R. (2017). Chiral gold nanoclusters: Atomic level origins of chirality, *Chem. Asian J.*, **12**, pp. 1839–1850.

195. Knoppe, S. and Bürgi, T. (2013). Chiroptical properties of intrinsically chiral thiolate-protected gold clusters, *Chimia*, **67**, pp. 236–239.

196. Schaaff, T. G., Knight, G., Shafigullin, M. N., Borkman, R. F. and Whetten, R. L. (1998). Isolation and selected properties of a 10.4 kDa gold: Glutathione cluster compound, *J. Phys. Chem. B*, **102**, pp. 10643–10646.

197. Kitaev, V. (2008). Chiral nanoscale building blocks—From understanding to applications, *J. Mater. Chem.*, **18**, pp. 4745–4749.

198. Ben-Moshe, A., Maoz, B. M., Govorov, A. O. and Markovich, G. (2013). Chirality and chiroptical effects in inorganic nanocrystal systems with plasmon and exciton resonances, *Chem. Soc. Rev.*, **42**, pp. 7028–7041.

199. Knoppe, S., Azoulay, R., Dass, A. and Bürgi, T. (2012). In situ reaction monitoring reveals a diastereoselective ligand exchange reaction between the intrinsically chiral $Au_{38}(SR)_{24}$ and chiral thiols, *J. Am. Chem. Soc.*, **134**, pp. 20302–20305.

200. Lopez-Acevedo, O., Tsunoyama, H., Tsukuda, T., Hakkinen, H. and Aikens, C. M. (2010). Chirality and electronic structure of the thiolate-protected Au_{38} nanocluster, *J. Am. Chem. Soc.*, **132**, pp. 8210–8218.

201. Garzón, I., Beltrán, M., González, G., Gutıerrez-González, I., Michaelian, K., Reyes-Nava, J. and Rodrıguez-Hernández, J. (2003). Chirality, defects, and disorder in gold clusters, *Eur. Phys. J. D*, **24**, pp. 105–109.

202. Garzón, I. L., Reyes-Nava, J. A., Rodriguez-Hernandez, J. I., Sigal, I., Beltrán, M. and Michaelian, K. (2002). Chirality in bare and passivated gold nanoclusters, *Phys. Rev. B*, **66**, p. 073403.

203. Qian, H., Eckenhoff, W. T., Zhu, Y., Pintauer, T. and Jin, R. (2010). Total structure determination of thiolate-protected Au38 nanoparticles, *J. Am. Chem. Soc.*, **132**, pp. 8280–8281.

204. Dolamic, I., Knoppe, S., Dass, A. and Bürgi, T. (2012). First enantioseparation and circular dichroism spectra of Au_{38} clusters protected by achiral ligands, *Nat. Commun.*, **3**, p. 798.

205. Zhu, M., Qian, H., Meng, X., Jin, S., Wu, Z. and Jin, R. (2011). Chiral Au_{25} nanospheres and nanorods: Synthesis and insight into the origin of chirality, *Nano Lett.*, **11**, pp. 3963–3969.

206. Dolamic, I., Varnholt, B. and Bürgi, T. (2015). Chirality transfer from gold nanocluster to adsorbate evidenced by vibrational circular dichroism, *Nat. Commun.*, **6**, p. 7117.

207. Riccardi, L., De Biasi, F., De Vivo, M., Burgi, T., Rastrelli, F. and Salassa, G. (2019). The dynamic origin of chirality transfer between chiral surface and achiral ligand in Au_{38} clusters, *ACS Nano*, **13**, pp. 7127–7134.

208. Knoppe, S., Wong, O. A., Malola, S., Häkkinen, H., Bürgi, T., Verbiest, T. and Ackerson, C. J. (2014). Chiral phase transfer and enantioenrichment of thiolate-protected Au_{102} clusters, *J. Am. Chem. Soc.*, **136**, pp. 4129–4132.

209. Malola, S., Lehtovaara, L., Knoppe, S., Hu, K.-J., Palmer, R. E., Bürgi, T. and Häkkinen, H. (2012). $Au_{40}(SR)_{24}$ cluster as a chiral dimer of 8-electron superatoms: Structure and optical properties, *J. Am. Chem. Soc.*, **134**, pp. 19560–19563.

210. Knoppe, S., Dolamic, I., Dass, A. and Bürgi, T. (2012). Separation of enantiomers and CD spectra of $Au_{40}(SCH_2CH_2Ph)_{24}$: Spectroscopic evidence for intrinsic chirality, *Angew. Chem. Int. Ed.*, **51**, pp. 7589–7591.

211. Zhu, Y., Wang, H., Wan, K., Guo, J., He, C., Yu, Y., Zhao, L., Zhang, Y., Lv, J. and Shi, L. (2018). Enantioseparation of $Au_{20}(PP_3)_4Cl_4$ clusters with intrinsically chiral cores, *Angew. Chem. Int. Ed.*, **57**, pp. 9059–9063.

212. Yang, Y., Zhang, Q., Guan, Z.-J., Nan, Z.-A., Wang, J.-Q., Jia, T. and Zhan, W.-W. (2019). Enantioselective synthesis of homochiral Au_{13} nanoclusters and their chiroptical activities, *Inorg. Chem.*, **58**, pp. 3670–3675.

213. Malola, S. and Häkkinen, H. (2019). Chiral inversion of thiolate-protected gold nanoclusters via core reconstruction without breaking a Au–S bond, *J. Am. Chem. Soc.*, **141**, pp. 6006–6012.

Part III
Catalytic Application of Nanoclusters

Chapter 8

Catalytic Application of Well-Defined Au Nanoparticles: Oxidation, Hydrogenation, and Coupling Reactions

Quanquan Shi,[a] Youhai Cao,[b] Zhaoxian Qin,[b] and Gao Li[b]

[a]*College of Science, Inner Mongolia Agricultural University, Hohhot 010018, China*

[b]*State Key Laboratory of Catalysis, Dalian Institute of Chemical Physics, Chinese Academy of Sciences, Dalian 116023, China*

gaoli@dicp.ac.cn

Gold clusters (1–3 nm size) have been recently documented as a novel nanogold material, especially in the catalysis. Atomically precise Au clusters (Au_nL_m, n and m stand for the number of gold atoms and organic ligands, respectively) possess unique crystal and electronic structures, thereby correlating the intrinsic catalytic performance with the clusters' structure and allowing to investigate the catalytic pathways (e.g., adsorption and activation processes). Further, the activity of Au clusters can be distinctly enhanced via the tailored removal of surface ligands (i.e., "L") to eliminate the steric hindrance during catalytic reactions. This chapter summarizes the homogeneous and heterogeneous reactions catalyzed over intact

Atomically Precise Nanoclusters
Edited by Yan Zhu and Rongchao Jin

ISBN 978-981-4877-08-4 (Hardcover), 978-1-003-11990-6 (eBook)
www.jennystanford.com

gold clusters and defective clusters (i.e., partial ligands are detached), which include the selective aerobic oxidation and hydrogenation as well as one-pot coupling reactions of furfural with ethanol to give biofuel.

8.1 Introduction

Metal nanoparticles (NPs) occupy an important niche in chemical transformations [1–5]. Since the 1980s, TiO_2-supported gold NPs of 3–6 nm size have been found to have good activity in the CO oxidation at low reaction temperature [6]; therefore, the investigation on the catalysis over AuNPs has sharply increased. Although gold NPs are more inactive than Pt- or PdNPs in many chemical transformations, the AuNPs exhibit much better selectivity toward target products [7, 8]. However, the polydispersity of the conventional NPs creates a major issue; that is, the size-dependent catalytic activities of NPs are averaged out in polydisperse catalysts. The preparation of ultrasmall gold particles (the so-called gold clusters) is another big challenge. To overcome these obstacles, the synthesis of well-defined gold clusters (1–3 nm size with monodispersity) catalysts is highly desirable.

Very recently, remarkable advances have been achieved in the synthesis of small-sized clusters in solution phase [9]. These specific-size gold clusters Au_nL_m are composed of a precise number of Au atoms (n) and ligands (m), with n ranging up to a few hundred atoms (e.g., $n < 300$, equivalent size range from 1 to 3 nm). The gold clusters exhibit strong electron energy quantization [9], which is opposed to the continuous conduction band in metallic gold nanoparticles or bulk gold. Thus, clusters become semiconductors and possess a sizable bandgap [e.g., ~1.6 eV for $Au_{38}S_2(SR)_{20}$] [10]. In addition, Au clusters can be controllably doped with foreign atoms (e.g., Pd, Pt, Ag, and Cu) [11–13], which is of interest to nanocatalysis, as the catalytic performance of gold clusters can be turned on atom-by-atom basis.

In this chapter, we introduce well-defined Au clusters as a novel promising nanocatalyst. The Au cluster-catalyzed homogeneous and heterogeneous reactions are well summarized, for example, CO oxidation, sulfoxidation, hydrogenation of aldehydes, semihydrogenation of alkynes, aerobic oxidation of amines into imines, and one-pot conversion of furfural with ethanol to give biofuel (Fig. 8.1).

Moreover, the catalytic mechanism and the reaction pathway are thoroughly discussed.

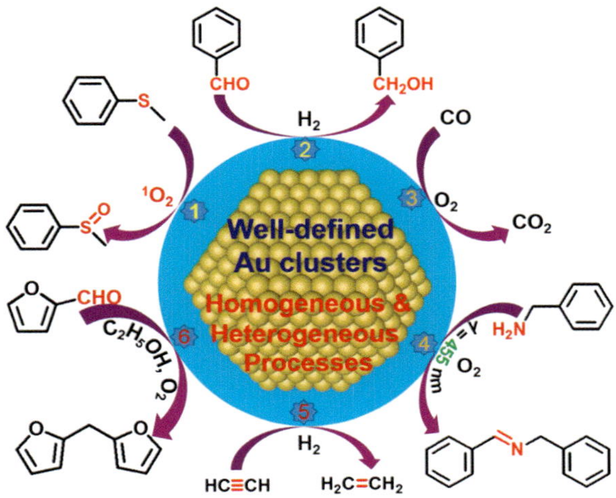

Figure 8.1 Gold cluster-catalyzed homogeneous and heterogeneous reactions.

8.2 Homogeneous Catalysis

These gold nanoclusters show high thermal stability in the solution phase, as some of the Au clusters are obtained via "size-focusing" processes under harsh conditions (e.g., at 80 °C) [9]. For example, the robust $Au_{99}(SPh)_{42}$ nanocluster can survive under 150 °C annealing in the presence of N_2, evidenced by the analyses of UV-Vis and matrix-assisted laser desorption ionization mass spectrometry (MALDI-MS) [14, 15]. On the other hand, these ligand-protected clusters usually exhibit good capability in electronic transitions [16–20]. Hence, these clusters can well be explored in the homogeneous catalysis under mild reaction conditions (<120 °C, in the absence of strong oxidants such as peroxide).

8.2.1 Hydrogenation of Aldehyde

The activation of H_2 gas over gold nanoparticles is deemed as the pivotal step during the hydrogenation process. The cluster itself shows very weak capacity to dissociate the H–H bond; however, gold nanoclusters are capable of activating H_2 with the aid of organic bases, for example, pyridine [21, 22]. Thus, the phenthiolate-capped gold clusters were first studied in the chemoselective hydrogenation of 4-nitrobenzaldehyde. The catalytic hydrogenation reaction was carried out under mild conditions (80 °C for 4 h) in the presence of pyridine and using toluene as the solvent [23]. A size dependence was found to be following the order of $Au_{25}(SPh)_{18}$ (13.5% nitrobenzaldehyde conversion) > $Au_{36}(SPh)_{24}$ (13.0%) > $Au_{99}(SPh)_{42}$ (8.9%). It is worthy to note that the chemoselectivity for reducing aldehyde group (i.e., the 4-nitrobenzyl alcohol product) was ~100% and the clusters were sufficiently robust and remained intact after the reactions. These catalytic results clearly demonstrated that gold clusters are associated with active sites during reactions.

Further, the aqueous soluble of $Au_n(SR)_m$ nanoclusters [containing glutathione-capped $Au_{15}(SG)_{13}$, $Au_{18}(SG)_{14}$, $Au_{25}(SG)_{18}$, and $Au_{38}(SG)_{24}$ and captopril-capped $Au_{25}(Capt)_{18}$] was examined in the hydrogenation to study the size dependence and the ligand effect [24]. The free clusters, except $Au_{18}(SG)_{14}$, remained intact after the catalytic hydrogenation, supported by the analyses of UV-Vis and polyacrylamide gel electrophoresis (PAGE). Drastic size dependence and steric effect of protecting organic ligands were observed in the reactions. The catalytic activity (based on the conversion) follows the order of $Au_{38}(SG)_{24}$ (20.6% conversion) > $Au_{25}(SG)_{18}$ (16.0%) > $Au_{18}(SG)_{14}$ (10.2%) > $Au_{15}(SG)_{13}$ (7.8%), and $Au_{25}(Capt)_{18}$ (23.3%) > $Au_{25}(SG)_{18}$ [42]. Considering the catalytic results over the Au_{15} to Au_{99}, a volcano-like performance over gold nanoclusters is clearly found.

Next, DFT (density functional theory) simulation was applied to study the mechanism. It showed that both –CHO and –NO_2 groups of the reactant are in close interaction with the surface staples of clusters (RS–Au–SR) (Fig. 8.2a). The average adsorption energies of nitrobenzaldehyde on different sized Au clusters are moderately strong and similar in strength [$Au_{38}(SCH_3)_{24}$: –0.86 eV, $Au_{25}(SCH_3)_{18}$: –1.03 eV, $Au_{18}(SG)_{14}$: –0.92 eV, and $Au_{15}(SG)_{13}$: –0.88 eV]. Therefore, the catalytic activity is primarily associated with the surface area of Au clusters, consistent with the observed trend of the conversion versus the cluster size (Fig. 8.2b). Hydrogen is dissociated over gold clusters with the help of pyridine and then the –CHO group is selectively hydrogenated via surface Au atoms [24].

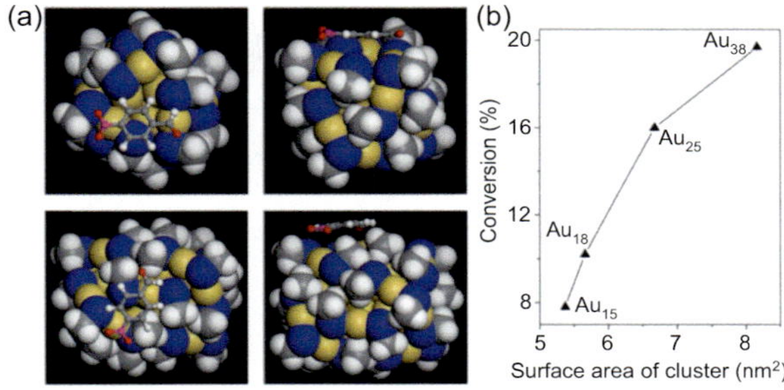

Figure 8.2 (a) Nitrobenzaldehyde molecule adsorbed on $Au_{25}(SCH_3)_{18}$ nanoclusters. Colour codes: Au, yellow; S, blue; C, grey; H, white; O, red; N, pink. (b) Conversion of nitrobenzaldehyde versus the surface area of Au clusters. Reprinted with permission from Ref. [24], Copyright 2014, American Chemical Society.

The well-defined Au nanorod capped by cetyltrimethylammonium, as the homogeneous catalyst, was also evaluated in the hydrogenation of $NO_2C_6H_4CHO$ in water [25]. The selectivity for the product of $NO_2C_6H_4CH_2OH$ is the same as that over the Au cluster catalyst (decent 100%). The Au nanorod catalysts showed an aspect ratio–dependent reactivity and performed much better than the Au spherical NPs. It is also found that the catalytic reaction prefers to occur on the Au{111} of gold nanorods.

8.2.2 Photo-oxidation

Recently, metal clusters have emerged as a novel photosensitizer for the formation of singlet oxygen ($^1O_2$) using visible light [26–29]. The energy difference between the triplet state and the ground state of the photosensitizer is expected to be larger than the activation energy of the triplet oxygen ($^3O_2 \rightarrow {}^1O_2$ with $\Delta E = 0.97$ eV, Scheme 8.1a) [30]. Of note, the formation of $^1O_2$ can be detected by chemical trapping probes (e.g., 1,3-diphenylisobenzofuran (DPBF) and diaminobenzidine) and the direct observation of the characteristic $^1O_2$ emission (ca. 1276 nm). The singlet oxygen is the activated oxygen species and more active than the oxygen molecule; therefore, $^1O_2$ can be exploited in mild selective aerobic oxidations. Initially, Hideya et al. [26] found that $Au_{25}(SR)_{18}$ (where H-SR: phenylethanethiol and captopril) with a HOMO–LUMO (highest occupied and lowest unoccupied molecular orbitals) gap of 1.3 eV can directly generate $^1O_2$ without using the conventional organic photosensitizers under visible/near-IR irradiation (e.g., 532, 650, and 808 nm). Further, the in situ generated $^1O_2$ was applied in the selective oxidation of sulfide at 25 °C and 55 °C using $CHCl_3$ as the solution. Relatively low conversion of 4–12% with completely 100% selectivity for the sulfoxide production was achieved. The $Au_{25}(SR)_{18}^-$ nanoclusters are intact during oxidation.

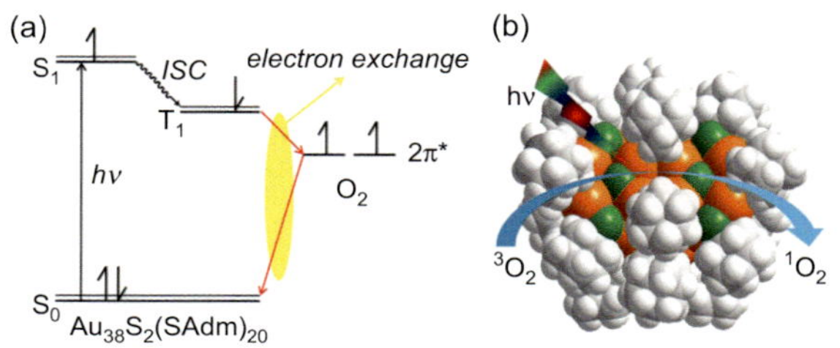

Scheme 8.1 (a) Mechanism of the Dexter-type electron exchange coupling between $Au_{38}S_2(SAdm)_{20}$ cluster and an oxygen molecule for the generation of singlet oxygen. (b) Illustration of the generation process on the open $Au_{38}S_2(SAdm)_{20}$ structure. Color codes: Au, orange; S, green; C, grey; H, white. Reprinted with permission from Ref. [28], Copyright 2017, American Chemical Society.

Next, Li and coworkers [28] observed that the $Au_{38}S_2(SAdm)_{20}$ clusters (–SAdm = 1-adamantanethiolate) of 1.57 eV HOMO–LUMO gap also exhibited as an excellent photosensitizer for the $^1O_2$ generation under visible/near-IR (infrared) (532 and 650 nm) irradiation. The $^1O_2$ quantum yield of the $Au_{38}S_2(SAdm)_{20}$ (Φ_{Au38}) was estimated to be 0.22, which is relative to anthracene standard ($\Phi_{anthracene}$ is ca. 0.70). The mechanism of singlet oxygen generation via the Au cluster photoexcitation is proposed to involve a Dexter-type electron exchange coupling between the excited photosensitizer and ground state, triplet oxygen molecules ($^3O_2$) (Scheme 8.1). The gold cluster exhibited a "pocket-like" surface structure, serving as the site for adsorption and activation of reactant molecules (Scheme 8.1b).

Further, photoexcited $Au_{38}S_2(SAdm)_{20}$ clusters are shown to have promising catalytic performance in the selective aerobic oxidation of sulfide into sulfoxide and amine into imine (Tables 8.1

and 8.2). The catalytic reactions were carried out in open quartz vessels with O_2 bubbling under irradiation (λ = 532 nm). Several aspects are noteworthy in the sulfoxidation: (i) No activity of $Au_{38}S_2(SAdm)_{20}$ nanoclusters was detected in dark. (ii) $Au_{25}(PET)_{20}$ and $Au_{38}S_2(SAdm)_{20}$ clusters gave 18% and 57% conversion, respectively, when the solution was exposed to exciting light of λ = 532 nm. Both clusters gave a decent 100% selectivity to produce sulfoxide. (iii) $Au_{38}S_2(SAdm)_{20}$ presented a better catalytic activity than $Au_{25}(PET)_{20}$, demonstrating the higher efficiency of $^1O_2$ photogeneration over the former cluster. UV-Vis analysis showed that the clusters were intact after completion of the reaction. The clusters exhibited good reusability; no appreciable loss of catalytic activity and selectivity was observed after three cycles. (iv) No catalytic reaction occurred in the absence of gold nanoclusters or oxygen, indicating that the sulfide was oxidized by the in situ generated $^1O_2$ species. The activity over the Au_{38} in the homogeneous catalytic process was more efficient than those over heterogeneous processes [31, 32].

Table 8.1 Comparison of catalytic performances of $Au_{25}(PET)_{18}$ (PET = 2-phenylethanethiolate) and $Au_{38}S_2(SAdm)_{20}$ in the selective oxidation of sulfide into sulfoxide under light irradiation with λ = 532 nm at room temperature[a]

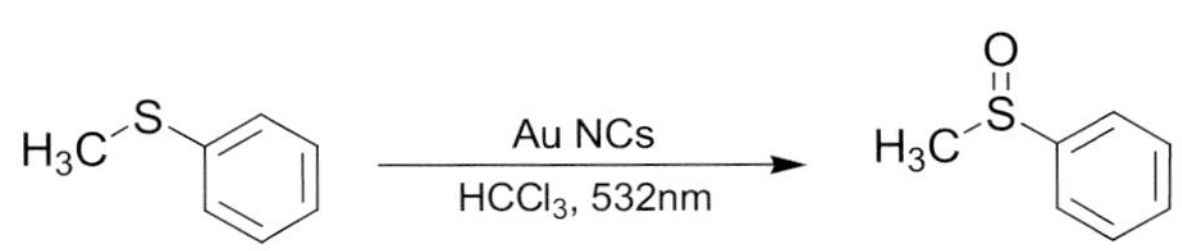

Au cluster	Gas	Light	Conversion	Selectivity
$Au_{38}S_2(SAdm)_{20}$	O_2	–	n.r.	–
$Au_{25}(PET)_{18}$	O_2	532 nm	18%	100%
$Au_{38}S_2(SAdm)_{20}$	O_2	532 nm	57%	100%
$Au_{38}S_2(SAdm)_{20}$[b]	O_2	532 nm	58%	100%
$Au_{38}S_2(SAdm)_{20}$[c]	O_2	532 nm	56%	100%
–	O_2	532 nm	n.r.	–
$Au_{38}S_2(SAdm)_{20}$	N_2	532 nm	n.r.	–

[a]Reaction conditions: 0.05 mmol sulfide, 0.2 μmol Au clusters for 12 h
[b]2nd and [c]3rd tests
n.r. = no reaction

Table 8.2 Photocatalytic oxidation of benzylamine over $Au_{38}S_2(SAdm)_{20}$ in the presence of O_2 under light-emitting diode (LED) irradiation ($\lambda \sim 455$ nm)

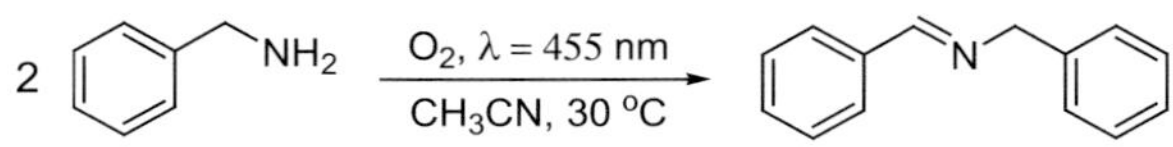

Catalyst	Gas	Light	Conversion	Selectivity
–	O_2	455 nm	–	–
$Au_{38}S_2(SAdm)_{20}$	O_2	Dark	–	–
$Au_{38}S_2(SAdm)_{20}$	N_2	455 nm	–	–
$Au_{38}S_2(SAdm)_{20}$	O_2	455 nm	99%	>99%
$Au_{38}(PET)_{24}$	O_2	455 nm	20%	>99%

Reaction conditions: benzylamine (0.2 mmol), dodecane (0.1 mmol), 0.5 μmol Au clusters at 30 °C for 30 min

Next, Au_{38} nanogold catalysts were applied in a one-step chemical transformation of benzylamine to *N*-(phenylmethylene) benzenemethanamine (Table 8.2). The amines are typically synthesized via direct oxidation of amines or condensation of an amine with an aldehyde. Ninety-nine percent conversion of benzylamine to amine was completed in 30 min with greater than 99% selectivity, which is more efficient than that catalyzed by gold nanoparticles (particle size: 6–12 nm, loaded on metal oxides) [33]. The turnover frequency of the catalytic oxidation, which is given by TOF = [reacted mole of amine]/[(mole of cluster) × (reaction time)], can reach to about 1584 h^{-1}. These clusters also were intact during the reaction, and part of the photocatalytic oxidation catalyzed over clusters should occur via a mechanism involving electron transfer and generation of benzylamine radical cation intermediates [34].

Zhang et al. [29] found that the $[Au_{13}(dppe)_5Cl_2]Cl_3$ cluster with a propeller-like structure can be employed as an efficient photosensitizer for the generation of singlet oxygen with a high quantum yield of $\Phi_{Au13} = 0.71$, which is considerably higher than that of anthracene (organic dye). The photosensitizer of organic dye is not stable during the generation of singlet oxygen. The HOMO–LUMO gap of the $[Au_{13}(dppe)_5Cl_2]^{3+}$ cluster was determined to be 1.9 eV via the ultraviolet photoemission spectroscopy (UPS) analysis. In all, these gold clusters with atomic precision showed excellent behavior and were found to be a novel efficient photosensitizer in the generation of singlet oxygen.

8.3 Heterogeneous Catalysis

Although intact gold clusters showed the catalytic activity during the reaction process, the catalytic efficiency is not promising, as the active sites (e.g., the gold atoms at the staple motif) are well covered by organic ligands. So, the partial removal of the surface ligands would keep the size of the particle (e.g., the number of gold atoms) and enhance the catalytic performance (including the activity and the product selectivity). These capping ligands (e.g., thiolate and phosphine) can be selectively detached with the aid of organic base (e.g., pyridine), water, Lewis acid [e.g., $M(OAc)_x$], ionic liquids, and so on [35–38].

8.3.1 Oxidation

O_2 activation is one of the steps during the selective oxidation process. Gold nanoclusters have been proved to have a good capacity for strong interactions with O_2 molecules in experiment results (e.g., emerging as the singlet oxygen photosensitizer).

8.3.1.1 CO oxidation

CO oxidation was first documented as a probe reaction over naked gold clusters [4]. Regarding ligand-capped gold particles, Nie et al. [39] first found that the $Au_{25}(SR)_{18}/CeO_2$ catalyst exhibits moderate activity in CO oxidations. However, they claimed that the $Au_{25}(SR)_{18}$ should be intact during the oxidation process. Further, Wu et al. [40] also investigated the nature of active sites of the $Au_{25}(SR)_{18}$ nanocluster for oxidation of CO. They found that CO molecules were not able to adsorb onto the surface of intact $Au_{25}(SR)_{18}$ clusters at room temperature. In situ Fourier transform infrared spectroscopy (FTIR) analysis showed that the protecting ligands are gradually removed when the reaction temperature increases to 150 °C. Of note, the gold clusters grew up at high temperature (e.g., > 200 °C) and almost all of the surface ligands were detached. The cationic $Au^{\delta+}$ sites (δ: 0 ~ +1) are found to play an important role in the low-temperature CO oxidation. Next, DFT calculations suggested that CO can adsorb favorably onto exposed gold atoms upon the removal of thiolate ligands. They speculated that the CO oxidation over the $Au_{25}(SR)_{18}/CeO_2$ catalyst proceeded predominantly via the redox mechanism, where O_2 molecules were activated on the CeO_2 surface and the CO reactant was activated on dethiolated gold sites, supported by isotopic labeling experiments.

Very recently, clusters with Ag and Cu dopants have been also explored in the oxidation [41, 42]. Of note, Ag and Cu dopants are preferentially located at the cluster kernel instead of the staple motif [11]. The X-ray photoelectron spectroscopy (XPS) analysis and DFT suggest that the electron was transferred from dopants to gold atoms and the electronic structure of clusters was modified by dopants. The gold clusters with copper dopants can improve the catalytic activity, following the order of

$$Cu_xAu_{25-x}(SR)_{18} > Au_{25}(SR)_{18} > Ag_xAu_{25-x}(SR)_{18}.$$

The $Cu_xAu_{25-x}(SR)_{18}$ catalyst shows excellent recyclability and durability (no appreciable loss of activity in the experiment of 133 h) [41]. TOF is found to be about 4 s^{-1}, which is slightly higher than that for the conventional gold NP catalysts. The FTIR analyses and DFT suggested that thiolate ligands were partially removed under reaction conditions ($T > 120$ °C). DFT indicated that dopant effects were more pronounced when dopants are located at the staple motif to interact directly with CO molecules.

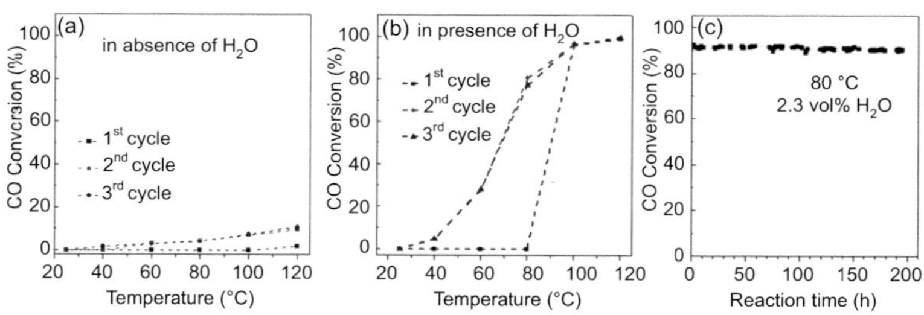

Figure 8.3 Temperature dependence of CO conversion over $Au_{25}(SC_{12}H_{25})_{18}/CeO_2$ catalysts (a) in the absence of water and (b) in the presence of water. (c) CO conversion as a function of reaction time. Reprinted with permission from Ref. [43], Copyright 2018, Royal Society of Chemistry.

Further, Li et al. [43] reported a discovery that water vapor provokes the mild removal of surface ligands from $Au_{25}(SC_{12}H_{25})_{18}$ clusters in a controlled manner. They found that the catalytic activity of cluster/CeO_2 is improved from nearly zero conversion of CO (in the absence of water) to 96.2% (in the presence of 2.3 vol% H_2O) at the same temperature (100 °C) (Fig. 8.3a,b). The TOF of the ceria-supported Au cluster catalyst improved twofold to 8.7 s^{-1}. The cluster exhibited high stability during the CO oxidation process under moisture conditions (up to 20 vol% water vapor, Fig. 8.3c).

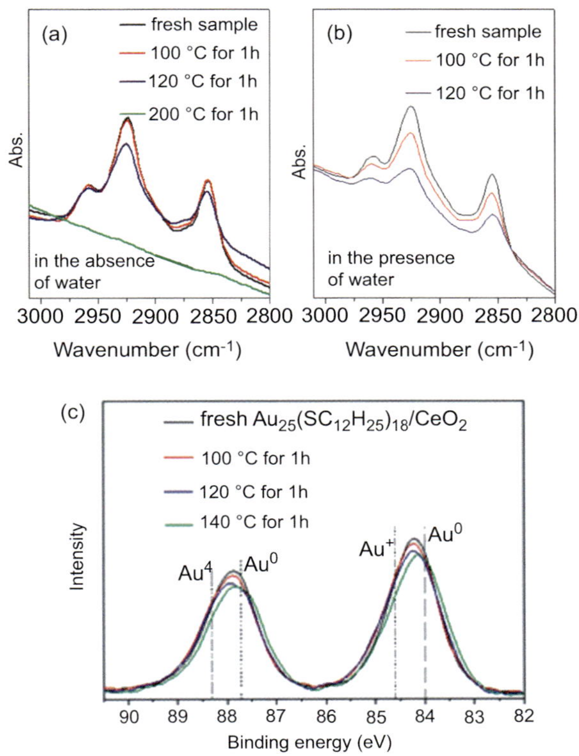

Figure 8.4 (a,b) FTIR spectra and (c) Au4f XPS analysis of $Au_{25}(SR)_{18}/CeO_2$ catalysts after the CO oxidation at different reaction temperatures under mixed gases (1.67% CO, 3.33% O_2, and 95% He) in the absence or presence of moisture. Reprinted with permission from Ref. [43], Copyright 2018, Royal Society of Chemistry.

To study the possibility of "ligand-off" clusters in the presence of water vapor, FTIR analysis was performed on the cluster catalyst. After 100 °C treatment in the absence of water, the FTIR spectrum was superimposable on that of the as-prepared catalyst (Fig. 8.4a), implying that all the thiolate ligands were still capped on the cluster surface and the Au_{25} should be intact. It was also confirmed by the TGA, as no weight loss was found up to 120 °C in a dry atmosphere. For comparison, the surface ligands on Au clusters were partially removed at 100 °C when water vapor was present (Fig. 8.4b), indicating that water vapor only can assist the removal of thiolate ligands. Further, DFT results on $Au_{25}(SCH_3)_{18}$ also showed that the "$-SCH_3$" removal in the presence of water is more favorable (by 45.9 kcal/mol) than that in the absence of water. Interestingly, gold atoms are reduced when thiolate ligands are removed, as the binding energies (BEs) of Au $4f_{5/2}$ and $4f_{7/2}$ were downshift (Fig. 8.4c). In all, the dominantly cationic $Au^{\delta+}$ ($0 < \delta < 1$) species on the partial naked clusters were found to be the catalytic active sites for the CO activation.

8.3.1.2 Photo-oxidation of amines to imines

$Au_{25}(PPh_3)_{10}Cl_2(SC_3H_6SiO_3)_5/TiO_2$ was applied as an efficient photocatalyst for the selective oxidation of amines to imines under visible light irradiation [44]. The diffuse reflectance UV-Vis spectrum of $[Au_{25}]$/P25 showed similar optical peaks (especially the characteristic peak at 675 nm) with that of free clusters (Fig. 8.5a) [44, 45]. It also was confirmed by the TEM (transmission electron microscopy) analysis, as the cluster was still ~1.2 nm on the P25 surface (Fig. 8.5b–d). Only one single peak at $\delta = 33.2$ ppm was observed in the solid state ^{31}P magic angle spinning (MAS) NMR (nuclear magnetic resonance) spectroscopy (Fig. 8.5c), demonstrating the presence of PPh_3 ligands. These results indicated that $[Au_{25}]$ clusters were intact on the P25 surface.

Surprisingly, $[Au_{25}]$/P25 photocatalysts exhibited promising activity in the aerobic oxidation of amines to imines. The selectivity toward the target products imines was close to 100%. Next, catalytic oxidations were exploited in various amine derivatives (Table 8.3). The TOF of $[Au_{25}]$/P25 reached 1522 h^{-1} when 4-methylbenzlamine was used as the reactant, which is much higher than those catalyzed by AuNP and other metals catalysts. Of note, the detachment of phosphine from clusters were obviously observed during the reactions, implying the uncovered Au atoms were the catalytically active sites for the photocatalysis.

Table 8.3 Photo-oxidation of a range of amine substrates over $[Au_{25}]$/P25 catalyst

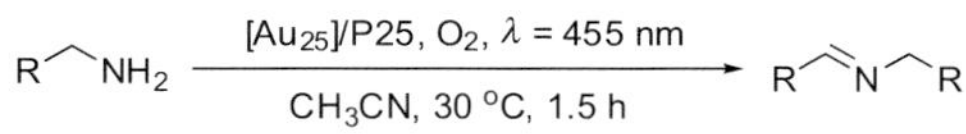

entry	substrate	product	conv. (%)	sel. (%)	TOF (h^{-1})
1	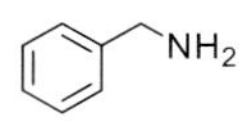	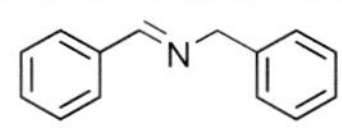	98	99	878
2	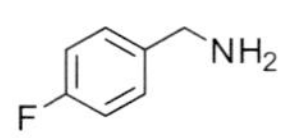	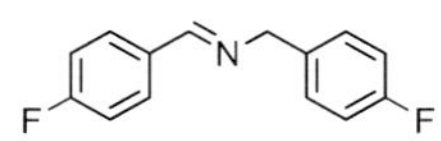	82	99	1049
3	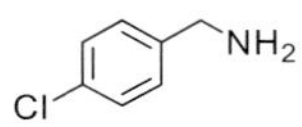	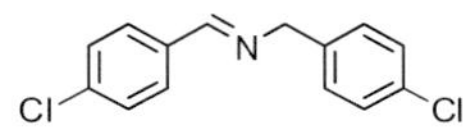	97	99	1184
4	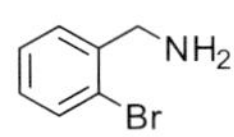	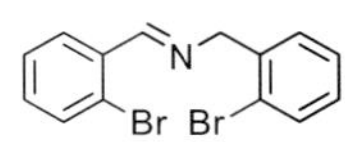	73	97	312
5	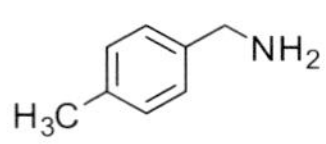	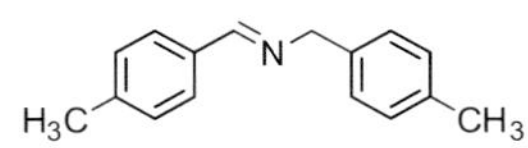	99	99	1522
6	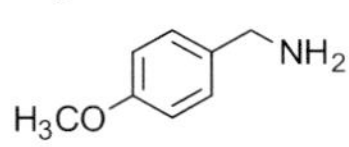	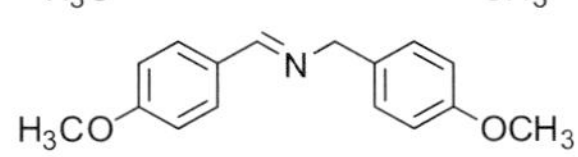	96	97	1025
7	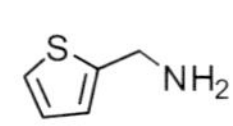	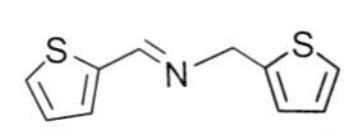	94	99	638
8	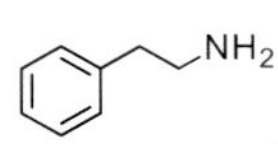	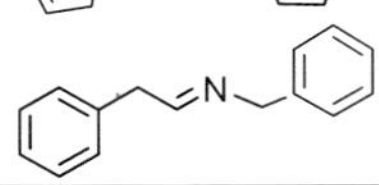	69	87	

To study the reaction pathway of the photo-oxidation, several controlled experiments were carried out. (i) The activity significantly decreased when $K_2S_2O_8$ (scavenger for electrons) and ammonium oxalate (scavenger for holes) were added into the reaction system, demonstrating the photogenerated holes and electrons were the initial drivers for the photocatalysis. (ii) The activity also distinctly decreased when TEMPO was added. Of note, TEMPO plays a similar role as oxygen, which can abstract hydrogen from the metal surface instead of from the substrate. It means the Au–H species were indeed generated during the photocatalytic reaction. Based on these catalytic results, the reaction mechanism can be rationalized, as illustrated in Scheme 8.2.

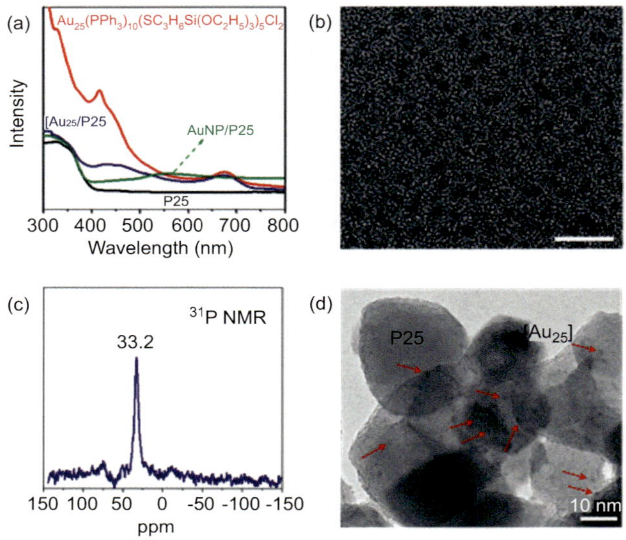

Figure 8.5 (a) UV-Vis of Au_{25} nanoclusters and diffuse reflectance UV-Vis spectra of $[Au_{25}]$/P25, AuNP/P25, and P25 support. (b) TEM image of free Au_{25} clusters. (c) Solid state ^{31}P MAS NMR spectrum and (d) TEM image of $[Au_{25}]$/P25. Reprinted with permission from Ref. [44], Copyright 2017, American Chemical Society.

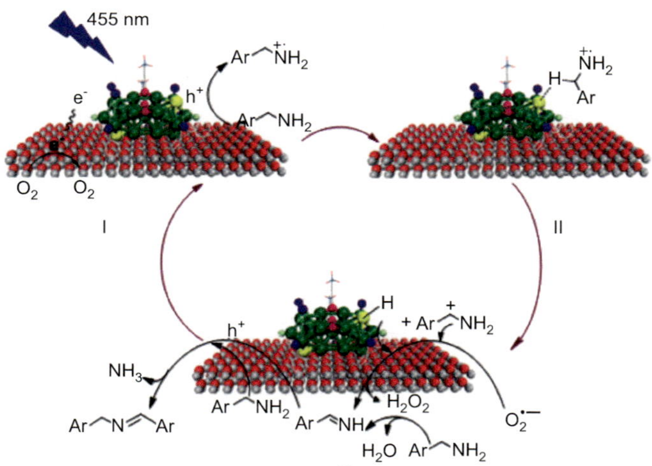

Scheme 8.2 Proposed mechanism for photo-oxidation of benzylamine catalyzed by TiO_2-suppported $[Au_{25}]$ clusters. Color codes: Au, green; exposed Au, yellow; P, purple; S, magenta; Cl, light green; Ti, gray; O, pink. Reprinted with permission from Ref. [44], Copyright 2017, American Chemical Society.

8.3.2 Hydrogenation

8.3.2.1 Hydrogenation of aldehydes

Supported cluster catalysts were applied in the selective hydrogenation of aldehydes to alcohols [46, 47]. Catalytic reactions were carried out in water. It was found that the activity of the clusters was largely influenced by the oxide supports; the aldehyde conversion was improved from 27.9% [using $Au_{99}(SPh)_{42}/SiO_2$ as the catalyst] to 69.8% [$Au_{99}(SPh)_{42}/TiO_2$] and then to 93.1% [$Au_{99}(SPh)_{42}/CeO_2$] [23]. Next, the $Au_{99}(SPh)_{42}/CeO_2$ catalyst was investigated for a range of substrates with aldehyde and nitro groups (Table 8.4). The effects of the nitro group at the ortho-, meta-, and para-position were examined, and the nitro group at the meta-position (*m*-nitrobenzaldehyde) gave the best activity. Further, the effect of the electronic factor (including methyl, hydroxyl, and chloro groups) of substrates also was explored; the hydroxyl group on the substrate largely affected the catalytic activity. However, the $Au_{99}(SPh)_{42}/CeO_2$ catalysts had limited capacity to activate other aldehyde reactants (e.g., benzaldehyde), as most of the protecting ligands were capped on the surface of the particles.

Thus, Lewis acid [such as $Cr(NO_3)_3$, $Co(OAc)_2$, $NiCl_2$, $Cu(OAc)_1$, $Cu(OAc)_2$, and $Cu(NO_3)_2$] was introduced to the reaction system to peel the surface ligands [38]. The partial removal of the "–Au(SR)–" unit from the parent $Au_{25}(SR)_{18}$ nanoclusters was clearly supported by ESI-MS (electrospray ionization mass spectrometry) analysis and simulated by DFT calculation {i.e., $[Au_{25}(SCH_3)_{18}]^- + n\ M(NH_3)_4{}^{z+} + n\ NH_3 \rightarrow [Au_{25\text{-}n}(SCH_3)_{18\text{-}n}]^- + n\ [M(NH_3)_3(SCH_3)]^{(z-1)+} + n\ Au(NH_3)_2{}^+$, where n = 1–4}. It is very interesting that the catalytic activity of $Au_{25}(SR)_{18}/CeO_2$ was largely improved in the presence of Lewis acid at relatively low temperature (Table 8.5). The degree of promotion of the catalytic activity of $Au_{25}(SR)_{18}/CeO_2$ by M^{z+} is in the order $Co^{2+} > Ni^{2+} \approx Cr^{3+} > Cu^+ > Cu^{2+}$. Of note, the anions acted merely as a spectator in the catalytic reactions. Further, $Au_{25}(SR)_{18}/CeO_2$ with $Co(OAc)_2$ was applied in the hydrogenation of benzaldehyde and its derivatives and it showed good recyclability.

Table 8.4 Chemoselective hydrogenation of a range of substrates with nitro and aldehyde groups over $Au_{99}(SPh)_{42}/CeO_2$

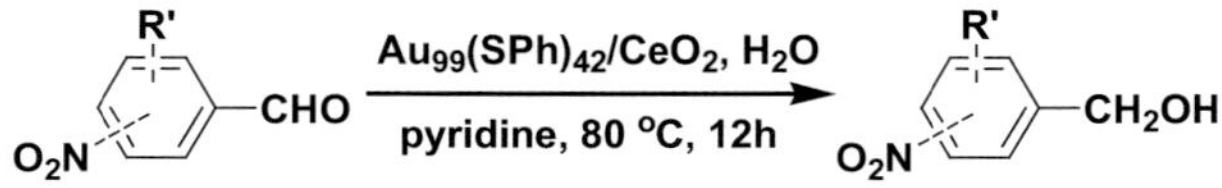

Entry	Substrate	Product	Conv. (%)
1	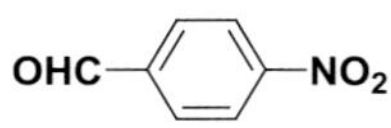	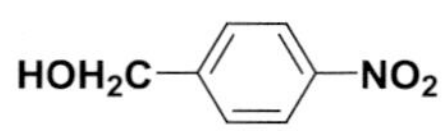	93.1
2	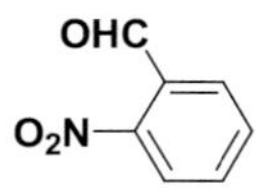	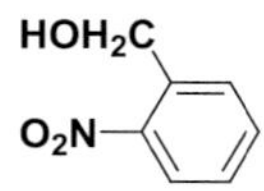	73.2
3	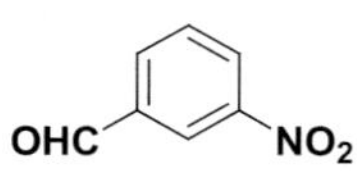	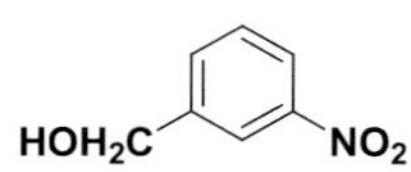	98.5
4	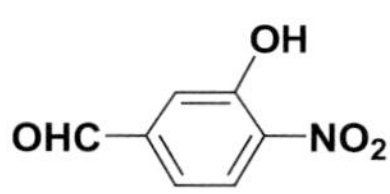	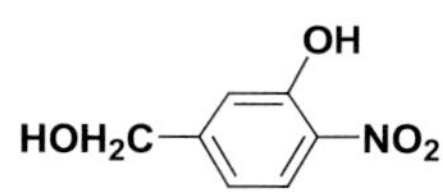	84.1
5	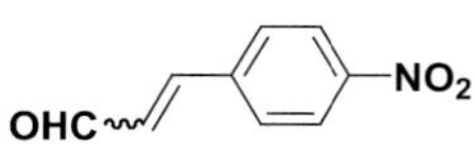	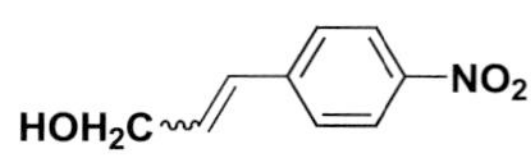	90.0
6	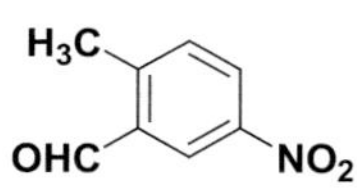	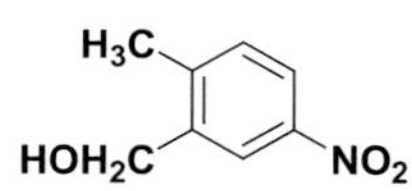	98.9
7	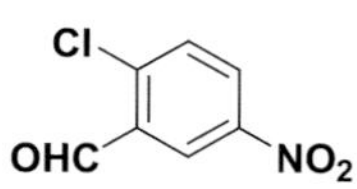	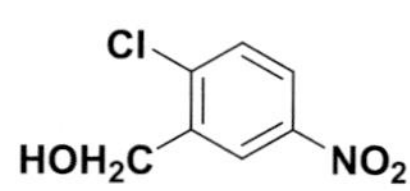	99.5

Source: Adapted with permission from Ref. [38], Copyright 2015, American Chemical Society.

Table 8.5 Influence of different Lewis acids ($M^{z+}X_z$) on the selective hydrogenation of 4-nitrobenzaldehyde catalyzed by $Au_{25}(SR)_{18}/CeO_2$

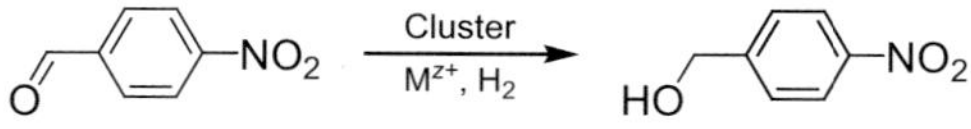

Table 8.5 (*continued*)

Cluster	Lewis acids	Conversion	Selectivity to alcohol
$Au_{25}(SR)_{18}$	None	13.4%	100%
$Au_{25}(SR)_{18}$	$Co(OAc)_2$	90.1%	100%
$Au_{25}(SR)_{18}$	$Cr(NO_3)_3$	75.8%	100%
$Au_{25}(SR)_{18}$	$NiCl_2$	78.2%	100%
$Au_{25}(SR)_{18}$	$Cu(OAc)_1$	61.1%	100%
$Au_{25}(SR)_{18}$	$Cu(OAc)_2$	50.0%	100%
$Au_{25}(SR)_{18}$	$Cu(NO_3)_2$	50.8%	100%
$Au_{25}(SR)_{18}$	$CoCl_2$	90.8%	100%
None	$Co(OAc)_2$	n.r.	–

n.r. = no reaction

8.3.2.2 Semihydrogenation

Next, spherical $Au_{25}(SC_2H_4Ph)_{18}$ (noted as Au_{25} sphere) and rod-shaped $Au_{25}(PPh_3)_{10}(C{\equiv}CPh)_5Cl_2$ clusters (noted as Au_{25} rod) also showed good catalytic activity in the semihydrogenation of alkynes [48, 49]. The semihydrogenations catalyzed over the "ligand-on" Au_{25} were examined in a combination of solvents, bases, and temperatures. The ethanol/H_2O as solvent, pyridine as base, and reaction temperature of 100 °C were chosen as the optimized conditions for the catalytic semihydrogenation. Next, the cluster catalysts were examined in various terminal alkynes and it was found that they gave rise to high conversion (up to >99%) and ~100% selectivity for alkenes (Table 8.5). Notably, the "ligand-on" Au_{25} catalysts were silent in the case of internal alkynes (<1% conversion). However, over the "ligand-off" Au_{25} catalysts (Au_{25} clusters were thermal treated at 300 °C), the internal alkynes can undergo semihydrogenation to give *Z*-alkenes (Table 8.6), which is similar to the conventional gold nanoparticle catalysts. Based on these catalytic results, a unique activation pathway of terminal alkynes by "ligand-on" gold nanoclusters was identified, which should follow a deprotonation activation pathway via a $R'{-}C{\equiv}C{-}[Au_nL_m]$, in contrast with the activation mechanism on the conventional gold nanocatalysts. The activation pathway was strongly supported by the observation of $R'{-}C{\equiv}C{-}[Au_n(SR)_m]$ via FTIR and by observing the deprotonated $-C{\equiv}CPh$ as the protecting ligand on rod-shaped $Au_{25}(PPh_3)_{10}(C{\equiv}CPh)_5X_2$ nanoclusters [49].

Table 8.6 Comparison of catalytic performance in semihydrogenation of the alkynes over the "ligand-on" and "ligand-off" cluster catalysts

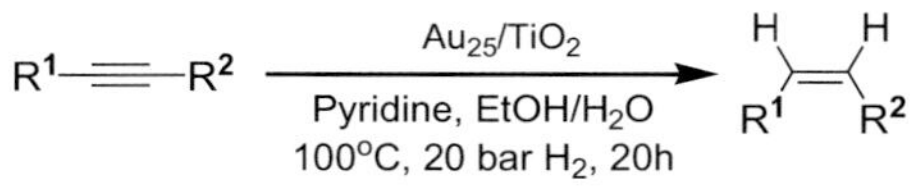

Catalyst	Alkyne	Conversion (%)	Selectivity (%)
"Ligand-on" Au_{25} sphere	PhC_2H_4–C≡CH	>99	–
	Ph–C≡C–CH_3	<1	–
	Ph–C≡C–Ph	<1	–
	C_6H_{13}–C≡C–CO_2CH_3	<1	–
"Ligand-off" Au_{25}	PhC_2H_4–C≡CH	95.6	–
	Ph–C≡C–CH_3	52.8	97
	Ph–C≡C–Ph	59.7	>99
	C_6H_{13}–C≡C–CO_2CH_3	52.6	99
"Ligand-on" Au_{25} rod	PhC_2H_4–C≡CH	99.8	–
	Ph–C≡C–CH_3	<1	–
	Ph–C≡C–Ph	<1	–
	C_6H_{13}–C≡C–CO_2CH_3	<1	–

Finally, the tentative mechanism of semihydrogenation was proposed: a deprotonation activation of terminal alkyne and bridging adsorption of R–C≡C on the catalytic site should be similar in "ligand-on" Au_{25} clusters (Fig. 8.6). The reactants (i.e., hydrogen and alkyne) should be absorbed on the triangular Au_3 site of $Au_{25}(SR)_{18}$ and the "waist" site of $Au_{25}(PPh_3)_{10}(C{\equiv}CPh)_5X_2$. Then both H– bond and –C≡CH triple bond would be activated in the presence of a base (e.g., pyridine). Finally, the alkene products should be formed after the addition process with the aid of gold atoms of active sites.

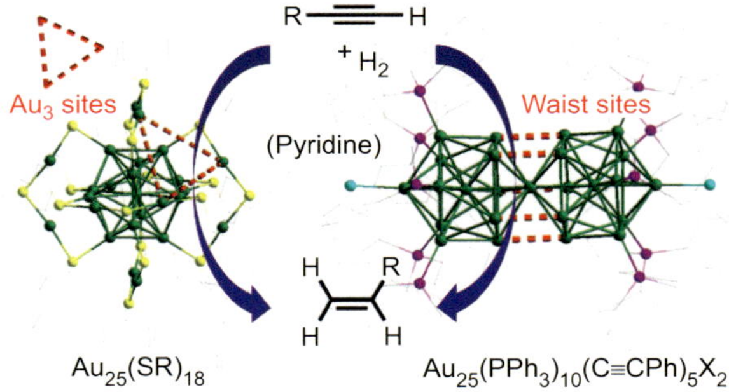

Figure 8.6 Proposed mechanism for the "ligand-on" Au_{25} cluster-catalyzed semihydrogenation of terminal alkynes to alkenes using H_2 as the reducing agent. Left panel: $Au_{25}(SR)_{18}$ nanoclusters. Right panel: $Au_{25}(PPh_3)_{10}(C{\equiv}CPh)_5X_2$ (X = Cl/Br) nanoclusters. Color codes: Au, green; S, yellow; C, grey; P, pink; X, cyan. The areas marked in orange are the Au_3 active sites (left panel) and the waist active sites (right panel) in the alkyne semihydrogenation reactions. Reprinted with permission from Ref. [49], Copyright 2014, American Chemical Society.

8.3.3 One-Pot Cascade Coupling

The above discussions showed that ultrasmall clusters exhibited high activity in the chemical transformation when partial ligands were removed. Of note, the small particles will grow to big ones when all the thiolate/phosphine ligands are detached during the annealing process (e.g., 300 °C). Thus, the direct synthesis of ultrasmall clusters, supported on oxides, may largely improve the catalytic performance [50–52]. Recently, the preparation of Au/oxide catalysts via a wet synthetic process of Au:PVA colloids (PVA: polyvinyl alcohol) and oxides was achieved. The PVA ligands could be easier to be removed at relative temperatures (e.g., 200 °C in air) and the gold clusters would not be aggregated.

Fang et al. [51] synthesized gold clusters of 2–3 nm (using NiO as support) and applied it in the one-pot conversion of furfural with cellulosic ethanol to generate C_5–C_7 hydrocarbon, which is a high-energy density fuel. The one-pot reactions were performed at 130 °C under 1 MPa air, stirring at 600 rpm for 6 h using ethanol, K_2CO_3, and a variety of aldehyde derivatives. First time furyl chemicals via the one-pot cross-aldol condensation was obtained, which was mainly due to the natural basic property of NiO supports (evidenced by the CO_2-TPD). Further, it was found that the substitution of hydrogen at the position 5 of furfural with a methyl or a nitro group exerts a considerable influence on the aldehyde conversion (Table 8.7).

Table 8.7 Au/NiO catalysts for aldehydes conversion

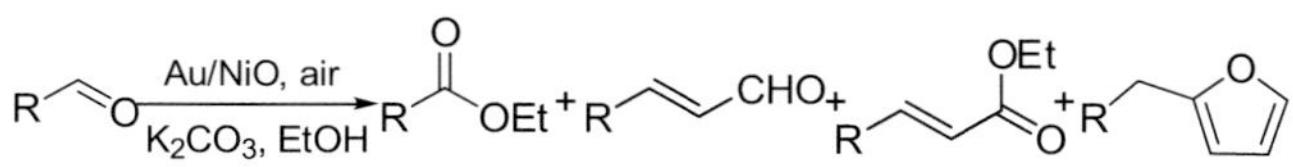

A B C D

Entry	Aldehyde	Conversion (%)	Selectivity (%)			
			A	B	C	D
1	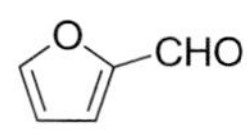	96	7	55	4	34
2	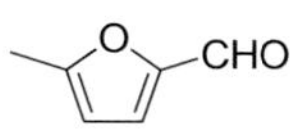	73	9	74	1	16
3	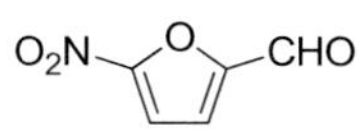	54	50	46	2	n.d.
4	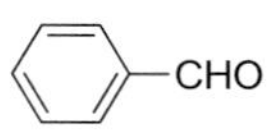	65	55	28	6	11
5	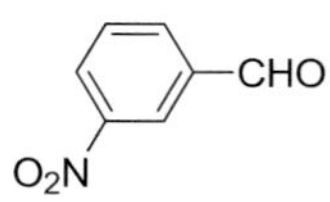	93	54	29	17	n.d.

Reaction conditions: 10 mg Au/NiO catalysts; 0.42 mmol aldehydes; 10 mg K_2CO_3; 3 mL ethanol; 130 °C for 6 h under 1 MP of air; n.d. = not detected.

Finally, products of B and D can be fully (i.e., 100%) converted to C_nH_{2n+2} or C_nH_{2n-6} hydrocarbons when the atmosphere of the reaction changes to H_2 (1 MPa) after the completion of the hydrogenolysis process (Scheme 8.3). This hydrogenolysis should occur on the synergistic effect of gold particles and Ni^0 species on the surface of NiO supports.

Further, the one-pot cascade coupling was investigated in the ethanol conversion with cinnamaldehyde and it finally gave rise to the C_{11}–C_{13} hydrocarbon (liquid biofuels) [52]. A 70% selectivity for C_{11}–C_{13} hydrocarbon is achieved over Au/NiO. Chemo-adsorption and in situ infrared spectroscopy investigations showed that the cascade reactions should process a CH_3CHO^* intermediate, which reacts with cinnamaldehyde at the interfacial perimeter of Au/NiO composite to give the final products (Fig. 8.7).

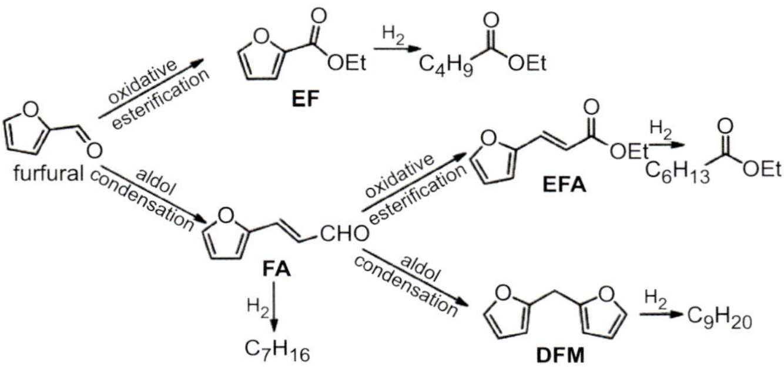

Scheme 8.3 The conversion pathway of furfural to C_7 and C_9 hydrocarbons via the oxidative esterification and cross-aldol condensation with ethanol catalyzed over Au/NiO catalysts in the presence of O_2 and K_2CO_3, followed by a hydrogenolysis process using H_2. Reprinted with permission from Ref. [51], Copyright 2019, American Chemical Society.

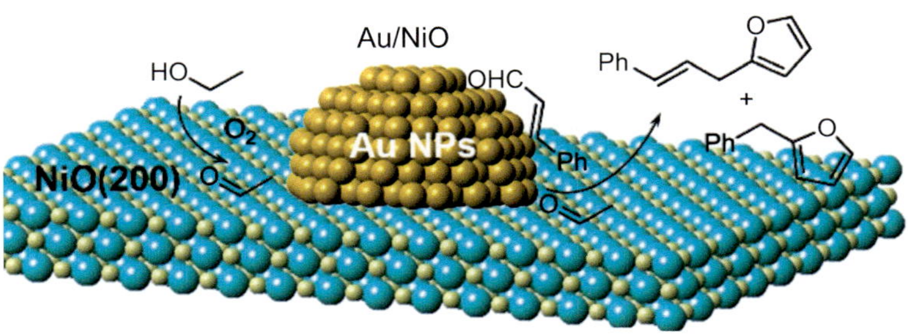

Figure 8.7 The catalytic pathway of the cascade aldol condensation and Michael addition of ethanol and cinnamaldehyde in the presence of oxygen. Color codes: Au, orange; O, light yellow; Ni, cyan. Reprinted with permission from Ref. [52], Copyright 2019, Royal Society of Chemistry.

8.4 Conclusions

Over the decades, the outstanding applications of gold clusters have been well explored and developed in the homogeneous and heterogeneous catalytic processes [53], leading to a novel perspective for nanocatalysis, especially in CO oxidation to CO_2, selective sulfoxidation to sulfoxide, hydrogenation of aldehydes to alcohol, semihydrogenation of alkynes into alkenes, aerobic oxidation of amines into imines, and one-pot conversion of furfural with ethanol to produce biofuel (e.g., C_9–C_{13}). These well-defined gold clusters provided new opportunities for achieving fundamental understanding of nanocatalysis at the atomic level, such as the insight into the size dependence and deep understanding of the molecular activation, catalytic active sites, and catalytic mechanism by correlation with the structures of gold nanoclusters. Moreover, these cluster catalysts are further expected to bridge the gap between model catalysts and realistic catalytic applications, although much work remains to be carried out in the future.

Acknowledgments

The work is financially supported by Liaoning Revitalization Talents Program (XLYC1807121) and the Program of Higher-Level Talents of Inner Mongolia Agricultural University (NDYB2016-03).

References

1. Corma, A. and Garcia, H. (2008). Supported gold nanoparticles as catalysts for organic reactions, *Chem. Soc. Rev.*, **37**, pp. 2096–2126.
2. Li, G. and Jin, R. (2013). Atomically precise gold nanoclusters as new model catalysts, *Acc. Chem. Res.*, **46**, pp. 1749–1758.
3. Yamazoe, S., Koyasu, K. and Tsukuda, T. (2014). Nonscalable oxidation catalysis of gold clusters, *Acc. Chem. Res.*, **47**, pp. 816–824.
4. Haruta, M. (2002). Catalysis of gold nanoparticles deposited on metal oxides, *Cattech*, **6**, pp. 102–115.
5. Li, Z., Abroshan, H., Liu, C. and Li, G. (2017). A critical review on the catalytic applications of non-metallic gold nanoclusters: Selective oxidation, hydrogenation, and coupling reactions, *Curr. Org. Chem.*, **21**, pp. 476–488.
6. Haruta, M., Kobayashi, T., Sano, H. and Yamada, N. (1987). Novel gold catalysts for the oxidation of carbon-monoxide at a temperature far below 0-degrees-C, *Chem. Lett.*, **2**, pp. 405–408.
7. Hammer, B. and Norskov, J. K. (1995). Why gold is the noblest of all the metals, *Nature*, **376**, pp. 238–240.

8. Zhang, G., Wang, R. and Li, G. (2018). Non-metallic gold nanoclusters for oxygen activation and aerobic oxidation, *Chin. Chem. Lett.*, **29**, pp. 687–693.
9. Jin, R., Zeng, C., Zhou, M. and Chen, Y. (2016). Atomically precise colloidal metal nanoclusters and nanoparticles: Fundamentals and opportunities, *Chem. Rev.*, **116**, pp. 10346–10413.
10. Liu, C., Li, T., Li, G., Nobusada, K., Zeng, C., Pang, G., Rosi, N. and Jin, R. (2015). Observation of body-centered cubic gold nanocluster, *Angew. Chem. Int. Ed.*, **54**, pp. 9826–9829.
11. Jin, R. and Nobusada, K. (2014). Doping and alloying in atomically precise gold nanoparticles, *Nano Res.*, **7**, pp. 285–300.
12. Zhang, C., Chen, Y., Wang, H., Li, Z., Zheng, K., Li, S. and Li, G. (2018). Transition-metal-mediated catalytic properties of CeO_2-supported gold clusters in aerobic alcohol oxidation, *Nano Res.*, **11**, pp. 2139–2148.
13. Li, Z., Yang, X., Liu, C., Wang, J. and Li, G. (2016). Effects of doping in 25-atom bimetallic nanocluster catalysts for carbon–carbon coupling reaction of iodoanisole and phenylacetylene, *Prog. Nat. Sci. Mater. Int.*, **26**, pp. 477–482.
14. Liu, C., Lin, J., Shi, Y. and Li, G. (2015). Efficient synthesis of $Au_{99}(SPh)_{42}$ nanoclusters, *Nanoscale*, **7**, pp. 5987–5990.
15. Chen, Y., Wang, J., Liu, C., Li, Z. and Li, G. (2016). Kinetically controlled synthesis of $Au_{102}(SPh)_{44}$ nanoclusters and catalytic application, *Nanoscale*, **8**, pp. 10059–10065.
16. Li, G., Abroshan, H., Liu, C., Zhuo, S., Li, Z., Xie, Y., Kim, H., Rosi, N. and Jin, R. (2016). Tailoring the electronic and catalytic properties of Au_{25} nanoclusters via ligand engineering, *ACS Nano*, **10**, pp. 7998–8005.
17. Zhou, M., Higaki, T., Hu, G., Sfeir, M. Y., Chen, Y. and Jiang, D. (2019). Three-orders-of-magnitude variation of carrier lifetimes with crystal phase of gold nanoclusters, *Science*, **364**, pp. 279–280.
18. Yan, C., Liu, C., Abroshan, H., Li, Z., Qiu, R. and Li, G. (2016). Surface modification of adamantane-terminated gold nanoclusters using cyclodextrins, *Phy. Chem. Chem. Phy.*, **18**, pp. 23358–23364.
19. Qin, Z., Zhao, D., Zhao, L., Xiao, Q., Wu, T., Zhang, J., Wan, C. and Li, G. (2019). Tailoring the stability, photocatalysis and photoluminescence properties of Au_{11} nanocluster via doping engineering, *Nanoscale Adv.*, **1**, pp. 2529–2536.

20. Zheng, K., Zhang, J., Zhao, D., Yang, Y., Li, Z. and Li, G. (2019). Motif mediated $Au_{25}(SPh)_5(PPh_3)_{10}X_2$ nanorod of conjugated electron delocalization, *Nano Res.*, **12**, pp. 501–507.

21. Yan, M., Jin, T., Ishikawa, Y., Minato, T., Fujita, T., Chen, L., Bao, M., Asao, N., Chen, M. and Yamamoto, Y. (2012). Nanoporous gold catalyst for highly selective semihydrogenation of alkynes: Remarkable effect of amine additives, *J. Am. Chem. Soc.*, **134**, pp. 17536–17542.

22. Ren, D., He, L., Yu, L., Ding, R., Liu, Y., Cao, Y., He, H. and Fan, K. (2012). An unusual chemoselective hydrogenation of quinoline compounds using supported gold catalysts, *J. Am. Chem. Soc.*, **134**, pp. 17592–17598.

23. Li, G., Zeng, C. and Jin, R. (2014). Thermally robust $Au_{99}(SPh)_{42}$ nanoclusters for chemoselective hydrogenation of nitrobenzaldehyde derivatives in water, *J. Am. Chem. Soc.*, **136**, pp. 3673–3679.

24. Li, G., Jiang, D., Kumar, S., Chen, Y. and Jin, R. (2014). Size dependence of atomically precise gold nanoclusters in chemoselective hydrogenation and active site structure, *ACS Catal.*, **4**, pp. 2463–2469.

25. Li, G., Zeng, C. and Jin, R. (2015). Chemoselective hydrogenation of nitrobenzaldehyde to nitrobenzyl alcohol with unsupported Au nanorod catalysts in water, *J. Phys. Chem. C*, **119**, pp. 11143–11147.

26. Kawasaki, H., Kumar, S., Li, G., Zeng, C., Kauffman, D., Yoshimoto, J., Iwasaki, Y. and Jin, R. (2014). Generation of singlet oxygen by photoexcited $Au_{25}(SR)_{18}$ clusters, *Chem. Mater.*, **26**, pp. 2777–2788.

27. Chen, Y., Zhang, C., Yang, C., Zhang, J., Zheng, K., Fang, Q. and Li, G. (2017). Waugh type $[CoMo_9O_{32}]^{6-}$ cluster with atomically dispersed Co^{3+} derives from Anderson type $[CoMo_6O_{24}]^{3-}$ anion for photocatalytic oxygen molecule activation, *Nanoscale*, **9**, pp. 15332–15339.

28. Li, Z., Liu, C., Abroshan, H., Kauffman, D. and Li, G. (2017). $Au_{38}S_2(SAdm)_{20}$ photocatalyst for one-step selective aerobic oxidations, *ACS Catal.*, **7**, pp. 3368–3374.

29. Zhang, J., Zhou, Y., Zheng, K., Abroshan, H., Kauffman, D., Sun, J. and Li, G. (2018). Diphosphine-induced chiral propeller arrangement of gold nanoclusters for singlet oxygen photogeneration, *Nano Res.*, **11**, pp. 5787–5798.

30. DeRosa, M. and Crutchley, R. (2002). Photosensitized singlet oxygen and its applications, *Coord. Chem. Rev.*, **233**, pp. 351–371.

31. Liu, C., Yan, C., Lin, J., Yu, C., Huang, J. and Li, G. (2015). One-pot synthesis of $Au_{144}(SCH_2Ph)_{60}$ nanoclusters and catalytic application, *J. Mater. Chem. A*, **3**, pp. 20167–20173.

32. Li, G., Qian, H. and Jin, R. (2012). Gold nanocluster-catalyzed selective oxidation of sulfide to sulfoxide, *Nanoscale*, **4**, pp. 6714–6717.

33. Wang, C. and Astruc, D. (2014). Nanogold plasmonic photocatalysis for organic synthesis and clean energy conversion, *Chem. Soc. Rev.*, **43**, pp. 7188–7216.

34. Naya, S., Kimura, K. and Tada, H. (2013). One-step selective aerobic oxidation of amines to imines by gold nanoparticle-loaded rutile titanium(IV) oxide plasmon photocatalyst, *ACS Catal.*, **3**, pp. 10–13.

35. Abroshan, H., Li, G., Lin, J., Kim, H. and Jin, R. (2016). Molecular mechanism on the activation of $Au_{25}(SCH_2CH_2Ph)_{18}$ nanoclusters by imidazolium-based ionic liquids for catalysis, *J. Catal.*, **337**, pp. 72–79.

36. Liu, C., Abroshan, H., Yan, C., Li, G. and Haruta, M. (2016). One-pot synthesis of $Au_{11}(PPh_2Py)_7Br_3$ for the highly chemoselective hydrogenation of nitrobenzadehyde, *ACS Catal.*, **6**, pp. 92–99.

37. Lin, J., Li, W., Liu, C., Huang, P., Zhu, M., Ge, Q. and Li, G. (2015). One-phase controlled synthesis of Au_{25} nanospheres and nanorods from 1.3 nm Au:PPh_3 nanoparticles: The ligand effects, *Nanoscale*, **7**, pp. 13663–13670.

38. Li, G., Abroshan, H., Chen, Y., Jin, R. and Kim, H. J. (2015). Experimental and mechanistic understanding of aldehyde hydrogenation using Au_{25} nanoclusters with Lewis acids: Unique sites for catalytic reactions, *J. Am. Chem. Soc.*, **137**, pp. 14295–14304.

39. Nie, X., Qian, H., Ge, Q., Xu, H. and Jin, R. (2012). CO oxidation catalyzed by oxide-supported $Au_{25}(SR)_{18}$ nanoclusters and identification of perimeter sites as active centers, *ACS Nano*, **6**, pp. 6014–6022.

40. Wu, Z., Jiang, D., Mann, A. K. P., Mullins, D. R., Qiao, Z., Allard, L. F., Zeng, C., Jin, R. and Overbury, S. H. (2014). Thiolate ligands as a double-edged sword for CO oxidation on CeO_2 supported $Au_{25}(SCH_2CH_2Ph)_{18}$ nanoclusters, *J. Am. Chem. Soc.*, **136**, pp. 6111–6122.

41. Li, W., Liu, C., Abroshan, H., Ge, Q., Yang, X., Xu, H. and Li, G. (2016). Catalytic CO oxidation using bimetallic M_xAu_{25-x} clusters: A combined experimental and computational study on doping effects, *J. Phys. Chem. C*, **120**, pp. 10261–10267.

42. Li, G. and Jin, R. (2016). Atomic level tuning of the catalytic properties: Doping effects of 25-atom bimetallic nanoclusters on styrene oxidation, *Catal. Today*, **278**, pp. 187–191.

43. Li, Z., Li, W., Abroshan, H., Ge, Q., Zhou, Y., Zhang, C., Li, G. and Jin, R. (2018). Dual effects of water vapor over ceria-supported gold clusters, *Nanoscale*, **10**, pp. 6558–6565.

44. Chen, H., Liu, C., Wang, M., Zhang, C., Luo, N., Wang, Y., Abroshan, H., Li, G. and Wang, F. (2017). Visible light gold nanocluster photocatalyst: Selective aerobic oxidation of amines to imines, *ACS Catal.*, **7**, pp. 3632–3638.

45. Chen, H., Liu, C., Wang, M., Zhang, C., Li, G. and Wang, F. (2016). Thermally robust silica-enclosed Au_{25} nanocluster and its catalysis, *Chin. J. Catal.*, **37**, pp. 1787–1793.

46. Zhang, Y., Yang, X., Zhou, Y., Li, G., Li, Z., Liu, C., Bao, M. and Shen, W. (2016). Selective hydrogenation of C=C bond in α,β-unsaturated aldehydes and ketones over Pd-Au clusters of 1.8 nm, *Nanoscale*, **8**, pp. 18626–18629.

47. Wen, Z., Li, Z., Ge, Q., Zhou, Y., Sun, J. and Li, G. (2018). Robust nickel cluster@Mes-HZSM-5 composite nanostructure with enhanced catalytic activity in the DTG reaction, *J. Catal.*, **363**, pp. 26–33.

48. Chen, H., Li, Z., Qin, Z., Kim, H., Abroshan, H. and Li, G. (2019). Silica-encapsulated gold nanoclusters for efficient acetylene hydrogenation to ethylene, *ACS Appl. Nano Mater.*, **2**, pp. 2999–3006.

49. Li, G. and Jin, R. (2014). Gold nanocluster-catalyzed semihydrogenation: A unique activation pathway of terminal alkynes, *J. Am. Chem. Soc.*, **136**, pp. 11347–11354.

50. Shao, B., Zhao, W., Shi, Q., Miao, S., Huang, J., Wang, L., Li, G. and Shen, W. (2019). Facet-dependent of TiO_2 nanocrystallites for gold nanoparticle in CO oxidation, *Chin. J. Catal.*, **40**, pp. 1534–1539.

51. Fang, Q., Qin, Z., Shi, Y., Liu, F., Barkaoui, S., Abroshan, H. and Li, G. (2019). Au/NiO composite: A catalyst for one-pot cascade conversion of furfural, *ACS Appl. Energy Mater.*, **2**, pp. 2654–2661.

52. Shi, Y., Tian, S., Shi, Q., Waheed, A., Cao, Y. and Li, G. (2019). Cascade aldol condensation of aldehyde via the aerobic oxidation of ethanol over Au/NiO composite, *Nanoscale Adv.*, **1**(9), pp. 3654–3659.

53. Panigrahi, S., Basu, S. and Praharaj, S. (2007). Synthesis and size-selective catalysis by supported gold nanoparticles: Study on heterogeneous and homogeneous catalytic process, *J. Phys. Chem. C*, **111**, pp. 4596–4605.

Chapter 9

Catalytic Application of Atomically Precise Metal Nanoclusters as Heterogeneous Catalysts in Industrially Important Chemical Reactions

Yongnan Sun and Yan Zhu
School of Chemistry and Chemical Engineering, Nanjing University, Nanjing 210093, China
zhuyan@nju.edu.cn

Metal catalysts are one type of the most important catalysts for the majority of current chemical processes. The emergence of atomically precise metal nanoclusters has brought site-specific catalysis of the nanocatalysts into bloom, which has constituted one of the most exciting topics. On the basis of their atomic structures, these nanoclusters allow for identification of the nature of catalysts. These nanoclusters have exhibited unexpected catalytic properties for a series of industrially important catalytic reactions.

Atomically Precise Nanoclusters
Edited by Yan Zhu and Rongchao Jin

ISBN 978-981-4877-08-4 (Hardcover), 978-1-003-11990-6 (eBook)
www.jennystanford.com

9.1 Introduction

Metal nanoparticles as heterogeneous catalysts are widely applied in industrial chemical reactions, such as the conversion of fossil, the conversion of biomass, fuel generation, environmental catalysis and electrocatalysis, and so on. However, the conventional metal catalysts have some major issues—in particular, the inherent size polydispersity and precise size control at the atomic level. These major limitations preclude fundamental investigations of the precise structure–catalytic activity relationships. Due to the size dispersity of nanoparticles, it is not possible to achieve an in-depth understanding of the origin of the size dependence of nanocatalysts; moreover, it is impossible to identify catalytically active species in nanoparticle catalysis. Almost all of the current studies give rise to only an ensemble average of the catalytic performance due to the structural polydispersity and heterogeneity of the conventional nanocatalysts. Although significant efforts have been invested in preparing well-defined nanoparticles (e.g., model catalysts), fundamental nanocatalysis research still significantly lags behind. Further, it is impossible to explore new properties in the catalytic field.

Therefore, it is of paramount importance to synthesize atomically precise metal nanoparticles and use such nanoparticles as well-defined catalysts. By solving the atomic structure of nanoparticles, one will be able to precisely correlate the catalytic properties with the exact atomic structure of nanoparticles and learn what control the surface activation, surface active site structure, and catalytic mechanism. By learning these fundamental principles, one will ultimately be able to design new types of highly active and highly selective nanocatalysts for a variety of catalytic processes.

Atomically monodisperse metal nanoclusters are ideally composed of an exact number of metal atoms, for example, ranging from a dozen to hundreds (equivalent diameter from subnanometer to about 3 nm) [1]. These nanoclusters are unique and vastly different from their larger counterparts—metal nanocrystals (typically 3–100 nm). Small metal nanoclusters (e.g., $n < 100$) behave like molecules and exhibit strong quantum confinement effects; relatively larger ones ($100 < n < 1000$) exhibit intermediate properties between molecular behavior and metallic properties.

Overall, the nonmetallic behavior of metal nanoclusters in both size regimes is particularly important for nanocatalysis. More important, on the basis of their atom packing structures and unique electronic properties, these metal nanoclusters will permit a precise correlation of the particle structure with catalytic properties and also allow for identification of catalytically active sites on the metal nanoparticle. Solving these fundamental questions will not only provide excited opportunities for unraveling catalysis at an atomic level, but also promote the exploration of new or challenging chemical processes. The atomically precise metal nanocluster catalysts will ultimately open an avenue for a new type of metal catalysts with high efficiency for industrially important chemical reactions.

9.2 Catalysis of Surface Active Sites

9.2.1 Selective Oxidation

Selective oxidation reactions, such as methane oxidation, olefin oxidation, cyclohexane oxidation, and VOCs (volatile organic compounds) oxidation, are important for petroleum industry. Turner and co-workers reported very small gold entities derived from 55-atom gold clusters supported on inert SiO_2 that were efficient and robust catalysts for the selective oxidation of styrene by dioxygen toward benzaldehyde, styrene epoxide, and acetophenone (Fig. 9.1) [2]. A sharp size threshold in catalytic activity was found; that is, ~2 nm is the boundary between activity and non-activity. Ultrasmall gold nanoclusters have high catalytic activity due to their altered electronic structures that can dissociatively chemisorb oxygen and induce selective oxidation of styrene. This study stimulates investigations into catalysis of atomically precise metal nanoclusters. Zhu and co-workers further studied the size effect catalysis using Au_{25}, Au_{38}, and Au_{144} nanoclusters as catalysts for styrene oxidation [3]. As listed in Table 9.1, the catalytic activity of three Au_n nanocluster catalysts showed a strong dependence on size: the smaller Au_n nanoclusters had higher activity. The activity was presented in a descending sequence: $Au_{25} > Au_{38} > Au_{144}$. Meanwhile, it was found that the catalytic properties of Au_n nanoclusters are mainly determined by metal rather than by ligands, which is in contrast with organometallic catalysts that show a large dependence on the ligands bonded to metal centers. The Au_n nanoclusters also exhibited good stability during the oxidation reaction, which is of particular importance in order to put it to real-world use of such catalysts.

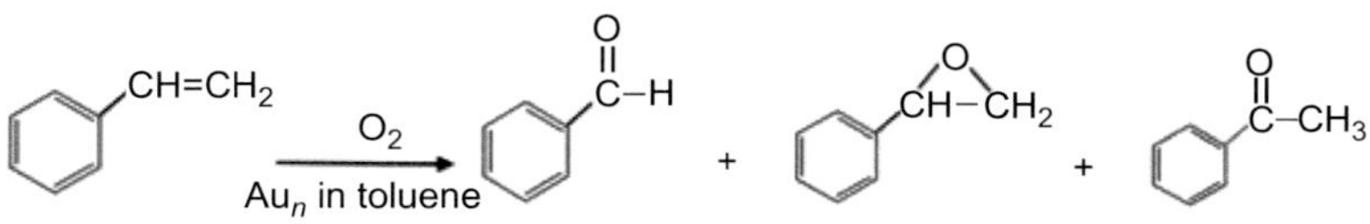

Figure 9.1 The styrene oxidation catalyzed by Au_n nanoclusters. Reprinted with permission from Ref. [3], Copyright 2010, John Wiley & Sons, Inc.

Table 9.1 Catalytic performance of thiolate capped $Au_n(SR)_m$ cluster catalysts for selective oxidation of styrene with O_2

Catalysts	Conversion [%]	Selectivity [%]		
		1	2	3
1. $Au_{25}(SCH_2CH_2Ph)_{18}$	27 ± 1.0	70	24	6
2. $Au_{25}(SC_6H_{13})_{18}$	25 ± 0.8	69	26	5
3. $Au_{38}(SCH_2CH_2Ph)_{24}$	14 ± 1.2	72	24	4
4. $Au_{38}(SC_{12}H_{25})_{24}$	15 ± 1.2	71	25	4
5. $Au_{144}(SCH_2CH_2Ph)_{60}$	12 ± 0.5	80	20	trace
6. $Au_{144}(SC_{12}C_{25})_{60}$	11 ± 0.8	84	16	trace
7. ~3 nm Au NC[a]	6 ± 1.2	82	18	trace

[a]These particles are stablized by octanethiolates.

Note: Products 1–3 are benzaldehyde, styrene epoxide, and acetophenone, respectively.

Source: Reprinted with permission from Ref. [3], Copyright 2010, John Wiley & Sons, Inc.

The reaction mechanism of styrene oxidation over Au_n nanoclusters was explored in terms of identification of active oxygen species [4]. Tiron (disodium 4,5-dihydroxy-1,3-benzenedisulfonate) was used as a superoxide trapping agent, DMPO (5,5′-dimethylpyrroline N-oxide) was employed to detect hydroxyl radicals, and TEMP (2,2,6,6-tetramethylpiperidine) trapped the single oxygen. As shown in Fig. 9.2, the electron paramagnetic resonance (EPR) studies showed that singlet oxygen was detected in the styrene oxidation over the Au_{10} catalyst when TEMP was added to the reaction system. When the singlet oxygen inhibitor azide was added in the reaction, the conversion of styrene dropped remarkably. The observation unambiguously confirmed that the single oxygen was the active oxygen species in the styrene oxidation over the Au_{10} catalyst.

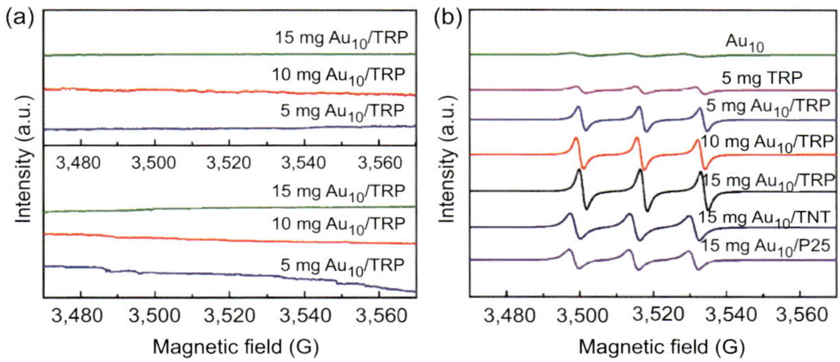

Figure 9.2 (a) EPR profiles of the Au_{10} catalysts using DMPO and Tiron as trapping agents. (b) EPR profiles of the Au_{10} catalysts with TEMP as the trapping agent. Reprinted with permission from Ref. [4], Copyright 2016, Springer Nature.

Another important oxidation reaction is the alcohol oxidation to produce aldehydes or acids. Tsukuda and co-workers found that more amounts of residual thiolates on Au_{25} gave rise to higher selectivity for benzaldehyde from selective oxidation of benzyl alcohol, but lower activity [5]. The thiolate ligands not only reduced the oxidation ability of Au_{25}, but also inhibited the esterification reaction on the surface of the Au_{25} nanocluster by site isolation (Fig. 9.3). They also found that single Pd atom doping into Au_{33} and Au_{43} formed $PdAu_{33}$ and $PdAu_{43}$ that improved the catalytic activity for selective oxidation of benzyl alcohol [6]. Especially, Pd_1Au_{24} supported on carbon nanotubes showed distinct catalytic activity for the oxidation of benzyl alcohol, mainly because the electron transfer from Pd to Au led to more negative Au sites [7].

CO oxidation is a key reaction in automotive emission abatement and the reaction can be catalyzed by novel metal nanoparticles such as Au, Pt, Pd supported on oxides. Atomically precise Au_n nanoclusters supported on CeO_2 are capable of catalyzing CO oxidation under wild conditions. Jin's group reported interface engineering of gold nanoclusters for CO oxidation catalysis and performed two comparison experiments (Fig. 9.4) [8]. The first one ruled out

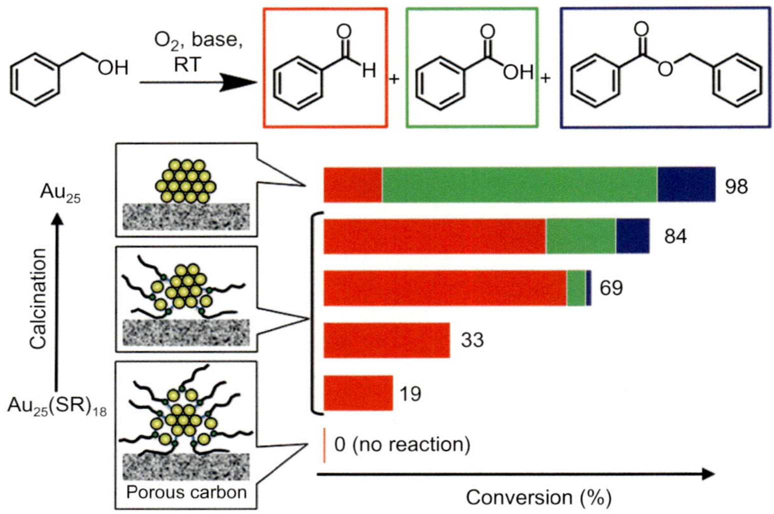

Figure 9.3 The selective oxidation of benzyl alcohol over the Au_{25} catalysts pretreated at different temperatures to remove ligands. Reprinted with permission from Ref. [5], Copyright 2014, American Chemical Society.

the commonly thought barrier that the steric hindrance of ligands at the interface between the thiolate, gold, and support inhibits CO adsorption onto gold sites. The second one supported the view that the crystal facets of Au_n nanoclusters play an important role in the oxidation of CO, and the icosahedral structure of Au_{38} showed better activity than the face-centered cubic (fcc) structure of Au_{36}. Ligand-protected Au_n nanoclusters have little activity due to the metal surface passivated by organic ligands and no sites available on the surface of Au_n for adsorption and conversion of reactants. Wu and coworkers studied the effect of thiolate ligands on Au_n nanoclusters on CO oxidation [9]. They found that the passivated $Au_{25}(SR)_{18}$ showed no activity for CO oxidation and the thiolate ligands were partially removed to chemically adsorb CO. It suggests that in the heterogeneous reactions, the ligands are rigidly bound to the surface of Au_n nanoclusters, thus preventing reactants from adsorbing and further reaction.

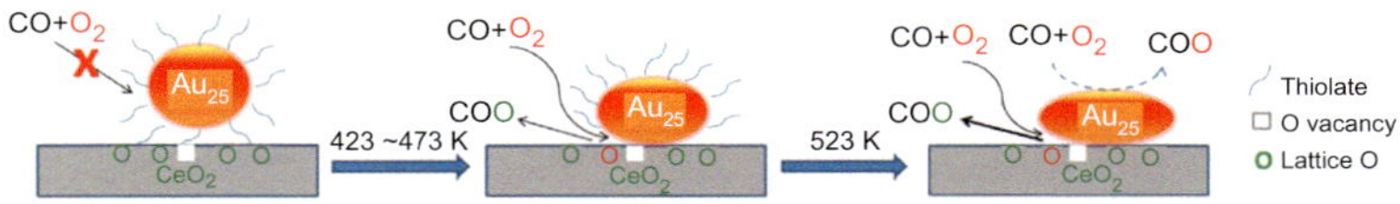

Figure 9.4 CO oxidation mechanism on intact, partially, and fully dethiolated $Au_{25}(SR)_{18}/CeO_2$. Reprinted with permission from Ref. [9], Copyright 2014, American Chemical Society.

Li and Ge reported a discovery that the presence of water can improve the activity of CO oxidation from nearly no conversion of CO in the absence of water to 96% in the presence of 2.4% water due to the removal of surface ligands from gold nanoclusters under water vapor atmosphere [10]. They found that the peroxide species were the active oxygen species in CO oxidation and indicated that the perimeter sites of the interface of Au_{25}/CeO_2 catalyst were likely to be the active sites for catalytic CO oxidation. Li et al. also compared the activity of bimetal nanoclusters such as Ag_xAu_{25-x}, Cu_xAu_{25-x}, and Au_{25} for CO oxidation and observed a trend in the activity: Cu_xAu_{25-x} > Au_{25} > Ag_xAu_{25-x} (Fig. 9.5). The preferable adsorption of CO over the three nanoclusters was in the order: $Cu_xAu_{25-x} > Au_{25} > Ag_xAu_{25-x}$, which accounted for the activity trend.

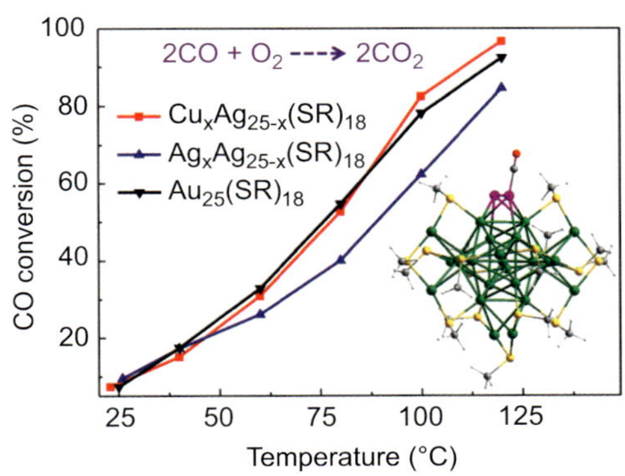

Figure 9.5 Catalytic oxidation of CO over the Cu_xAu_{25-x}, Au_{25}, and Ag_xAu_{25-x} nanoclusters. Reprinted with permission from Ref. [10], Copyright 2016, American Chemical Society.

Selective oxidation of cyclohexane can produce cyclohexanol and cyclohexanone that are important intermediates in industry to make nylon-related materials. Zhu and coworkers reported structure-specific catalytic oxidation of cyclohexane by two isomeric $Au_{28}(SR)_{20}$ nanoclusters with the same core structure but different surface atom structures [11]. The $Au_{28}(TBBT)_{20}$ (TBBT = 4-tert-butylthiophenol) nanoclusters showed higher activity for the oxidation of cyclohexane than the $Au_{28}(SC_6H_{11})_{20}$ nanoclusters, which was ascribed to more electron-deficient oxygen species on the surface of $Au_{28}(TBBT)_{20}$ as well as stronger interaction of reagent molecules and $Au_{28}(SC_6H_{11})_{20}$. Tsukuda and co-workers studied the size effect of Au_n nanoclusters supported on hydroxyapatite for oxidation of cyclohexane and observed volcano-like behavior that peaked at Au_{39} [12]. The phenomenon is very interesting and cannot be explained merely by surface area and low-coordination environment, and thus further work is in progress.

9.2.2 Selective Hydrogenation

Selective hydrogenation is one type of most important catalytic reactions in the petroleum industry. Atomically precise metal nanoclusters are also well-defined model catalysts to catalyze a series of important chemical processes. One typical example is the selective hydrogenation of α,β-unsaturated ketones, whose products are valuable intermediates for the production of perfumes and flavors. Zhu and co-workers reported an interesting work that Au_{25} is capable of transforming α,β-unsaturated ketones with H_2 to α,β-unsaturated alcohols with 100% selectivity, that is, no side products such as unsaturated alcohols and unsaturated ketones (Fig. 9.6) [13].

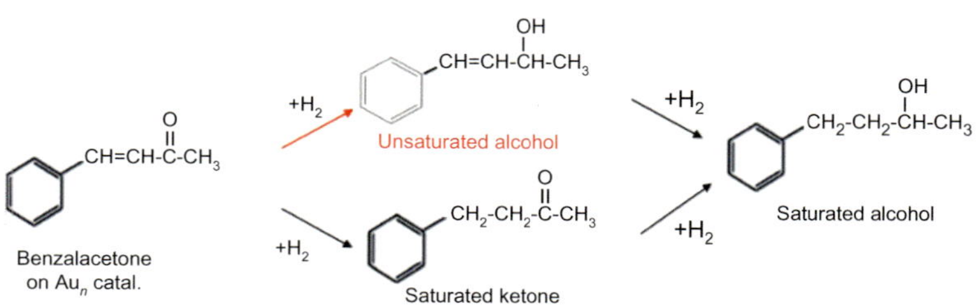

Figure 9.6 The reaction processes of benzalacetone hydrogenation.

Zhu et al. reported the first example of stereoselective hydrogenation of bicyclic ketone using $Au_{25}(SR)_{18}$ nanoclusters [14]. The electron-rich Au_{13} core activates the C=O group, and the

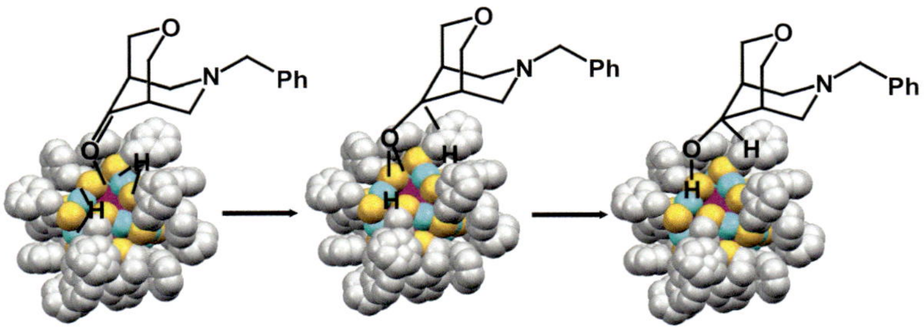

Figure 9.7 The proposed mechanism of stereoselective hydrogenation of bicyclic ketone on the $Au_{25}(SR)_{18}$. Reprinted with permission from Ref. [14], Copyright 2010, Elsevier.

electron-deficient Au_{12} shell provides sites for H_2 adsorption and dissociation. The stereoselective hydrogenation is due to the spatial restriction imposed by $Au_{25}(SR)_{18}$ nanocluster and the activated geometry of ketones. The configuration of bicyclic ketone adsorbed on the $Au_{25}(SR)_{18}$ is backbone rigidity, limiting the conformational degrees of freedom, and the bicyclic ketone tumbling is restricted in the atomic structure of $Au_{25}(SR)_{18}$, and hence the product is toward one of the stereoisomers (Fig. 9.7).

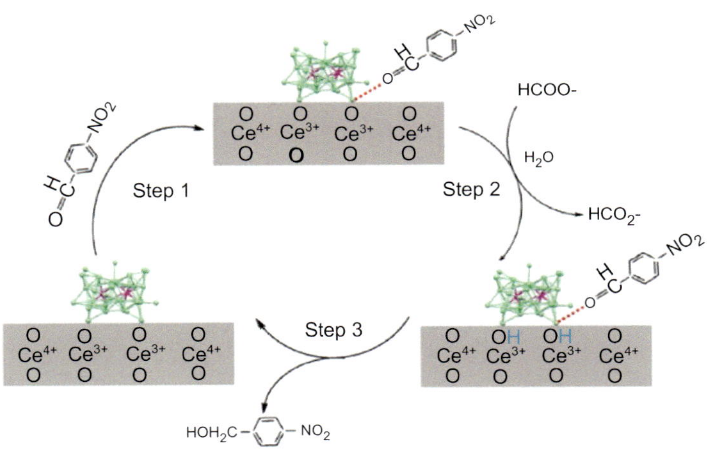

Figure 9.8 Mechanism of transfer hydrogenation of nitrobenzaldehyde on $Au_{38}(SR)_{24}/CeO_2$. Reprinted with permission from Ref. [15], Copyright 2018, American Chemical Society.

Zhao and co-workers exploited a transfer hydrogenation strategy to study the selective hydrogenation of aldehyde groups using Au_{25}, Au_{38}, Au_{52}, and Au_{144} nanoclusters as catalysts [15]. A distinct size-dependence catalysis was observed and among the four catalysts, $Au_{38}(SR)_{24}$ nanoclusters gave the highest catalytic activity. Dethiolate nanoclusters can give a contrast selectivity toward nitro hydrogenation compared to thiolate-protected nanoclusters. Interestingly, $Au_{38}(SR)_{24}$ nanoclusters were found to have the reversible chemoselectivity to this transfer hydrogenation reaction, and its selectivity can be restored by treating the ligand-off nanoclusters with thiolate (Fig. 9.8).

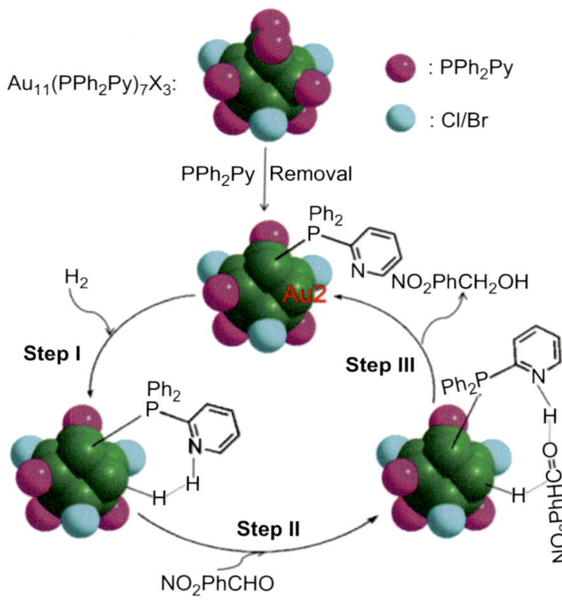

Figure 9.9 Mechanism of nitrobenzaldehyde hydrogenation on $Au_{11}(PPh_2Py)_7Br_3$ in the absence of base. Reprinted with permission from Ref. [17], Copyright 2016, American Chemical Society.

Li and co-workers studied the catalytic activity of $Au_{25}(SR)_{18}$ nanoclusters for the hydrogenation of aldehydes with the introduction of a base and transition metal ions as Lewis acid [16]. The addition of Lewis acid enhanced the activity of $Au_{25}(SR)_{18}$ nanoclusters. The study found new species like $Au_{25-x}(SR)_{18-x}$ ($x = 1–4$) in the reaction. The corresponding calculations suggested that the –Au–SR– could be removed from $Au_{25}(SR)_{18}$ to produce $Au_{24}(SR)_{17}$ with the help of Lewis acids and ammonia, which in turn exposed the Au_{13} core as the active site to reactants and then hydrogenation was carried out. Li and co-workers also studied chemoselective hydrogenation of nitrobenzaldehyde catalyzed by $Au_{11}(PPh_2Py)_7Br_3$ (PPh_2Py = diphenyl-2-pyridylphosphine) nanoclusters [17]. The $Au_{11}(PPh_3)Cl_3$ nanoclusters are active in the chemoselective hydrogenation of nitrobenzaldehyde in the presence of a base like pyridine, whereas the $Au_{11}(PPh_2Py)_7Br_3$ nanoclusters also can achieve 100% chemoselectivity to nitrobenzyl alcohol in the absence of the base. The removal of one ligand from the Au_{11} nanocluster led to the exposure of the gold core and provided the active site for the reaction. The H–H bond of H_2 can be activated with the aid of the free PPh_2Py ligand (Fig. 9.9).

Selective conversion of acetylene to ethylene is a major step in processing chemical feedstock in industry. Li and co-workers synthesized a robust catalyst of gold nanoclusters encapsulated in microporous silica, which exhibited 98% selectivity toward ethylene with 99.5% conversion efficiency in the selective hydrogenation of acetylene [18]. The small size effect and positively charged surface gold induced by the electronegative species in the silica pores are the main reasons to improve the activation of acetylene. In addition, it has been found that the doping of Cd in $Au_{23}(SR)_{20}$ can improve the selectivity for the semihydrogenation of acetylene in an ethylene-rich stream, suggesting that the Cd alloying with gold nanoclusters is a useful way to enhance the activity of gold nanocluster catalysts in heterogeneous reactions [19].

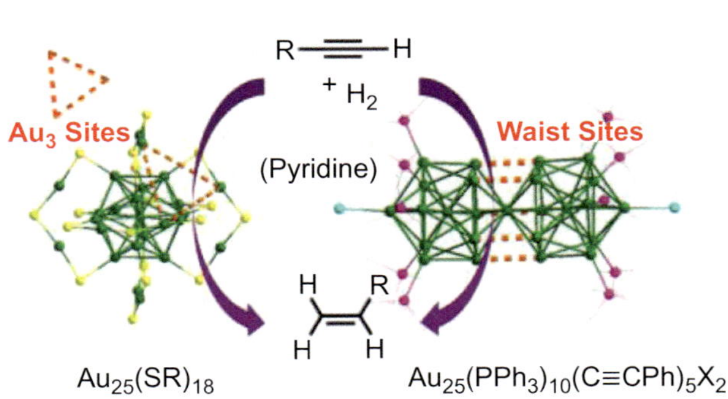

Figure 9.10 Mechanism of semihydrogenation of terminal alkynes over $Au_{25}(SR)_{18}$ and $Au_{25}(PPh_3)_{10}(C{\equiv}CPh)_5X_2$ nanoclusters. Reprinted with permission from Ref. [20], Copyright 2014, American Chemical Society.

The semihydrogenation of terminal alkynes into alkenes can be also catalyzed by atomically precise Au_n nanoclusters. It is observed that both $Au_{25}(SR)_{18}$ and $Au_{25}(PPh_3)_{10}(C{\equiv}CPh)_5X_2$ nanoclusters can achieve 100% selectivity to the semihydrogenation product with 100% conversion (Fig. 9.10). The open triangular Au_3 facet on the $Au_{25}(SR)_{18}$ and the waist sites of the $Au_{25}(PPh_3)_{10}(C{\equiv}CPh)_5X_2$ are thought to be active sites in the reaction [20].

Wang and co-workers prepared two isostructural $[Au_{38}(L)_{20}(Ph_3P)_4]_2^+$ (L = alkynyl or thiolate) with fcc-type Au_{34} core [21]. The only difference in the two Au_{38} nanoclusters is the L ligands. The two Au_{38} nanoclusters act as catalysts to catalyze the semihydrogenation of alkynes, where the alkynyl-protected Au_{38} were very active with 97% conversion, including terminal and internal alkynes toward alkenes, while the thiolated Au_{38} nanoclusters exhibited a very low conversion, that is, <2%. The results not only show that Au_{38} protected by alkynyl can give rise to high semihydrogenation of alkynes, but also demonstrate that the ligands should not be ignored and may play a vital role in some catalytic reactions. This study shows a novel example of understanding the ligand effect on the catalytic performance of gold nanocluster catalysts.

Yan and co-workers explored the support effect on the catalytic properties of Au_{25} nanoclusters as precatalysts [22]. They found that there was no significant size growth for Au_{25} supported on HAP (hydroxyapatite) and P_{25} after thermal treatment to remove ligands, owing to the strong interaction between the gold and the supports. In contrast, dramatic aggregations of Au_{25} on AC (activated carbon), PGO (pyrolyzed graphene oxide), and SiO_2 supports were observed. The surface thiolate ligands deactivated the nitrobenzene hydrogenation and the removal of ligands from Au_{25} enabled the hydrogenation of nitrobenzene. Yan and Xie et al. also found that protective ligands can modulate the catalytic active sites of metal nanoclusters in solution and hence can induce the reaction pathway toward desirable products [23]. They found that the thiolate ligands, that is, *p*-mercaptobenzoic acid, in Au_{25} nanoclusters can direct the catalytic hydrogenation of 4-nitrophenol in solution to form azobenzene intermediates. These results indicate that protective ligands impact the activation of H_2 and hydrogenation pathway (Fig. 9.11).

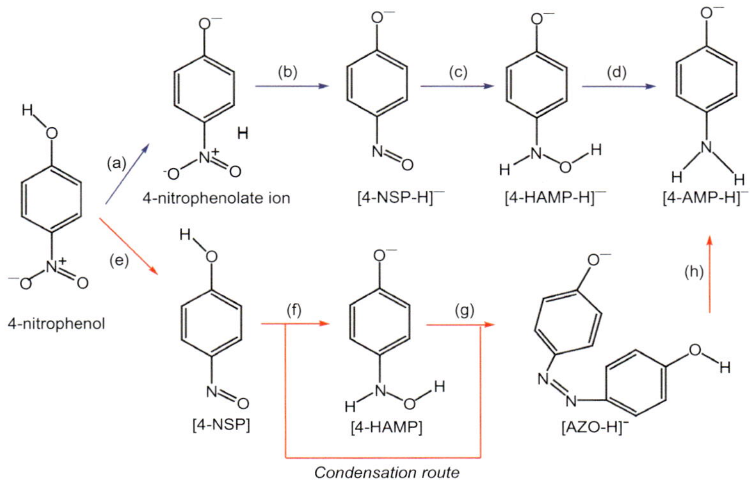

Figure 9.11 The reaction pathways for the hydrogenation of 4-nitrophenol on Au_{25}(*p*-MBA)$_{18}$ in solution. Reprinted with permission from Ref. [23], Copyright 2018, John Wiley & Sons, Inc.

Tsukuda et al. reported the catalytic dehydrogenation of ammonia borane using $[Ag_{44}(SC_6H_4F)_{30}]^{4-}$ nanoclusters as catalysts [24]. The $[Ag_{44}(SC_6H_4F)_{30}]^{4-}$ had different mechanism of dehydrogenation of NH_3BH_3 from other nanoparticles: the NH_3BH_3 and O_2 reaction to produce H_2 on $[Ag_{44}(SC_6H_4F)_{30}]^{4-}$ and the NH_3BH_3 and H_2O reaction to produce H_2 on other nanoparticles.

Zhu and co-workers focused on the ligand-shielding effect, and of particular interest are studies that involve triggering the catalytic ability of metal nanoclusters without the need to remove any ligand via thermal treatments that destroy their precise structures [25]. They synthesized six-atom Ni protected by 12 thiolates, where $Ni_6(SR)_{12}$ had no activity for hydrogenation of nitriles. Interestingly, when NH_3 was added in the reaction system, there was >99% yield toward primary amines in the hydrogenation of nitriles on $Ni_6(SR)_{12}$ nanoclusters (Table 9.2). Their studies indicated that the NH_3 is crucial to remain the open NiS_3 site of the activated $Ni_6(SR)_{12}$ by preventing the deactivation of $Ni_6(SR)_{12}$ caused by the recombination reaction. NH_3 reduces the rate-limiting bridge-bond cleavage steps in catalytic cycles and suppresses the thiolate-shielding effects.

Table 9.2 Catalytic hydrogenation of 4-cyanotetrahydropyran on different catalysts with/without NH_3

Entry	Catalyst	Additive	Conversion (%)	PA Selectivity (%)
1	$Ni_6(SC_2H_4Ph)_{12}$	NH_3	100	>99
2	$Ni_6(SC_2H_4Ph)_{12}$[a]	No	9	>99
3	$Ni_6(SC_6H_{11})_{12}$	NH_3	100	>99
4	$Ni_6(SC_6H_{11})_{12}$[a]	No	8	>99
5	$Ni_6(SC_6H_{13})_{12}$	NH_3	100	>99
6	$Ni_6(SC_6H_{13})_{12}$[a]	No	9	>99
7	NiB	NH_3	73	88
8	NiB	No	71	78
9	Ni nanoparticles	NH_3	18	61
10	Ni nanoparticles	No	18	62
11	NiS	NH_3	n.d.	n.d.
12	NiS	No	n.d.	n.d.

Source: Reprinted with permission from Ref. [25], Copyright 2018, Royal Society of Chemistry.

Reaction conditions: 0.07 mmol catalyst, 0.08 mmol nitrile, 20 ml solvent, 10 h, 4.5 °C, 6 atm H_2 pressure.

[a]Reaction time: 20 h

PA = primary amine

Zhu et al. reported three-atom Pd nanoclusters to catalyze the hydrogenation of nitrobenzene and benzaldehyde and the Pd_3 catalyst showed highly efficient selectivity to benzalaniline [26]. The result is obviously in contrast to Pd nanoparticles that gave the completely hydrogenated product benzylaniline. The different adsorption modes on Pd_3 and Pd nanoparticles are thought to be responsible for the different selectivity. Density functional theory (DFT) calculations found that the distinct selectivity in Pd_3 nanoclusters and Pd nanoparticles is due to the presence of ligands in Pd_3 nanoclusters, which can inhibit the complete hydrogenation process (Fig. 9.12). The study aids to explore atomically precise metal nanoclusters for the synthesis of fine chemicals.

Hydrodesulfurization is an industrial catalytic process for sulfur removal from hydrocarbons in crude petroleum to gain ultraclean fuel. Molybdenum sulfides with nickel or cobalt are typical catalysts for industrial hydrodesulfurization, yet the catalysts have high activity but low active site densities. Zhu and co-workers reported $Ni_6(SR)_{12}$ nanoclusters to be used as the catalyst for hydrodesulfurization of thiophene [27]. The $Ni_6(SR)_{12}$ supported on MoS_2 exhibited better catalytic activity for hydrodesulfurization of thiophene than the commercial $NiMoS/Al_2O_3$: $Ni_6(SR)_{12}/MoS_2$ catalyst gave a complete conversion of thiophene at 250 °C, while commercial catalyst gave the complete thiophene conversion at 300 °C (Fig. 9.13). Especially, the on-and-off dynamics of thiophene on $Ni_6(SR)_{12}$ nanoclusters enabled efficient hydrodesulfurization processes while maintaining the nanocluster atomicity. The study highlights the potential for pursuing practical applications of precise transition metal nanoclusters for the chemical industry.

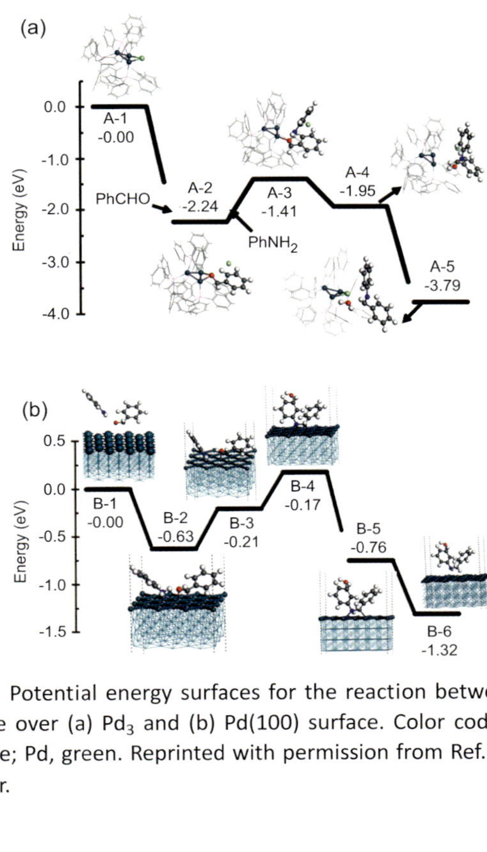

Figure 9.12 Potential energy surfaces for the reaction between aniline and benzaldehyde over (a) Pd_3 and (b) Pd(100) surface. Color codes: C, black; H, white; N, blue; Pd, green. Reprinted with permission from Ref. [26], Copyright 2019, Elsevier.

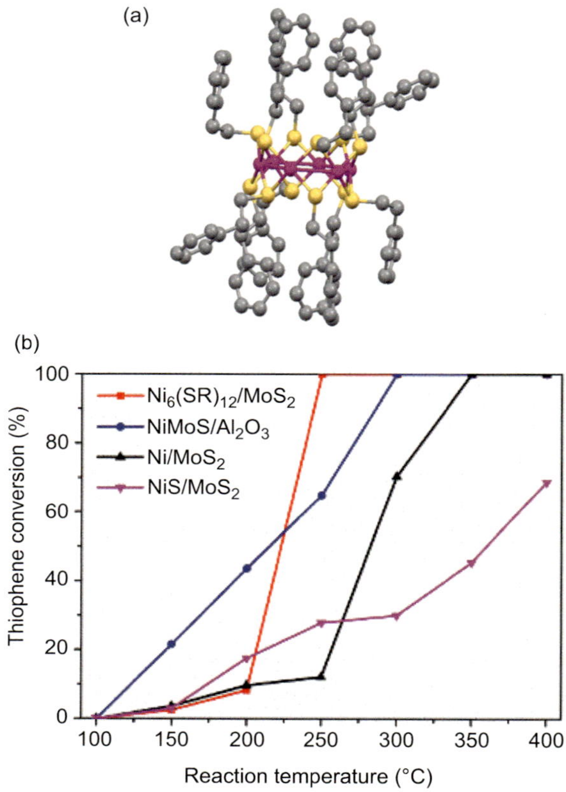

Figure 9.13 (a) Structure of $Ni_6(SR)_{12}$. (b) Catalytic results of different Ni-based catalysts for hydrodesulfurization of thiophene. Reprinted with permission from Ref. [27], Copyright 2019, Royal Society of Chemistry.

Selective hydrogenation of CO_2 into liquid fuels and chemicals is of significant importance for mitigating global warming and energy supply problems. Zhu and co-workers reported an interesting work where the isomers in the $Au_{28}(SR)_{20}$ nanoclusters with the same fcc Au_{20} core but different surface structure can direct the hydrogenation of CO_2 toward different paths [28]. The conversion of CO_2 to methanol as the major product was detected on $Au_{28}(TBBT)_{20}$ nanoclusters, while both methanol and methyl formate (MF) were obtained with a molar ratio of 2:1 on $Au_{28}(SC_6H_{11})_{20}$ nanoclusters. Based on theoretical studies, one mechanism proposed that the formation of MF over $Au_{28}(SC_6H_{11})_{20}$ nanoclusters is attributed to the formation of the Au_3 active site. The Au_3 active site forms only between the kernel and a $Au(SR)_2$ staple of the gold cluster, involving the metallic kernel Au and the ionic staple Au. DFT results reveal that the MF formed on the Au_3 active site is exclusively MF. Approximately, the count of the methanol-favoring sites in the $Au_{28}(SC_6H_{11})_{20}$ nanocluster is about two-thirds of that for $Au_{28}(TBBT)_{20}$; this value is consistent with CH_3OH ratio between $Au_{28}(TBBT)_{20}$ and $Au_{28}(SC_6H_{11})_{20}$ (Fig. 9.14).

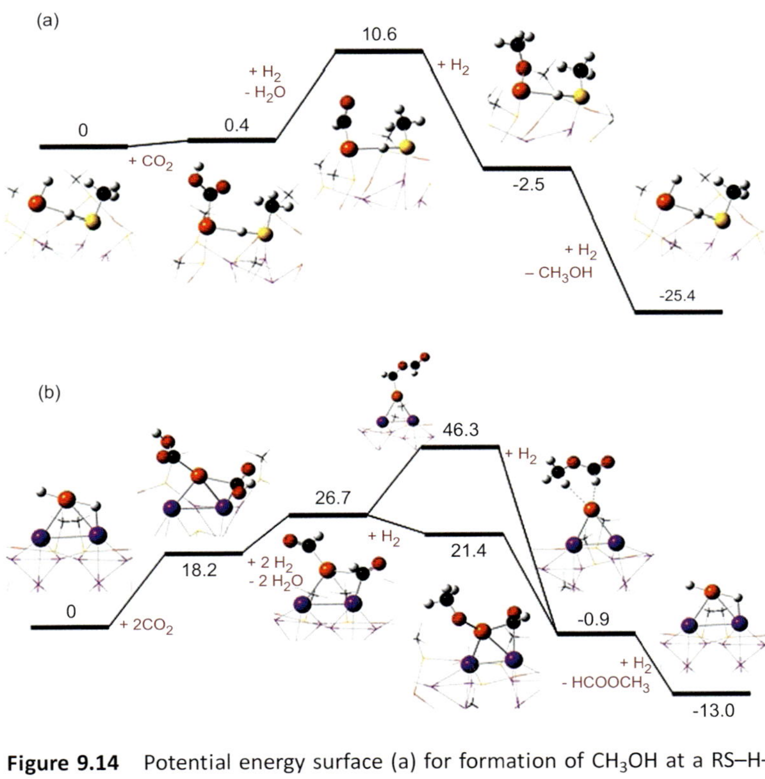

Figure 9.14 Potential energy surface (a) for formation of CH_3OH at a RS–H–Au site of hydrogenated $Au_{28}(TBBT)_{20}$ and (b) for formation of MF at a Au_3H_2 site of hydrogenated $Au_{28}(SC_6H_{11})_{20}$. Reprinted with permission from Ref. [28], Copyright 2019, John Wiley & Sons, Inc.

9.2.3 Other Catalytic Reactions

Besides transformation of CO_2 with H_2 to industrially relevant high-value added chemicals, the electrocatalytic reduction of CO_2 into useful products is another promising way to get renewable energy. On the basis of the strong quantum confinement effect and solved atomic structures of metal nanoclusters, Jin et al. studied the CO_2 electroreduction on metal nanoclusters at an atomic scale [29, 30]. First, they studied the interaction between CO_2 and $Au_{25}(SR)_{18}$ and found a CO_2-induced redistribution of charge within the Au_{25} nanoclusters. $Au_{25}(SR)_{18}$ can promote the CO_2 to CO reaction within 90 mV of the formal potential, and in peak CO production at −1.0 V versus RHE, the rate of CO_2 conversion was 7–700 times higher than that for Au nanoparticles with large size and 10–100 times higher than those for current state-of-the-art processes. This is the first example to show that $Au_{25}(SR)_{18}$ nanoclusters can achieve unexpected electrochemical conversion of CO_2 into CO (Fig. 9.15). The optimized systems can gain high turnover numbers (TON) approaching 6×10^6 mol CO_2/mol catalyst during total 36 h. Daytime photovoltaic-powered CO_2 conversion was tested for 12 h and low light or nighttime for 24 h with a solar rechargeable battery was also studied (Fig. 9.16). This study provided initial performance data for scalability and technical viability of CO_2 electroreduction technologies.

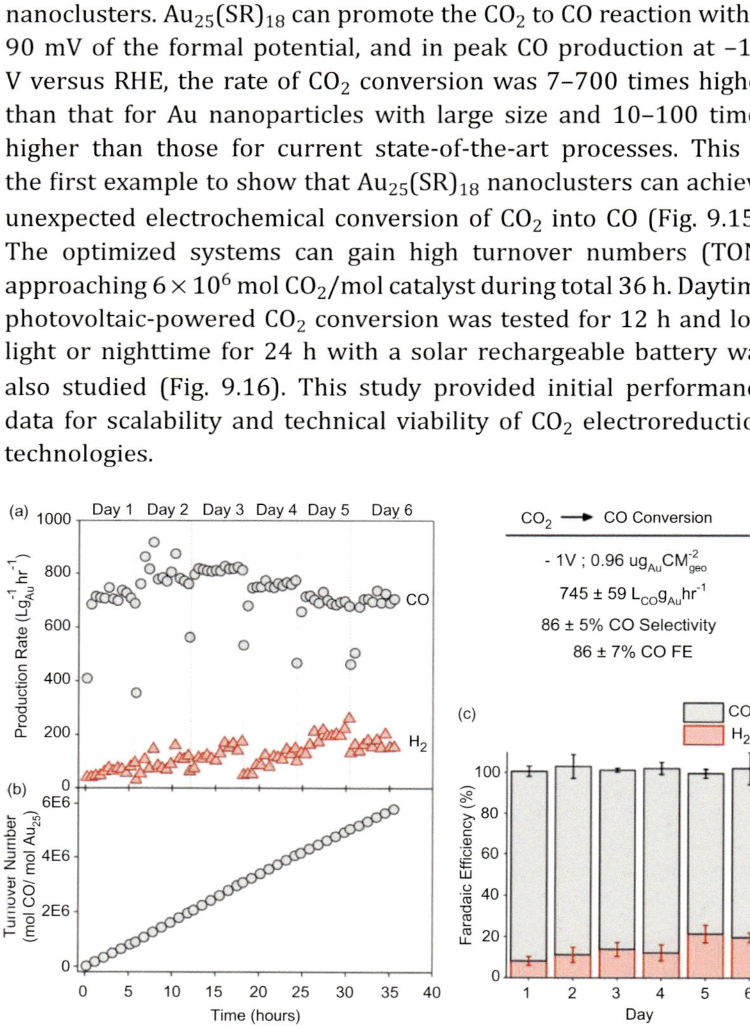

Figure 9.15 (a) Product formation rate. (b) Cumulative turnover number. (c) Faradaic efficiency. Reprinted with permission from Ref. [29], Copyright 2015, American Chemical Society.

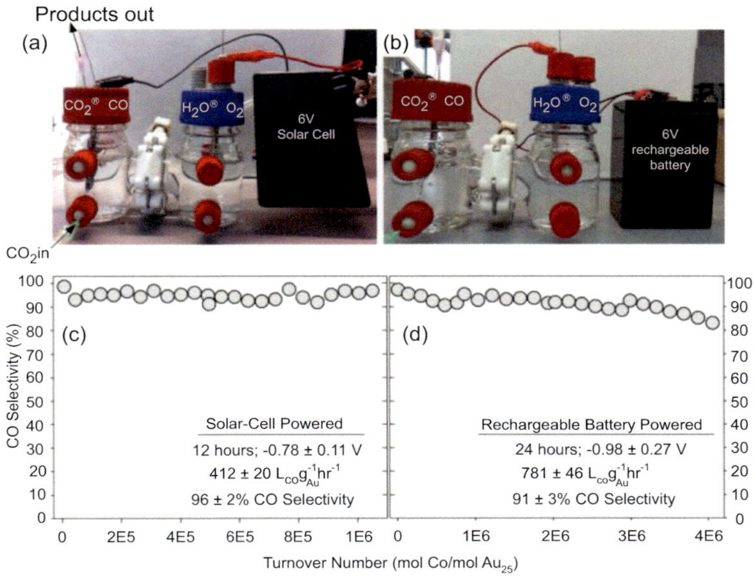

Figure 9.16 (a,b) An electrochemical CO_2 reactor powered by solar panels and a solar rechargeable battery. (c,d) $CO_2 \rightarrow CO$ selectivity as a function of TON. Reprinted with permission from Ref. [29], Copyright 2015, American Chemical Society.

Photocatalysis of atomically precise metal nanoclusters is also promising to enhance catalytic activity with the help of visible light or sunlight. Jin and co-workers dispersed $Au_{25}(SR)_{18}$ nanoclusters on TiO_2 and under visible light the $Au_{25}(SR)_{18}$ can have a small bandgap semiconductor behavior to absorb visible light and enable electron–hole separation as well as produce singlet oxygen, rather than as an efficient electron trap to facilitate the electron–hole separation [31]. It is demonstrated that the potential of $Au_{25}(SR)_{18}$ is more negative than the potential of TiO_2 and it facilitates the transfer of the photogenerated electrons from $Au_{25}(SR)_{18}$ to the conduction band. The efficient separation of electrons and holes and the reduction in recombination of electron–hole lead to the high photocatalysis of $Au_{25}(SR)_{18}$ for several catalytic reactions, including oxidation of sulfide to sulfoxide, degradation of methyl orange, and gentrification coupling of benzylamine to form *N*-benzylidenebenzylamine (Fig. 9.17). Among these catalytic oxidation reactions, the active oxygen species are detected to be single oxygen, which can be efficiently formed through the direct photosensitization by $Au_{25}(SR)_{18}$ nanoclusters [32].

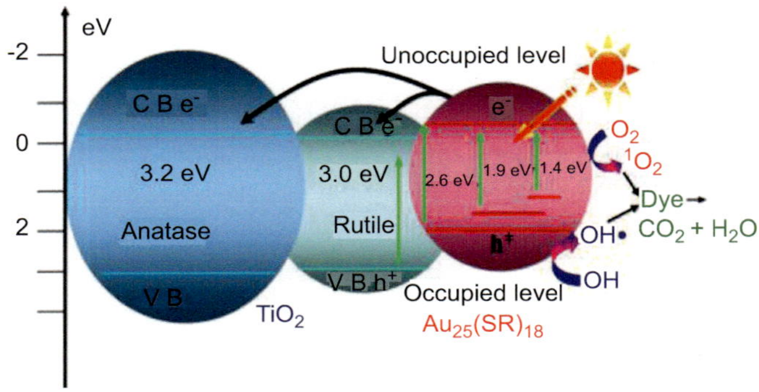

Figure 9.17 The photocatalytic mechanism of $Au_{25}(SR)_{18}/TiO_2$ under visible light irradiation. Reprinted with permission from Ref. [31], Copyright 2013, American Chemical Society.

9.3 Catalysis of Non-surface Active Sites

9.3.1 Central Atom Doped by a Foreign Atom

Jin et al. reported the crystal structure and catalytic properties of Pt_1Au_{24} nanocluster where Pt replaces the central Au atom to form a Pt-centered icosahedral core capped by six $Au_2(SC_2H_4Ph)_3$ staples [33]. The $Pt_1Au_{24}(SR)_{18}$ nanoclusters supported on TiO_2 as catalysts catalyzed the styrene oxidation and exhibited 90.8% conversion of styrene with 89.9% selectivity for benzaldehyde, while $Au_{25}(SR)_{18}$ gave rise to only 58.9% conversion of styrene with 54.0% selectivity to benzaldehyde under identical conditions. The results indicate that the central doping of Pt in the Au_{25} nanocluster can enhance the catalytic activity of origin Au_{25}. They also found that Pd can be doped into the central core of Au_{25} nanoclusters and $PdAu_{24}$ showed higher activity for hydrogenation of α,β-unsaturated ketone compared to Au_{25} (42% conversion vs 40%) [34].

Lee and co-workers reported the catalytic activity of the Pt_1Au_{24} nanocluster (note that Pt is also doped into the central core) for hydrogen production, which is higher than the benchmarking Pt catalyst [35]. Studies showed that H binds preferably on the hollow site of the surface of $PtAu_{12}$ core and the adsorbed H forms Pt–H bond. It was found that the H atom spontaneously moved into the subsurface to interact with Pt. The strong interaction of Pt–H accounts for the higher activity in the hydrogen evolution reaction (HER) (Fig. 9.18).

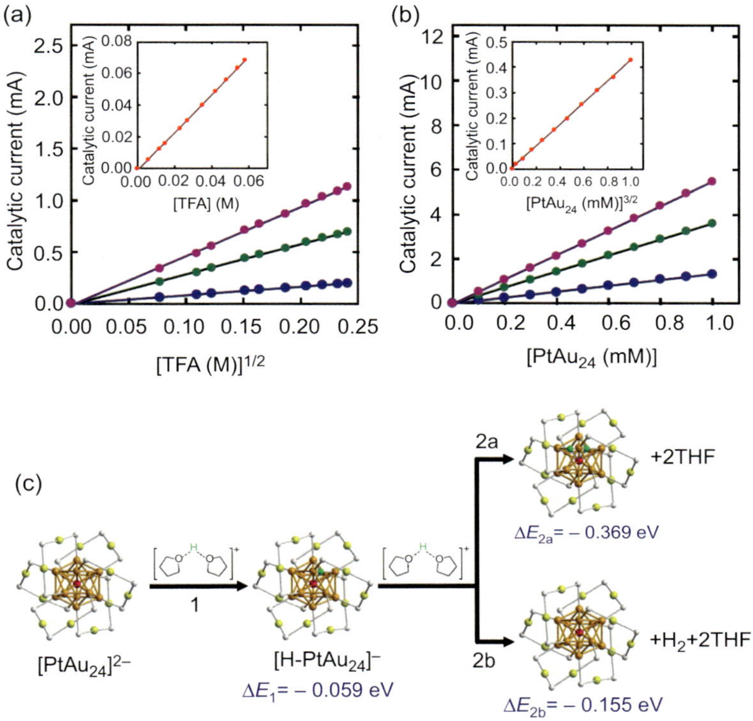

Figure 9.18 Dependence of the catalytic current, *I*c, (a) on the concentration of trifluoroacetic acid in the presence of $PtAu_{24}$ and (b) on the concentration of $PtAu_{24}$ in trifluoroacetic acid solution. (c) Calculated reaction energies for H_2 evolution on $PtAu_{24}$. Reprinted with permission from Ref. [35], Copyright 2017, Nature Publishing Group.

Zhu et al. studied the effects of single Au, Pd, and Pt into the core of Ag_{25} on the catalytic activity for the carboxylation of CO_2 with alkynes [36]. The catalytic activity can be shown in a descending order: Au@Ag_{24} > Pd@Ag_{24} ≈ Pt@Ag_{24} > Ag@Ag_{24}. The catalytic stability is also presented in the same order: Au@Ag_{24} > Pd@Ag_{24} ≈ Pt@Ag_{24} > Ag@Ag_{24} (Fig. 9.19). The Fourier transform infrared spectroscopy (FTIR) of terminal alkyne adsorbed on the four nanoclusters indicated that the carboxylation of CO_2 with terminal alkyne on the four nanoclusters proceeded via the deprotonation-induced activation pathway. IR and XPS analysis revealed that the transfer of electrons happened from external silver atoms to central dopants. The DFT calculations showed that the core is heavily negatively charged in Pd@Ag_{24} and

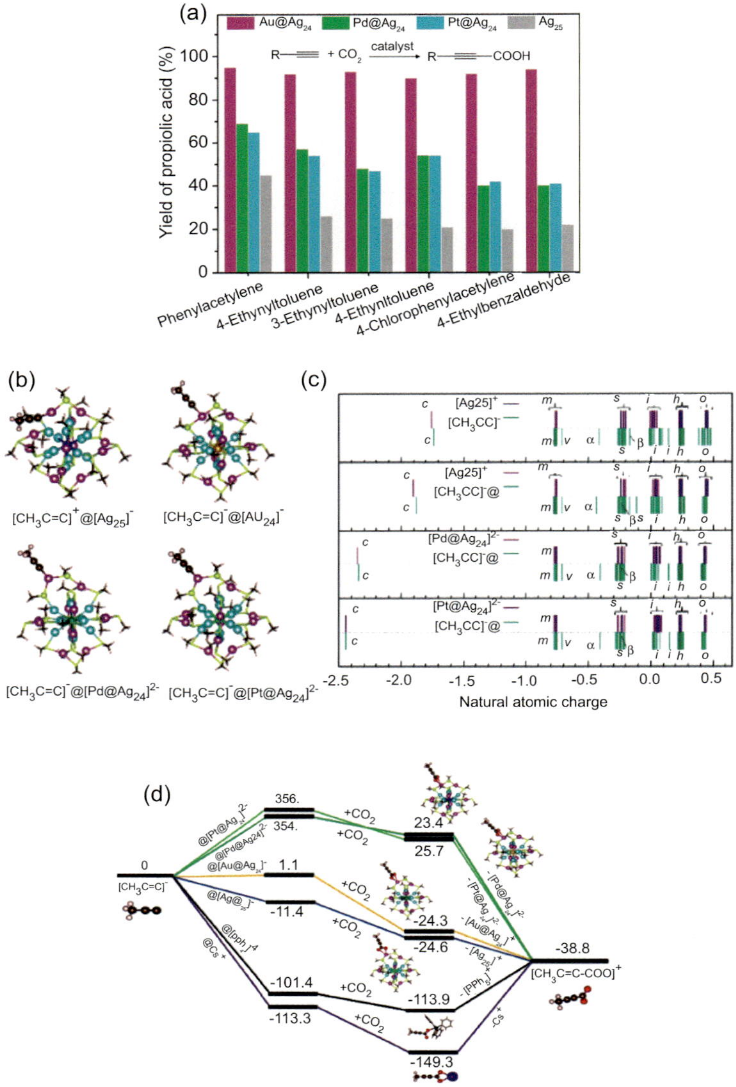

Figure 9.19 (a) The catalytic performances of the Au@Ag_{24}, Pd@Ag_{24}, Pt@Ag_{24}, and Ag@Ag_{24} nanoclusters for the carboxylation of CO_2 with alkynes. (b) DFT-optimized geometries for the adsorption products of $CH_3C\equiv C$ onto Au@Ag_{24}, Pd@Ag_{24}, Pt@Ag_{24}, and Ag@Ag_{24} nanoclusters. (c) Natural atomic charge spectra for the adsorption of $CH_3C\equiv C$ on Au@Ag_{24}, Pd@Ag_{24}, Pt@Ag_{24}, and Ag@Ag_{24} nanoclusters. (d) Potential energy surface for $CH_3C\equiv C$ + CO_2 to $CH_3C\equiv C$–COO on Au@Ag_{24}, Pd@Ag_{24}, Pt@Ag_{24}, and Ag@Ag_{24} nanoclusters. Reprinted with permission from Ref. [36], Copyright 2018, John Wiley & Sons, Inc.

Pt@Ag_{24}, it can repel C and cause the deformation of structure and the core is median negatively charged as found in Au@Ag_{24}, and the electrostatic interactions are reached a balance, leading to high stability and activity. The work indeed supports that the foreign core atoms of nanoclusters contribute to the catalytic performance, even though the atoms are surrounded by external atoms and away from surface sites.

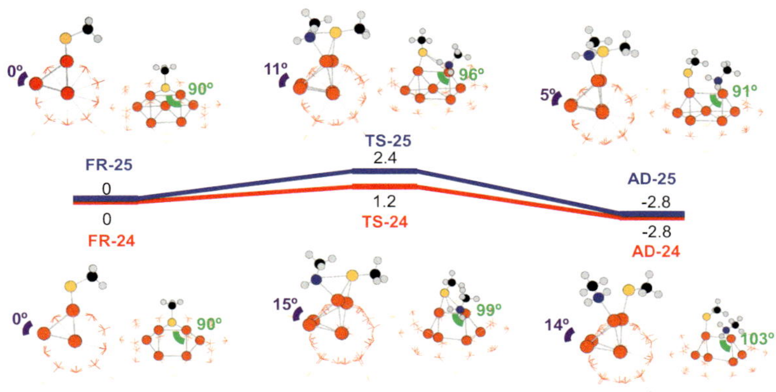

Figure 9.20 Activation of Au_{24} and Au_{25} upon the adsorption of CH_3NH_2 via bridge cleavage. Green and red arrows indicate exothermic and endothermic processes. Color codes: Au, orange; S, yellow; C, black; N, blue; H, grey; Cl, green; P, pink. Reprinted with permission from Ref. [37], Copyright 2019, Royal Society of Chemistry.

9.3.2 Appearance and Disappearance of Central Atom

It is extremely challenging to distinguish the contribution of every atom on different sites in a catalyst to the overall performances. Zhu et al. reported the first case of the loss in the central atom that enhanced the catalytic activity in homogeneous reactions, such as the intramolecular hydroamination of alkynes [37]. The $[Au_{24}(PPh_3)_{10}(SR)_5Cl_2]^+$ and $[Au_{25}(PPh_3)_{10}(SR)_5Cl_2]^{2+}$ nanoclusters were chosen as catalysts, where the only difference in the atomic structure was that Au_{24} had no core atom. At the initial reaction time, the Au_{24} without core atom always showed the higher reaction rate than the Au_{25} with core atom. DFT calculations found that the Au_{25} activation involves a forward twist and a backward twist between two Au_{12}, while for Au_{24} only a forward twist is required. The barrier height difference resulted in the higher reaction rate in the Au_{24} (Fig. 9.20).

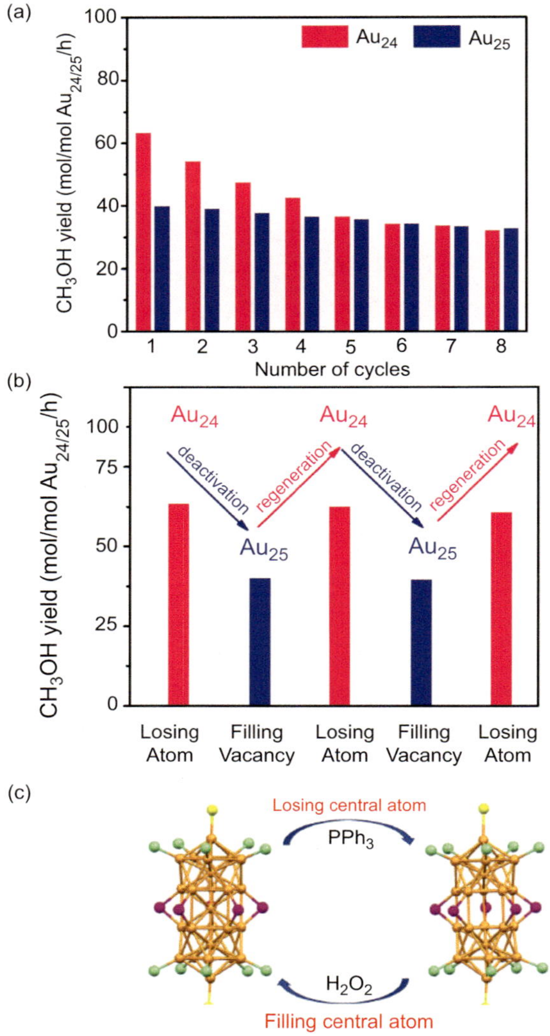

Figure 9.21 (a) Recyclability of Au_{24} and Au_{25} for methane with H_2O_2 reaction. (b) Activity reversible switching of Au_{24} and Au_{25}. (c) Schematic diagram of reversible conversion between Au_{24} and Au_{25}. Reprinted with permission from Ref. [38], Copyright 2019, John Wiley & Sons, Inc.

More interestingly, deactivation and regeneration can be reversibly switched in Au_{24} and Au_{25} nanoclusters. First, the Au_{24} nanoclusters gave rise to higher activity for oxidation of methane with H_2O_2 than the Au_{25} nanoclusters [38]. Under the exitance of H_2O_2, the Au_{24} underwent a structure transformation into Au_{25}, which is attributed to the decrease in activity through the reaction of methane with H_2O_2. A structure regeneration of Au_{24} can be realized by the destroyed Au_{24} reacting with the excess PPh_3. The structural conversion between Au_{24} and Au_{25} was reversible, and along with this the activity can be recovered (Fig. 9.21).

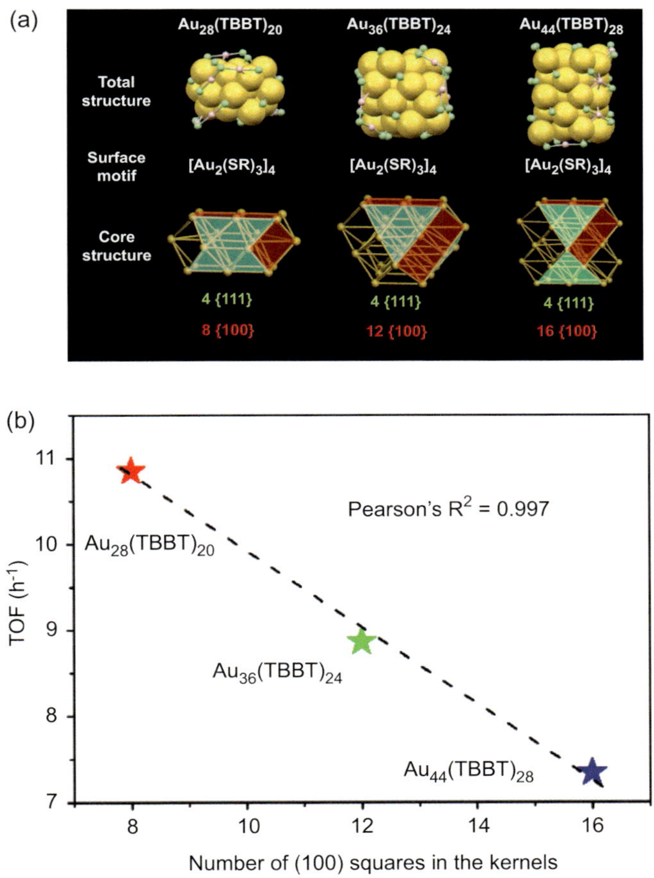

Figure 9.22 (a) Crystal structures and models of the kernels of $Au_{28}(SR)_{20}$, $Au_{36}(SR)_{24}$, and $Au_{44}(SR)_{28}$. Color codes: Yellow for Au atoms in the kernel, magenta for Au in dimeric staples, and green for S. (b) Correlation between the {100} square numbers in kernels and the catalytic activity of Au clusters. Reprinted with permission from Ref. [39], Copyright 2019, John Wiley & Sons, Inc.

Zhu and co-workers demonstrated that the evolution in catalytic activity can be derived by periodic transformation in the inner kernels in $Au_{8n+4}(SR)_{4n+8}$ (n = 3, 4, 5) nanoclusters, where their surface motifs are the same but kernels share a periodic structure [39]. In particular, the kernels of $Au_{8n+4}(SR)_{4n+8}$ (n = 3, 4, 5) nanoclusters have four {111} facets and a different number of {100} facets: 8 for $Au_{28}(SR)_{20}$, 12 for $Au_{36}(SR)_{24}$, and 16 for $Au_{44}(SR)_{28}$. The catalytic activity showed a descending order: $Au_{28}(TBBT)_{20}$ > $Au_{36}(TBBT)_{24}$ > $Au_{44}(TBBT)_{28}$. They ruled out the size-dependent activity and found a very interesting phenomenon that the catalytic activity is inversely proportional to the total number of {100} squares in the kernels of $Au_{8n+4}(SR)_{4n+8}$ (n = 3, 4, 5) nanoclusters. $Au_{28}(SR)_{20}$ with a minimum of {100} squares showed the highest activity and $Au_{44}(SR)_{28}$ with the most {100} squares exhibited the lowest activity. Further, the chemical adsorption capability of the $Au_{8n+4}(SR)_{4n+8}$ was in the order $Au_{28}(TBBT)_{20}$ > $Au_{36}(TBBT)_{24}$ > $Au_{44}(TBBT)_{28}$, which is in accordance with the activity sites in the three nanoclusters. Besides, the affinity for alkyne was also in a descending sequence: $Au_{28}(TBBT)_{20}$ > $Au_{36}(TBBT)_{24}$ > $Au_{44}(TBBT)_{28}$, which is in accordance with the activity as well (Fig. 9.22). These observations reveal that surface sites can be affected by the changes in inner structures and can further impact the catalytic properties.

In summary, the above examples of catalysts of atomically precise metal nanoclusters demonstrate the huge power of atomically precise, semiconducting metal catalysts for achieving super catalytic performance and atomically precise structure–property relationships. Such nanoclusters will not only provide new opportunities for unraveling catalysis at an atomic level, but also promote the exploration of new chemical processes with metal nanoclusters as well-defined, highly efficient catalysts. Atomically precise metal nanocluster catalysts will ultimately bring metal catalysis to an exciting new level.

Acknowledgments

We thank the support from Nanjing University.

References

1. Jin, R. C., Zeng, C. J., Zhou, M. and Chen, Y. X. (2016). Atomically precise colloidal metal nanoclusters and nanoparticles: Fundamentals and opportunities, *Chem. Rev.*, **116**, pp. 10346–10413.
2. Turner, M., Golovko, V. B., Vaughan, O. P. H., Abdulkin, P., Berenguer-Murcia, A., Tikhov, M. S., Johnson, B. F. G. and Lambert, R. M. (2008). Selective oxidation with dioxygen by gold nanoparticle catalysts derived from 55-atom clusters, *Nature*, **454**, pp. 981–983.
3. Zhu, Y., Qian, H. F., Zhu, M. Z. and Jin, R. C. (2010). Thiolate-protected Au_n nanoclusters as catalysts for selective oxidation and hydrogenation processes, *Adv. Mater.*, **22**, pp. 1915–1920.
4. Li, L. X., Huang, S. S., Song, J. J., Yang, N. T., Liu, J. W., Chen, Y. Y., Sun, Y. H., Jin, R. C. and Zhu, Y. (2016). Ultrasmall Au_{10} clusters anchored on pyramid-capped rectangular TiO_2 for olefin oxidation, *Nano Res.*, **9**, pp. 1182–1192.
5. Yoskamtorn, T., Yamazoe, S., Takahata, R., Nishigaki, J., Thivasasith, A., Limtrakul, J. and Tsukuda, T. (2014). Thiolate-mediated selectivity control in aerobic alcohol oxidation by porous carbon-supported Au_{25} clusters, *ACS Catal.*, **4**, pp. 3696–3700.
6. Hayashi, S., Ishida, R., Hasegawa, S., Yamazoe, S. and Tsukuda, T. (2018). Doping a single palladium atom into gold superatoms stabilized by PVP: Emergence of hydrogenation catalysis, *Top. Catal.*, **61**, pp. 136–141.
7. Xie, S. H., Tsunoyama, H., Kurashige, W., Negishi, Y. and Tsukuda, T. (2012). Enhancement in aerobic alcohol oxidation catalysis of Au_{25} clusters by single Pd atom doping, *ACS Catal.*, **2**, pp. 1519–1523.
8. Li, Y. W., Chen, Y. X., House, S. D., Zhao, S., Wahab, Z., Yang, J. C. and Jin, R. C. (2018). Interface engineering of gold nanoclusters for CO oxidation catalysis, *ACS Appl. Mater. Interfaces*, **10**, pp. 29425–29434.
9. Wu, Z. L., Jiang, D. E., Mann, A. K. P., Mullins, D. R., Qiao, Z. A., Allard, L. F., Zeng, C. J., Jin, R. C. and Overbury, S. H. (2014). Thiolate ligands as a double-edged sword for CO oxidation on CeO_2 supported $Au_{25}(SCH_2CH_2Ph)_{18}$ nanoclusters, *J. Am. Chem. Soc.*, **136**, pp. 6111–6122.
10. Li, W. L., Liu, C., Abroshan, H., Ge, Q. J., Yang, X. J., Xu, H. Y. and Li, G. (2016). Catalytic CO oxidation using bimetallic M_xAu_{25-x} clusters: A combined experimental and computational study on doping effects, *J. Phys. Chem. C*, **120**, pp. 10261–10267.

11. Wang, L. Q., Chai, X. Q., Cheng, X. L. and Zhu, Y. (2018). Structure-specific catalytic oxidation with O_2 by isomers in $Au_{28}(SR)_{20}$ nanoclusters, *ChemistrySelect*, **3**, pp. 6165–6169.

12. Liu, T., Tsunoyama, H., Akita, T., Xie, S. and Tsukuda, T. (2011). Aerobic oxidation of cyclohexane catalyzed by size-controlled Au clusters on hydroxyapatite: Size effect in the sub 2 nm regime, *ACS Catal.*, **1**, pp. 2–6.

13. Zhu, Y., Qian, H. F., Drake, B. A. and Jin, R. C. (2010). Atomically precise $Au_{25}(SR)_{18}$ nanoparticles as catalysts for the selective hydrogenation of α,β-unsaturated ketones and aldehydes, *Angew. Chem. Int. Ed.*, **49**, pp. 1295–1298.

14. Zhu, Y., Wu, Z. K., Gayathri, C., Qian, H. F., Gil, R. R. and Jin, R. C. (2010). Exploring stereoselectivity of Au_{25} nanoparticle catalyst for hydrogenation of cyclic ketone, *J. Catal.*, **271,** pp. 155–160

15. Zhao, J. B., Li, Q., Zhuang, S. L., Song, Y. B., Morris, D. J., Zhou, M., Wu, Z. K., Zhang, P. and Jin, R. C. (2018). Reversible control of chemoselectivity in $Au_{38}(SR)_{24}$ nanocluster-catalyzed transfer hydrogenation of nitrobenzaldehyde derivatives, *J. Phys. Chem. Lett.*, **9**, pp. 7173–7179.

16. Li, G., Abroshan, H., Chen, Y. X., Jin, R. C. and Kim, H. J. (2015). Experimental and mechanistic understanding of aldehyde hydrogenation using Au_{25} nanoclusters with Lewis acids: Unique sites for catalytic reactions, *J. Am. Chem. Soc.*, **137**, pp. 14295–14304.

17. Liu, C., Abroshan, H., Yan, C. Y., Li, G. and Haruta, M. (2016). One-pot synthesis of $Au_{11}(PPh_2Py)_7Br_3$ for the highly chemoselective hydrogenation of nitrobenzaldehyde, *ACS Catal.*, **6**, pp. 92–99.

18. Chen, H. J., Li, Z. M., Qin, Z. X., Kim, H. J., Abroshan, H. and Li, G. (2019). Silica-encapsulated gold nanoclusters for efficient acetylene hydrogenation to ethylene, *ACS Appl. Nano Mater.*, **2**, pp. 2999–3006.

19. Zhu, M., Wang, P., Yan, N., Chai, X. Q., He, L. Z., Zhao, Y., Xia, N., Yao, C. H., Li, J., Deng, H. T., Zhu, Y., Pei, Y. and Wu, Z. K. (2018). The fourth alloying mode by way of anti-galvanic reaction, *Angew. Chem. Int. Ed.*, **57**, pp. 4500–4504.

20. Li, Gao and Jin, R. C. (2014). Gold nanocluster-catalyzed semihydrogenation: A unique activation pathway for terminal alkynes, *J. Am. Chem. Soc.*, **136**, pp. 11347–11354

21. Wan, X. K., Wang, J. Q., Nan, Z. A. and Wang, Q. M. (2017). Ligand effects in catalysis by atomically precise gold nanoclusters, *Sci. Adv.*, **3**, p. e1701823.

22. Fang, J., Li, J. G., Zhang, B., Yuan, X., Asakura, H., Tanaka, T., Teramura, K., Xie, J. P. and Yan, N. (2015). The support effect on the size and catalytic activity of thiolated Au_{25} nanoclusters as precatalysts, *Nanoscale*, **7**, pp. 6325–6333

23. Nasaruddin, R. R., Chen, T. K., Li, J. G., Goswami, N., Zhang, J. G., Yan, N. and Xie, J. P. (2018). Ligands modulate reaction pathway in the hydrogenation of 4-nitrophenol catalyzed by gold nanoclusters, *ChemCatChem*, **10**, pp. 395–402

24. Urushizaki, M., Kitazawa, H., Takano, S., Takahata, R., Yamazoe, S. and Tsukuda, T. (2015). Synthesis and catalytic application of Ag_{44} clusters supported on mesoporous carbon, *J. Phys. Chem. C*, **119**, pp. 27483–27488.

25. Chai, X. Q., Li, T., Chen, M. Y., Jin, R. C., Ding, W. P. and Zhu, Y. (2018). Suppressing the active site-blocking impact of ligands of $Ni_6(SR)_{12}$ clusters with the assistance of NH_3 on catalytic hydrogenation of nitriles, *Nanoscale*, **10**, pp. 19375–19382.

26. Bao, L. Q., Zhao, C. C., Li, S. G. and Zhu, Y. (2019). Benzalaniline from nitrobenzene and benzaldehyde catalyzed efficiently by an atomically precise palladium nanocluster, *Chin. J. Catal.*, **40**, pp. 1499–1504.

27. Cheng, X. L., Chai, X. Q., Hu, W. G., Li, S. G. and Zhu, Y. (2019). The on-and-off dynamics of thiophene on a nickel cluster enables efficient hydrodesulfurization and excellent stability at high temperatures, *Nanoscale*, **11**, pp. 4369–4375.

28. Xu, J. Y., Chen, M. Y. and Zhu, Y. (2019). Selective hydrogenation of CO_2 dictated by isomers in $Au_{28}(SR)_{20}$ nanoclusters: Which one is better? *Chem. Eur. J.*, **25**, pp. 9185–9190.

29. Kauffman, D. R., Thakkar, J., Siva, R., Matranga, C., Ohodnicki, P. R., Zeng, C. J. and Jin, R. C. (2015). Efficient electrochemical CO_2 conversion powered by renewable energy, *ACS Appl. Mater. Interfaces*, **7**, pp. 15626–15632.

30. Kauffman, D. R., Alfonso, D., Matranga, C., Qian, H. F. and Jin, R. C. (2012). Experimental and computational investigation of Au_{25} clusters and CO_2: A unique interaction and enhanced electrocatalytic activity, *J. Am. Chem. Soc.*, **134**, pp. 10237–10243.

31. Yu, C. L., Li, G., Kumar, S., Kawasaki, H. and Jin, R. C. (2013). Stable $Au_{25}(SR)_{18}/TiO_2$ composite nanostructure with enhanced visible light photocatalytic activity, *J. Phys. Chem. Lett.*, **4**, pp. 2847–2852.

32. Kawasaki, H., Kumar, S., Li, G., Zeng, C. J., Kauffman, D. R., Yoshimoto, J., Iwasaki, Y. and Jin, R. C. (2014). Generation of singlet oxygen by photoexcited $Au_{25}(SR)_{18}$ clusters, *Chem. Mater.*, **26**, pp. 2777–2788.

33. Qian, H. F., Jiang, D. E., Li, G., Gayathri, C., Das, A., Gil, R. R. and Jin, R. C. (2012). Monoplatinum doping of gold nanoclusters and catalytic application, *J. Am. Chem. Soc.*, **134**, pp. 16159–16162.

34. Qian, H. F., Barry, E., Zhu,Y. and Jin, R. C. (2011). Doping 25-atom and 38-atom gold nanoclusters with palladium, *Acta Phys. Chim. Sin.*, **27**, pp. 513–519.

35. Kwak, K., Choi, W., Tang, Q., Kim, M., Lee, Y., Jiang, D. E. and Lee, D. A. (2017). A molecule-like $PtAu_{24}(SC_6H_{13})_{18}$ nanocluster as an electrocatalyst for hydrogen production, *Nat. Commun.*, **8**, p. 14723.

36. Liu, Y. Y., Chai, X. Q., Cai, X., Chen, M. Y., Jin, R. C., Ding, W. P. and Zhu, Y. (2018). Central doping of a foreign atom into the silver cluster for catalytic conversion of CO_2 toward C–C bond formation, *Angew. Chem. Int. Ed.*, **57**, pp. 9775–9779.

37. Wang, L. Q., Shen, K. Q., Chen, M. Y. and Zhu, Y. (2019). One-core-atom loss in a gold nanocluster promotes hydroamination reaction of alkyne, *Nanoscale*, **11**, pp. 13767–13772.

38. Cai, X., Saranya, G., Shen, K. Q., Chen, M. Y., Si, R., Ding, W. P. and Zhu, Y. (2019). Reversible switching of catalytic activity by shuttling an atom into and out of gold nanoclusters, *Angew. Chem. Int. Ed.*, **58**, pp. 9964–9968.

39. Sun, Y. N., Wang, E. D., Ren, Y. J., Xiao, K., Liu, X., Yang, D., Gao, Y., Ding, W. P. and Zhu, Y. (2019). The evolution in catalytic activity driven by periodic transformation in the inner sites of gold clusters, *Adv. Funct. Mater.*, **29**, p. 1904242.

Chapter 10

Density Functional Theory Studies for Catalysis of Atomically Precise Metal Clusters

Mingyang Chen[a,b] and Shenggang Li[c,d]

[a]*Center for Green Innovation, School of Materials Science and Engineering, University of Science and Technology Beijing, Beijing 100083, China*
[b]*Beijing Computational Science Research Center, Beijing 100193, China*
[c]*CAS Key Laboratory of Low-Carbon Conversion Science and Technology, Shanghai Advanced Research Institute, Chinese Academy of Sciences, Shanghai 201210, China*
[d]*School of Physical Science and Technology, ShanghaiTech University, Shanghai 201210, China*
mychen@ustb.edu.cn

Atomically precise metal clusters can be used as model catalysts for facilely correlating the structures with catalytic properties. Density functional theory (DFT) can be used to accurately predict various properties of sizable chemical systems, such as structural, electronic, and optical properties, and to efficiently study the catalytic mechanisms of atomically precise metal clusters. Benefiting from their precise structures, DFT computations are ideal for establishing the structure–activity correlations for such catalysts. In particular,

Atomically Precise Nanoclusters
Edited by Yan Zhu and Rongchao Jin

ISBN 978-981-4877-08-4 (Hardcover), 978-1-003-11990-6 (eBook)
www.jennystanford.com

structurally well-defined clusters can be engineered experimentally and modelled computationally, from which the structure–activity relationships can be reliably derived. This chapter aims to give a brief introduction to the application of DFT methods in computational catalysis and summarizes recent progresses in using DFT methods to explore the structure–activity relationship in catalysis of atomically precise metal clusters. The effects of several structural design factors for atomically precise metal cluster catalysts are considered, which include charge state, point vacancy, surface structure, structural evolution, and protective ligand. Finally, major challenges and opportunities for future applications of DFT methods in studying catalysis by the atomically precise metal clusters are given.

10.1 Introduction

Metal clusters exhibit unique physical and chemical properties, such as high surface-to-volume ratio, high catalytic activity, and tunability, and thus show significant prospects in energy and environmental applications [1]. The utilization of metal clusters, however, is unfortunately limited by their intrinsic features. For example, metal clusters are too reactive to be separated and stored. It is also difficult to obtain a pure metal cluster with monodisperse molecular structure and consistent properties. Recently, advances in the synthesis of atomically precise metal clusters [2, 3] have shown promises to overcome some of these challenges. An atomically precise metal cluster has a well-defined crystal structure comprising a metal cluster core and a shell of protective organic ligands. The metal cluster core can be considered as a superatom, where the occupancy of valence electrons may follow general rules similar to the 8-electron rule for main group atoms and the 18-electron rule for transition metal atoms. Because of the occupation rules, the metal cluster core exhibits a preference to incorporate a certain number of protective ligands and hence achieves the atomically precise structure. On the other hand, the protective ligands can fully or partially passivate the metal cluster core, which allows for the separation, storage, processing, and utilization of metal clusters. The crystal structure for $[Ag_{25}(SPhMe_2)_{18}]^-$ [4], a thiolated silver cluster, is shown in Fig. 10.1. $[Ag_{25}(SPhMe_2)_{18}]^-$ consists of an Ag_{13} icosahedron core and six $SR(AgSR)_2$ staples at precise surface positions of Ag_{13}.

The atomically precise metal clusters have quickly caught attention in the field of catalysis and are proposed as a class of effective model catalysts [5]. The effectiveness comes from the large surface-to-volume ratio and unique surface sites, while the model role is owing to the precise structure and uniform active sites. As model catalysts, the atomically precise metal clusters exhibit catalytic properties that are "purer" than those of the conventional metal catalysts, and thus they provide ideal opportunities for gaining fundamental understandings on catalysis. Both experimental and computational efforts are important for fundamental catalysis studies, for example, the identification of active sites, elementary reaction steps, and intermediate species. Typically, experiments provide spectroscopic evidences, while computations provide atomistic explanations. This is due to the advantages and limitations of both the experimental and computational methods (Fig. 10.2). Even with today's state-of-the-art techniques, it remains difficult to experimentally characterize catalytic reactions at an angstrom and femtosecond (fs) resolution, while specialized computations can excel at such scales and provide deductive explanations for experimental observations at larger scales. On the other hand, experiments can be run for much longer duration and in much large quantity than computations and are also considered to be more convincible than the latter, due to the intrinsic difficulties in computational methods and models.

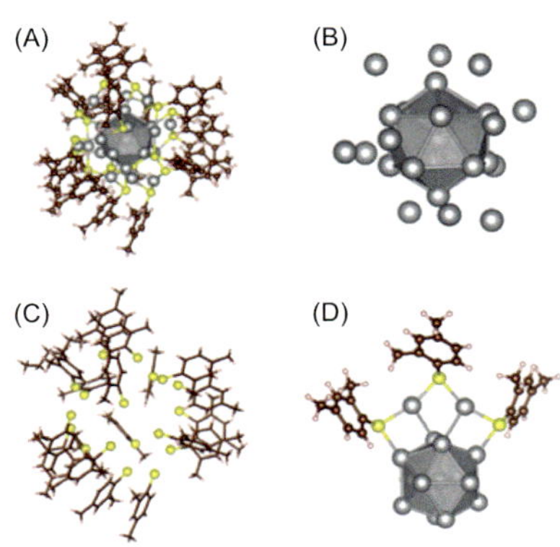

Figure 10.1 Crystal structure of $[Ag_{25}[SPhMe_2]_{18}]^-$. (A) Full atomistic structure with an Ag_{13} core and six $SR(AgSR)_2$ staples, (B) Ag_{13} core and 12 staple Ag atoms, (C) 18 $SPhMe_2$ thiolate protective ligands, and (D) anchoring of $SR(AgSR)_2$ on Ag_{13}. Color codes: Grey, Ag; yellow, S; black, C; white, H.

DFT [6, 7], a first-principles method derived under the quantum mechanics framework, is considered as the most efficient computational approach for solving atomistic systems that fit space scales in the range of Å to nm and timescales in the range of fs to picosecond (ps). In DFT, the many-body electronic structure of a molecular system is solved in the form of electron densities, and properties of the molecular system are expressed as functionals of electron densities. Therefore, the DFT methods are widely used to predict physical and chemical properties for molecules, clusters, bulk crystals, and slab surfaces. Due to its scope, resolution, accuracy, and versatility, DFT is the most extensively used computational method in catalysis. It can be used to accurately predict molecular structures, electronic structures, vibrational properties, optical properties, thermodynamics, kinetics, and so on, all of which are essential for the explanation of experimental observations as well as the understanding of reaction mechanisms in catalysis. The predictive power of the DFT methods also allows one to theoretically "synthesize" new catalysts and simulate new catalytic reactions due to the often great difficulty and high cost for the corresponding experimental studies, which can provide insights and guidance for experiments.

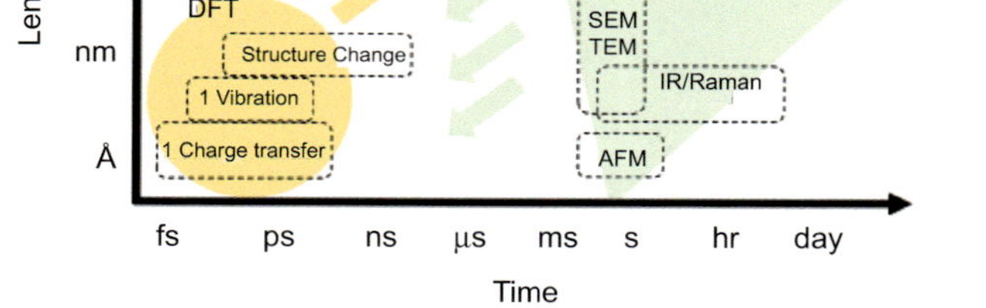

Figure 10.2 Schematic comparison and relation for the scope and resolution of the experimental techniques and the DFT computational methods.

In this chapter, a brief review on the progress of computational studies using the DFT methods for the catalysis of the atomically precise clusters will be given. The chapter is laid out in two major sections. In the first section, a brief overview of the theoretical background of the DFT approaches relevant to catalysis will be given. In the second section, applications of the DFT computations for the fundamental understanding of the catalysis of the atomically precise metal clusters are summarized. The chapter is concluded by outlining the current limitations and bottlenecks of DFT methods in catalysis of the atomically precise metal clusters and the future perspectives on improved utilizations of these DFT methods.

10.2 DFT-Related Methods for Cluster Catalysis

Modern DFT is built on the Hohenberg–Kohn theorems [8] for systems consisting electrons under the influence of an external potential $\upsilon_{\text{ext}}(\mathbf{r})$: (1) there is a one-to-one mapping between $\upsilon_{\text{ext}}(\mathbf{r})$ and ground-state electron density $n(\mathbf{r})$, and (2) the exact ground-state density minimizes the total energy. On the other hand, the ground-state density $n(\mathbf{r})$ is uniquely determined by the ground-state wavefunction Ψ. For a N-electron system with a normalized wavefunction Ψ,

$$n(\boldsymbol{r}) = \int d\boldsymbol{r}_2 \cdots d\boldsymbol{r}_N \psi^*(\mathbf{r},\mathbf{r}_2,\cdots,\boldsymbol{r}_N)\psi(\boldsymbol{r},\boldsymbol{r}_2,\ldots,\boldsymbol{r}_N) \tag{10.1}$$

The Hohenberg–Kohn theorems lead to a few important conclusions. For example, the wavefunction, the total energy, and the expectation value of a quantity for an electronic system are all functionals of $n(\mathbf{r})$; one can solve for the exact ground-state density if the exact form of the total energy as a functional of $n(\mathbf{r})$ is known; physical and chemical properties can be derived based on $n(\mathbf{r})$. Despite these promising prospects of the Hohenberg–Kohn theorems, the practical uses began only after Kohn and Sham proposed an explicit way to calculate the ground-state electron density for electronic systems [9]. With their proposed formulation, one can solve self-consistently the Kohn–Sham equation for the non-interactive electrons

$$\left(-\frac{\hbar^2}{2m}\nabla^2 + \upsilon_{\text{eff}}(\boldsymbol{r})\right)\phi_i^{KS}(\boldsymbol{r}) = \varepsilon_i\phi_i^{KS}(\boldsymbol{r}) \tag{10.2}$$

and obtain the density as

$$n(\boldsymbol{r}) = \sum_{i}^{N} \left|\phi_i^{KS}(\boldsymbol{r})\right|^2 \tag{10.3}$$

In Eq. 10.2, the Kohn–Sham effective potential is given by

$$\upsilon_{\text{eff}}(\boldsymbol{r}) = \int dr' \frac{n(r')}{|r - r'|} + \upsilon_{\text{XC}}(\boldsymbol{r}) + \upsilon_{\text{ext}}(\boldsymbol{r}) \tag{10.4}$$

where the first term is the Hartree term that describes the classical electron–electron repulsion and the second term υ_{XC} is exchange-correlation potential as the functional derivative of exchange-correlation energy (E_{XC}) with respect to $n(\boldsymbol{r})$. Equations 10.2–10.4 imply that knowledge of the explicit form of E_{XC}, a functional of $n(\boldsymbol{r})$, is required to iteratively solve for $n(\boldsymbol{r})$. Unfortunately, the exact form of E_{XC} with respect to $n(\boldsymbol{r})$ is unknown so far, and so many approximate forms have been derived, which are normally categorized as one of the LDA (local density approximation), GGA (generalized gradient approximation), meta-GGA, and hybrid functionals. With a chosen exchange-correlation functional, properties of a chemical system can be predicted either as the expectation value of a quantity for the ground state $n(\boldsymbol{r})$, from the perturbation of the $n(\boldsymbol{r})$ [10], or from the time evolution of $n(\boldsymbol{r})$ [11, 12].

The accuracy of DFT calculations are quite remarkable, given the relatively low computational complexity [$O(N^3)$ to $O(N^4)$] of DFT methods in comparison to higher level methods, such as the coupled cluster theory. Owing to error cancellation, DFT methods can predict catalytic reaction energies with small errors, typically a few kcal/mol, and are able to predict reasonable trends in catalysis. Among the DFT functionals, LDA is the simplest and least accurate, and thus it is not frequently used. DFT methods with GGA and hybrid functionals can accurately predict electronic structures and properties for various chemical systems, especially for chemical systems near equilibrium without strong electron correlations. Hybrid functionals are generally considered to be superior to GGA functionals, in terms of computational accuracy, but the trade-off is their much higher computational costs. PBE and B3LYP are the most widely used GGA and hybrid exchange-correlation functionals, respectively. The accuracy of DFT calculations is also dependent on other factors. A sufficiently large basis set that can better expand the wavefunction is necessary for accurate predictions. The inclusions of spin–orbit, relativistic, and dispersion corrections may also be critical for obtaining reliable prediction for more challenging chemical systems. Pseudopotentials are often required to reduce the size of basis sets and to include the relativistic effects in calculations. In the following, the commonly used DFT approaches in catalysis will be summarized.

10.2.1 Determination of Atomistic Structures for Catalysts

Determination of atomistic structures for catalysts can be done via molecular geometry optimization [13, 14]. In molecular geometry optimization, the total energy of the molecule is minimized with respect to the geometric parameters such as the bond length, bond angle, and even full Cartesian coordinates. Such a process is equivalent to locating a minimum on the potential energy surface (PES). For many regular chemical systems, DFT-level PESs exhibit good resemblance to the exact PESs (or PESs predicted at much higher computational levels), and DFT is ideal for determining equilibrium molecular geometries. Even LDA is able to predict reasonable local minimum structures with small basis sets, which can be used to predict atomistic structures for relatively large systems, such as atomically precise metal cluster with a large core and bulky ligands.

For the DFT modelling of atomically precise metal cluster catalysts, the first step is to obtain a structural model that is consistent with the experimental crystal structure. Usually one unit of the atomically precise metal cluster is extracted from the crystal structure as the initial structure for the geometry optimization of the fresh catalyst. The structures of catalytic reaction intermediates can be constructed from the modification of the optimized fresh catalyst structure on the basis of either spectroscopic evidences (e.g., infrared spectrum) or known reaction mechanisms for similar systems. If no basis is available for the initial guesses, the initial structures might be built based on chemical intuitions or by sampling multiple possible guesses. DFT methods can also be used to predict completely new atomically precise metal clusters [15].

The optimized structure for the fresh catalyst can be verified with the experimental single-crystal diffraction data. The validations of optimized structures for the reaction intermediates are less straightforward and rely on indirect evidences from spectroscopic measurements, for example, in situ and ex situ infrared and Raman spectroscopies. Extended X-ray absorption fine structure (EXAFS) provides radial distribution functions (RDFs) of atom pair distances in the catalyst sample as well as coordination numbers from data fitting, and it can be used to verify the optimized structures for the fresh catalyst, the persistent intermediate, and the used catalyst. EXAFS can be simulated for an atomistic structure [16]. The K-edge absorption of atom type *i* can be determined by

$$\chi_i(k)=\sum_j 4\pi\rho\int_0^{\infty} r^2\left(g_{ij}(r)-1\right)\chi_1(k,r)dr \qquad (10.5)$$

where $g_{ij}(r)$ is the RDF for atom type *j* with respect atom type *i*, and $\chi_1(k, r)$ represents the contribution by a single scattering atom (*j*-type) that can be determined ab initio. Reverse Monte Carlo [17] is handy to inversely guess the atomistic structures that match the EXAFS result, but it is not generally applicable for the structurally complex, atomically precise metal cluster catalysts with too many structural degrees of freedom. Another commonly used structure characterization technique X-ray powder diffraction (XRD) is not typically useful for characterizing atomically precise metal cluster catalysts, as XRD can indicate only short-range lattice spacings in the pulverized material, limited by the frequency range of X-ray and the large size of the unit cell.

The protective ligands of atomically precise metal cluster catalysts usually contain phenyl moieties and can significantly increase the DFT computational cost due to the $O(N^3)$–$O(N^4)$ scaling of DFT methods. For geometry optimization, large protective ligands with phenyl moieties can be replaced with smaller ligands [18]; for example, SC_2H_4Ph can be simplified into SCH_3 and PPh_3 can be simplified into PH_3. In general, simplified models can yield good optimized geometries and reasonable electronic structures for atomically precise metal cluster catalysts. Some catalytic reactions can be greatly affected by protective groups, for example, steric effects. To model these reactions, one can perform pre-optimizations with a simplified model and perform full-structure optimizations with the full model by substituting actual ligands for simplified ones.

We also note that although the molecular geometry optimization for a single atomically precise metal cluster containing over 100 transition metal atoms is possible [19], it is in general not feasible to study catalytic reactions of a large atomically precise metal cluster, especially the reactions for which the mechanisms are not clearly known. This is because to elucidate the catalytic mechanism, attempts such as guessing the reaction intermediates and transition state species are often made in a trial-and-error fashion for which a large number of geometry optimizations are required.

10.2.2 Determining Electronic Structures for Catalysts

Solving the electronic structure, or the ground-state electron density, of catalysts is essential for understanding the catalytic mechanism and is also important to predicting accurate theoretical spectra to compare with experimental observations. The correctness of the predicted electronic structure can be validated by several means, including optical, electrochemical, and magnetic characterizations. Atomically precise metal clusters with small metal cores (<2 nm) have quantized molecular orbitals and exhibit semiconductivity. The calculated HOMO–LUMO (highest occupied molecular orbital–lowest unoccupied molecular orbital) gap can be compared with the experimental energy gap from optical or electrochemical measurement. More comprehensive validations of the electronic structures can be done by directly matching the experimental and simulated absorption and emission spectra.

For atomically precise metal clusters, alteration of the magnetic moment can give rise to drastic change in electronic structure and molecular structure. One good example is distinction between the non-magnetic monoanionic singlet $^1[Au_{25}(SR)_{18}]^-$ and the magnetic neutral doublet $^2[Au_{25}(SR)_{18}]^0$ [20]. DFT calculations revealed that the absence of an electron in $^2[Au_{25}(SR)_{18}]^0$ with respect to $^1[Au_{25}(SR)_{18}]^-$ accounts for the resulting magnetic moment and also causes the Jahn–Teller distortion [21]. It has been proposed that styrene oxidation reaction over the $Au_{25}(SR)_{18}$ catalyst involves the state hopping between $^1[Au_{25}(SR)_{18}]^-$ and $^2[Au_{25}(SR)_{18}]^0$ [22].

Chemical shifts from X-ray photoelectron spectroscopy (XPS) can indicate the chemical environment for atoms in the catalyst. To explain XPS results, atomic charges and oxidation states can be estimated from DFT calculations via density of states [23] and population analyses. Commonly used population analyses include natural population analysis [24], Mulliken population analysis [25], Bader charge analysis [26], and Hirshfeld population analysis [27]. Comparison and discussion of the different population analysis methods can be found elsewhere [28].

10.2.3 Predicting Spectra for Catalysts

Characterization of atomically precise catalysts can involve various spectroscopic techniques. Infrared [29] and Raman [30] spectroscopies measure the vibrational frequencies of catalytic species and can be used to diagnose the local structures in the species. Vibrational frequencies can be determined theoretically, where the frequencies are related to the second derivatives of the total energy with respect to the normal coordinates and the intensities can be obtained using the density functional perturbation theory (DFPT) [10].

Photoabsorption and emission spectra such as UV-Vis [31] and fluorescence spectra [32] can be predicted using the linear response (LR) [33] or real-time (RT) [34] time-dependent density functional theory (TDDFT). On account of the computational cost, the LR-TDDFT is more feasible for atomically precise metal clusters. The photoelectron spectroscopies such as UPS and XPS (ultraviolet and X-ray photoelectron spectroscopies) can be indirectly compared with the DFT results or be directly emulated at the DFT level [35]. The binding energy for the electron may be predicted using the ΔSCF method [36], Slater–Janak transition state method [37], frozen orbital method [38], and so on.

Nuclear magnetic resonance (NMR) spectroscopy is another useful technique for structural characterization of catalysts [39]. NMR shielding tensors can be calculated using the gauge-independent atomic orbital (GIAO) method at the DFT level. The calculated relative chemical shifts from the GIAO method can be directly compared to the experimental chemical shifts. Electron paramagnetic resonance (EPR) spectroscopy is useful to detect catalytic species with unpaired electrons. EPR spectra are device dependent, and hence they cannot be directly emulated at the DFT level, but the hyperfine coupling constants are obtainable from DFT calculations [40].

10.2.4 Adiabatic and Non-adiabatic Molecular Dynamics

First-principles molecular dynamics (MD) methods can be used to study the nuclear motions (via adiabatic MD) and electronic motions (via non-adiabatic MD) for catalysts. The adiabatic MD has been used for semi-local geometry optimization for atomically precise metal clusters to determine their equilibrium geometries [41]. Compared to the gradient-driven local geometry optimization, the trajectories of MD simulations can overcome small energy barriers, and thus MD is useful for determining the orientation and packing of protective ligands. The environment factors such as solvent and medium can be explicitly included in MD simulations, although such practices lead to significant increases in the computational cost. Also, due to the high computational cost, adiabatic MD has not been used in elucidation of the mechanisms of realistic catalytic reactions over atomically precise metal clusters.

Non-adiabatic MD [42] deals with the electronic motions between adiabatic MD steps on the basis of TDDFT, where electrons can hop between several PESs. Surface hopping probability of electrons [43] from non-adiabatic MD can be used to predict the dynamic processes of electron photoexcitation, electron transfer, and electron–hole recombination, and it is therefore capable of modelling photocatalysis. Such capacities align with the great potential of atomically precise metal clusters in photocatalysis [44], as they often exhibit semiconductive properties, with the HOMO–LUMO gap suitable for absorption of solar energy. Non-adiabatic MD has been used to study the charge transfer interaction between the metal core and ligands of atomically precise metal clusters [45], but it has not been utilized in the corresponding photocatalysis applications to the best of our knowledge.

10.2.5 Transition State Theory and Microkinetics

Transition state theory [46] and microkinetics are powerful microscopic tools to express the catalytic reaction kinetics. Both generalized transition state theory and microkinetics employ the concept that the kinetics of a catalytic reaction may be reliably predicted with the knowledge of a few key stationary states of the catalyst complex. Transition state (TS), the quasi-equilibrium state that connects the reactant and the product for elementary reactions, is one of these key states. One essential function of the catalyst is to replace the transition state in the absence of the catalyst with a new one with a lower energy barrier [free energy of activation, $\Delta G^{\ddagger} = G(\text{TS}) - G\,(\text{reactant})$], so that faster reaction kinetics can be achieved. In the generalized transition state theory, the transition state is localized at the saddle point of the PES between the reactant and the product. Therefore, the transition state geometry can be determined by sampling a region of the PES or by performing gradient-based optimization searches on the PES. Computationally expensive force constant calculations (second derivatives on the PES) are required to determine the optimization search direction and verify the predicted transition state. Transition state theory has been applied in a few reaction mechanism studies for atomically precise metal cluster-catalyzed reactions, such as ligand exchange, isomerization, heterogeneous catalysis, and homogeneous catalysis [47–49], but its usage has been limited due to the high computational cost as compared to the wide applications of the transition state theory in traditional catalysis.

Microkinetics [50] is a method to solve the catalytic reaction rates based on a small set of crucial parameters determined from theories of reaction rates of the elementary steps, for example, transition state theory and collision theory. Microkinetics can predict turnover frequencies, and surface coverage for catalytic reactions, and can be used to predict product selectivity under different reaction conditions and microstructure–activity relationship for catalysts. It has not yet been applied for the catalysis study of atomically precise metal clusters, primarily because (1) the computational costs for the ab initio determination of microkinetic parameters are extremely high, (2) mechanisms for the catalysis of atomically precise metal clusters differ significantly from those in traditional catalysis, making it challenging for microkinetic modelling, and (3) the bases for estimating the necessary microkinetic parameters such as the pre-exponential factors are lacking. Thus, there are both opportunities and challenges for microkinetic simulations of catalytic reactions over atomically precise metal cluster catalysts. The applicability of microkinetics is apparent, given the resemblance between the atomically precise metal cluster catalysis and metal surface catalysis. The challenges partly come from the complex role played by the protective ligands, as the mechanisms of the protective ligands in the dynamic processes of catalyst activation and catalytic reaction are not clearly known yet.

10.3 Designing Factors for Atomically Precise Metal Cluster Catalysis from DFT Studies

The precise structures of atomically precise metal clusters determine their consistent chemical and physical properties, unique reaction mechanisms, and precise spectroscopic properties, all of which provide great opportunity for mechanistic studies of vital elementary catalytic reactions. Accurate prediction of the properties and spectra for a precise structure is one of the strengths of the first-principles DFT methods. In addition, owing to the highly monodisperse and highly passivated structures of the fresh catalyst, the structures of reaction intermediate complexes (formed from the catalyst and the precursors), despite usually being unknown, are also likely to be atomically precise, which makes it easy to come up with initial guesses and computationally model these structures. To reliably derive the reaction pathways and determine the catalytic mechanisms, one can take possible structure guesses, predict their spectra, and validate the structure guesses by direct comparison with the experimental spectra.

Because of the remarkable precision that can be achieved from both experiments and DFT computations for atomically precise metal clusters, it is possible to model cluster catalysts with subtle structural differences (such as atomic doping and vacancy) and unravel the contributions of the structural subtlety in the catalysis via a differential approach. Unlike the cases of bulk materials where the structural subtleties only statistically contribute to the exhibited properties of the catalyst that is an ensemble of various local catalytic structures, precise structural subtleties can be engineered experimentally and theoretically for the atomically precise cluster catalyst. If the introduction of a structural subtlety significantly changes the catalytic performance of the cluster catalyst, the structural subtlety is very likely to have an important role in the catalysis. Surgical analyses of subtlety–activity relationships are effective means, thanks to the uniqueness of atomically precise metal clusters as model catalysts, to resolve challenges in designing new functional catalysts. In this section, we introduce recent progresses in using DFT methods to elucidate the catalytic effects of natural and designed structural factors in atomically precise metal cluster catalysts.

10.3.1 Charge State of the Catalyst

Some atomically precise noble metal clusters can be considered as superatoms where the valence electron occupancy follows specific counting rules [51]. In a ligand-protected metal cluster with a spherical core, formulated as $[L_SA_NX_M]^z$, the valence electron count n for the metal cluster core is:

$$n = N^*V_A - M - z \tag{10.6}$$

where N is the number of metal atoms, V_A is the valency of atom A, M is the number of electron-deficient ligands X (e.g., thiolate), and z is the net charge of the superatom. L denotes a Lewis base ligand, which does not affect the electron count of the metal cluster core. It is indicated that analogue Aufbau principle exists for these spherical $[L_SA_NX_M]^z$ clusters, which is denoted by $1S^2$ | $1P^6$ | $1D^{10}$ | $2S^2$ $1F^{14}$ | $2P^6$ $1G^{18}$ | $2D^{10}$ $3S^2$ $1H^{22}$, where the capital letters indicate spherical harmonic projections of the valence electron orbitals with different angular momenta. Therefore, $[L_SA_NX_M]^z$ clusters with shell-closing configurations

$$n = 2, 8, 18, 34, 58, 92, 138, \ldots \tag{10.7}$$

exhibit superior stability. The superatom valence shell-closing rule is also found in other types of ligand-protected metal clusters; e.g., the rod-like poly(x)-icosahedral $Au_{12x+1}(PPh_3)_{10}(SC_2H_4Ph)_{5x-5}Cl_2$ [52] consisting of x vertex sharing Au_{13} icosahedral units may favor a shell-closing count of $8x$ and exhibit a net charge of $4 - x$. Equations 10.6 and 10.7 imply that the net charge of the superatom z could be tunable. In particular, altering a metal atom in the cluster core with a different V_A will lead to a different net charge z.

Liu et al. [53] reported that the catalytic properties of the cluster catalysts can be dramatically changed with the subtle alteration by a single atom in the metal cluster core. The effects of doping with a noble metal atom (Au, Pd, and Pt) into the core of an $[Ag_{25}SR_{18}]^-$ on the catalytic properties were explored (Fig. 10.3A), where the dopant was protected by 24 Ag atoms ($M@Ag_{24}$ = $Au@Ag_{24}$, $Pd@Ag_{24}$, and

Pt@Ag_{24}). The central doping of a single atom away from the active sites into the Ag_{25} cluster exhibits a substantial influence on the catalytic performance in the carboxylation reaction of CO_2 with terminal alkyne through C≡C bond formation to produce propiolic acid. The superatom valence electron count is 8 for $[Ag_{25}SR_{18}]^-$. For Ag and Au, V_A is 1, and for Pd and Pt, V_A is 0. Therefore, Au@Ag_{24} is monoanionic like Ag_{25}, whereas Pd@Ag_{24} and Pt@Ag_{24} are dianionic upon the retention of the 8-electron configuration.

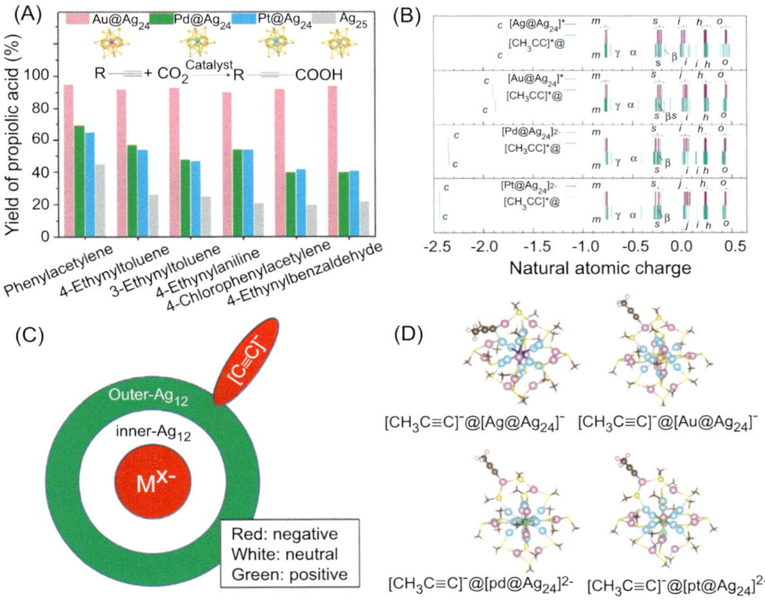

Figure 10.3 (A) Performances of M@Ag_{24} (M = Ag, Au, Pd, Pt) catalysts in promoting the carboxylation reactions of CO_2 with terminal alkynes. (B) Natural atomic charges calculated at the DFT level for the M@Ag_{24} catalyst with the $[CH_3C{\equiv}C]^-$ adsorbate. *c, o, m, i, s, h* denote core metal M, inner Ag_{12} shell, outer Ag_{12} shell, the methyl C, S, and H, respectively; α, β, and γ denote α-, β- and γ-C of $[CH_3C{\equiv}C]^-$. (C) Superanion–anion adsorption model for M@Ag_{24}. (D) Adsorption structures for $[CH_3C{\equiv}C]^-$ on M@Ag_{24} [53].

The catalytic performance and stability of M@Ag_{24} catalysts during the carboxylation reactions of CO_2 with terminal alkynes follow the order of Au > Pd = Pt > Ag. DFT calculations showed that M@Ag_{24} catalysts have similar molecular structures and near-gap electronic structures. The mechanism for the different catalytic behaviors of doped catalysts differ significantly from those for solid and single-atom catalysts. Infrared spectroscopy showed that the alkyne is adsorbed on $M@Ag_{24}$ in its deprotonated form. DFT geometry optimizations predicted that the adsorption of the deprotonated alkyne occurred at the outer-shell Ag with a linear configuration. The atomic charges from natural population analysis suggested that the core M, inner Ag_{12}, and outer Ag_{12} of $M@Ag_{24}$ are charge negative, neutral, and positive, respectively (Fig. 10.3B). The atomic charge on the core M is correlated with the net charge of the $M@Ag_{24}$ superanion in the order of Pd/Pt (−2.5) > Au (−1.9) > Ag (−1.7). In other words, although $M@Ag_{24}$ is anionic, its surface Ag sites are positively charged. A superanion–anion adsorption-related mechanism was proposed for the studied reactions based on the computational results (Fig. 10.3C). The adsorption of deprotonated alkyne (anion) on $M@Ag_{24}$ (superanion) is attributed to both outer Ag–C attraction and the electrostatic interaction between the core M and the Ag–C dipole. When the two forces are unbalanced, the superanion–anion adsorption leads to distorted or collapsed complex structures of $[RC{\equiv}C]^-$ $-[M@Ag_{24}]^{z-}$ for M = Ag, Pt, and Pd. A balance of the two forces is reached for $Au@Ag_{24}$, leading to the undistorted $[RC{\equiv}C]^-$–$[Au@Ag_{24}]^-$ adsorption complex. The proposed mechanism explains the different reactivities and stabilities of the central-atom doped Ag_{25} catalysts and reveals that the one-central-atom anionic doping can drastically change the outcomes of the catalytic reactions by controlling the net charge of the superanion. It is suggested from this study that (1) the anionic charge promotes deprotonation of the precursor in basic solution, (2) the superatom net charge can be controlled by atomic doping with different V_A, and (3) the atomic charge distribution, dependent on the net charge, significantly affects the stability and activity of the multi-layered spherical catalyst via Coulomb interactions.

Hu et al. [54] predicted the hydrogen evolution reaction (HER) activity of $M@Au_{24}$ catalysts for M = Hg, Cd, Ag, Au, Cu, Pd, and Pt (Fig. 10.4). The $M@Au_{24}$ structures were constructed from the central doping of the $[Au_{25}(SR)_{18}]^-$ with M. Upon hydrogen adsorption to form $H_xM@Au_{24}$, it was found that hydrogen adsorbate acted as metallic to fulfil the 8-electron shell-closing configuration of the $M@Au_{12}$ kernel. The adsorption sites were identified to be the Au atoms of the inner Au_{12} shell. Calculated H adsorption free energy with respect to a reversible hydrogen electrode ($H^+ + e^- \rightarrow$ ½ H_2) is shown in Fig. 10.4B for M@Au_{25}, which follows the order of Hg > Cd >> Ag > Au > Cu > Pd > Pt. Such an ordering is consistent with the expectations from the layered charge distribution model proposed in Ref. [53]. In order to fulfil the 8-electron shell-closing configuration, the groups IIB Hg and Cd at the central position are likely to be charge neutral, the groups IB Ag, Au, and Cu are monoanionic, and the groups VIII Pd and Pt are dianionic, which lead to a ordering of the predominant Coulomb attraction between central M and H^+: VIII > IB > IIB. This is another example benefiting from the facile tuning of the charge state for atomically precise metal cluster catalysts via atomic doping to affect the Coulombic surface–ligand adsorption.

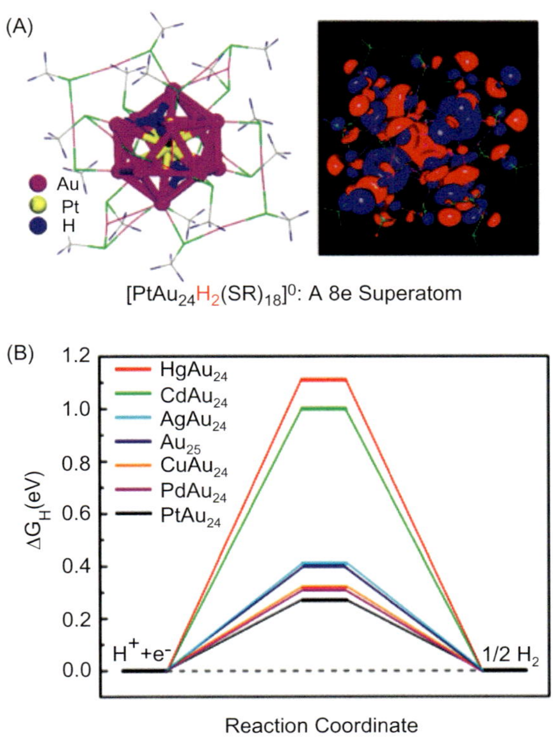

Figure 10.4 (A) Optimized structure and LUMO of $[PtAu_{24}H_2(SR)_{18}]^0$. (B) Hydrogen evolution reaction profiles for M@Au_{24} catalysts with M = Hg, Cd, Ag, Au, Cu, Pd, and Pt. Reprinted with permission from Ref. [54], Copyright 2017, American Chemical Society.

10.3.2 Point Vacancy

Removing the central Au atom from the bi-icosahedral $[Au_{25}(PPh_3)_{10}(SC_2H_4Ph)_5Cl_2]^{2+}$ [55] yields $[Au_{24}(PPh_3)_{10}(SC_2H_4Ph)_5Cl_2]^{+}$ [56] with a central point vacancy. Because almost every metal atom of the small atomically precise metal cluster catalyst contributes considerably to the catalytic reaction, the influences of such point vacancies to catalysis might be more significant than the influences of a point defect in traditional solid catalysts.

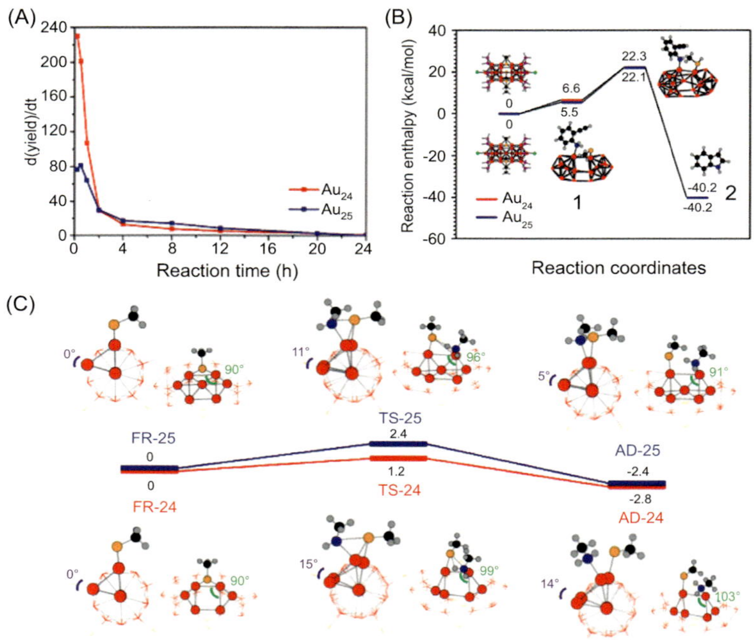

Figure 10.5 (A) Catalytic intramolecular cyclization of 2-ethnylaniline over the Au_{24} and Au_{25} supported on SiO_2. (B) Potential energy surface calculated at the DFT level for hydroamination of 2-ethnylaniline. (C) Activation of Au_{25} and Au_{24} complexes upon the adsorption of CH_3NH_2 via bridge cleavage. Energies in kcal/mol. Color codes: Au, orange; S, yellow; C, black; N, blue; H, grey; Cl, green; P, pink. Chemical species: **1**: 2-ethnylaniline, **2**: indole. The rhombus angle between two Au_{12} spheres is given in green text. The dihedral twist angle between two Au_{12} spheres is given in purple text [57].

Zhu and co-workers performed comparison studies of the bi-icosahedral Au_{24} and Au_{25} catalysts for intramolecular hydroamination of aminoalkynes (Fig. 10.5) [57] and methane-to-methanol conversion with H_2O_2 as the oxidant (Fig. 10.6) [48]. In both cases, the Au_{24} catalyst exhibited significantly higher catalytic performance than the Au_{25} catalyst, which led to the conclusion that the central point Au vacancy is key for the enhanced catalytic activity.

Goh et al. [58] predicted the molecular and electronic structures for the bi-icosahedral Au_{24} and Au_{25} clusters. The Au_{24} cluster has a larger HOMO–LUMO gap than that of the Au_{25} cluster, but their density of states spectra are alike, suggesting comparable electronic structures and similar molecular structures. A proton transfer mechanism was proposed for the intramolecular hydroamination of aminoalkynes over Au_{24} and Au_{25} [57], which follows the cleavage of a Au–S bond into a Au site for dative amine adsorption and a S site for proton binding. The predicted reaction energies for hydroamination of 2-ethynylaniline over Au_{24} and Au_{25} are comparable (Fig. 10.5B), which could not explain the experimental observation that reactions with Au_{24} are always twice as fast as those with the Au_{25} (and reactions over other spherical and rod-like Au clusters without point vacancy). In addition, the calculated natural atomic charge distributions are similar for Au_{24} and Au_{25}. So, what roles do the point defect and the central atom play in the different catalytic reactivities of Au_{24} and Au_{25}?

To address this question, the kinetic effects were considered by Zhu and co-workers [57]. In particular, during the activation of the catalyst via the Au–S cleavage upon adsorption or during the H^+ transfer between the dangling thiol and the organic adsorbate at the open Au site, the interaction between the adsorbate and the thiol is not negligible. This could lead to chemical changes in the two involved surface Au atoms (adsorbate Au and thiolate Au) that are separated by a vacancy in Au_{24} but are connected by a central Au in Au_{25}. DFT calculations were done with an abstract model, in which CH_3NH_2 was used to represent the generic amine (Fig. 10.5C). The activations of both Au_{24} and Au_{25} in the presence of amine have low energy barriers (1.2 and 2.4 kcal/mol, respectively). For such small barrier heights, energy becomes a poor indicator due to the intrinsic uncertanties and errors of the DFT method. Nevertheless, clear geometry differences were found for the two catalysts during the catalyst activation processes. For the Au_{24} activation, the fresh catalyst (FR-24, Fig. 10.5C) resembles a perfect pentagonal prism (considering the positions of the Au_5 rings near the equator), whereas the adsorption amino-complex (AD-24) and the activation transition state (TS-24) turn into antiprisms as two Au_{12} spheres are twisted about the axis of the rod. The dihedral angle twists for AD-24 and TS-24 are 14° and 15°, respectively, indicating that the twist due to the interaction between the two gold sites is barely hindered during the transition. As for the Au_{25} activation with amine, both FR-25 and AD-25 are in the shapes of approximate prisms, with the dihedral angle twist being 0° and 5°, respectively, while TS-25 is in the shape of antiprism with a dihedral angle twist of 9°. The constraints from the bonding between the central Au and the two near-equator Au_5 rings give rise to prism-like geometry of AD-25. This implies that the formation of TS-25 suffers from a greater strain (i.e., hindrance from the central Au–Au_5 bonds) than the formation of TS-24 does. The difference of 1.2 kcal/mol in the barrier height will lead to a rate constant ratio of 5.5:1 in favor of the Au_{24} activation reaction. Considering that the twist mechanism is likely involved in almost every reaction barrier of the hydroamination reactions where Au–amino and Au–thiol are coupled, this rate constant ratio may be applicable to the whole reaction, which explains the experimental value (ca. 3:1).

A different mechanism [48] was proposed to explain the different catalytic performances of the Au_{24} and Au_{25} catalysts in methane-to-methanol conversion with H_2O_2 (Fig. 10.6A,B). The corresponding reaction pathways were explored with DFT calculations. H_2O_2 first dissociatively adsorbs as two •OH at two activated Au sites. The adsorbed OH on the intact catalyst cannot initiate CH_4 activation. To react with CH_4, the Au–OH must be activated first by Au–S cleavage at Au. The Au_{24} can undergo isomerization via vacancy migration to form a M–Au_{24} isomer in the presence of OH (Fig. 10.6C), which introduces additional reaction pathways for Au_{24}. The reaction pathways for Au_{25} and original Au_{24} were considered (pathways in black in Fig. 10.6D,E). CH_4 is activated via a triangular TS **24d** for Au_{24} (and via a similar transition state for Au_{25}), with the barrier height $\Delta H^{\ddagger}$ of ~30 kcal/mol (Fig. 10.6E), which yields **24e** (Fig. 10.6D). The singly adsorbed Au–OH with broken Au–S bond can activate CH_4 and accept $•CH_3$ resulting from CH_4 activation. For CH_3

and OH in **24e** to recombine, OH must migrate to Au of Au–CH_3 since the large ∠Au–Au–C angle (125°) in HO–Au–Au–CH_3 forbids the two-site recombination. The OH migration is achieved via bending ∠Au–Au–O to form **24f**. A rigid scan of ∠Au–Au–O shows that the migration requires >40 kcal/mol to overcome the barrier (Fig. 10.6F). The successive one-site recombination of CH_3 and OH (**24f** → **24h** via **TS 24g**) has a low barrier of ~15 kcal/mol. The so-formed CH_3OH in **24h** is weakly adsorbed (E_{ads} = ~5 kcal/mol).

With the reaction pathways considered above, Au_{24} and Au_{25} would exhibit similar catalytic behaviors for the methane-to-methanol conversion, which contradicts the experimental results. DFT calculations revealed that the M–Au_{24} isomer, as the result of the central vacancy migration, allows for two additional shortcut pathways (red and blue paths in Fig. 10.6D,E) that enhance catalysis of Au_{24}. Both shortcut pathways avoid the rate-limiting OH migration barrier along **24e** → **24f**. The first shortcut (red in Fig. 10.6D,E) connects **24a** and **24f**, with the one-site CH_4 activation (**24b** → **24e**) and succeeding OH migration replaced by a two-site reaction (**24b*** → **24f** via **TS 24d*** with $\Delta H^{\ddagger}$ = 30 kcal/mol). The second shortcut (blue in Fig. 10.6D,E) connects **24e** and **24h**, with the one-site CH_3–OH recombination (**24f** → **24h**) and preceding OH migration replaced by a two-site CH_3–OH recombination (**24e*** → **24h** via **TS 24g*** with $\Delta H^{\ddagger}$ = 27 kcal/mol), directly leading to the CH_3OH product (**24h**). The suitable Au–adsorbate orientations at the central site of M–Au_{24} likely give rise to the two-site reactions. For example, in **24e***, ∠Au–Au–C of the HO–Au–Au–CH_3 site is 86°, and so bending ∠Au–Au–O directly yields **TS** of the two-site recombination reaction (Fig. 10.6F).

The isomerization of Au_{24} into M–Au_{24} also increases the mobility of the OH adsorbate and enhances the CH_3–OH recombination. For Au_{25}, the trajectory of OH migration connects two surface Au atoms of the same Au_5 ring and is always under the substantial steric effect of PPh_3 ligands, indicated by the barrier predicted at ∠Au–Au–OH = ~90° during the **25e** → **25f** transition (Fig. 10.6F). During the M–Au_{24} **24e*** → **24g*** transition, the trajectory of OH connects a Au of Au_5 ring and the central Au (at the position of the original vacancy), and so the steric effect from PPh_3 decreases as OH moves toward the central Au.

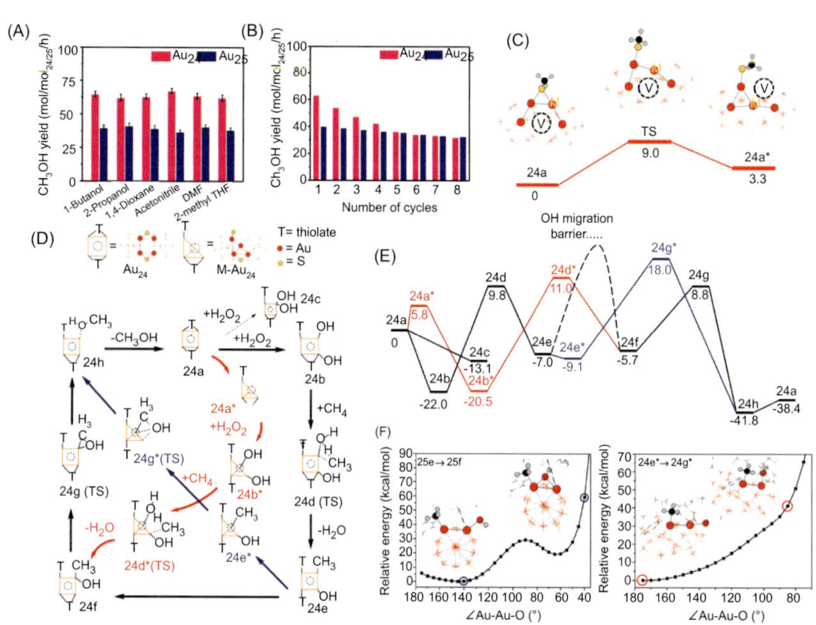

Figure 10.6 (A) Catalytic performance of methane oxidation over the Au_{24} and Au_{25} catalysts. (B) Recyclability of Au_{24} and Au_{25}. (C) Isomerization of Au_{24} via point vacancy migration ($Au_{24} \rightarrow$ M–Au_{24}). (D) Proposed catalytic mechanism and (E) predicted reaction enthalpy profiles for methane oxidation on Au_{24} at the DFT level. (F) Steric effects for the two-site CH_3–OH recombination at Au_{25} and M–Au_{24} catalysts. Reaction enthalpies are at 0 K in kcal/mol. Au_{24} complexes are labeled as 24**N**, **N** = a, b, c, and so on, and M–Au_{24} isomeric complexes are denoted as 24**N***. Black paths indicate the regular reaction pathways shared by Au_{24} and Au_{25} without involving the central vacancy, whereas the red and blue paths indicate the shortcut pathways that involve M–Au_{24} [48].

Different roles of the central vacancy in Au_{24} have been proposed for the hydroamination of aminoalkyne and methane-to-methanol conversion. This can be rationalized by the different surface modification effects by the reaction precursors in these two cases. The methane-to-methanol conversion is a two-site reaction where radical species form strong bonds with Au, whereas the hydroamination of aminoalkyne is essentially a one-site reaction (the thiolate is auxiliary for H^+ transfer) where the aminoalkyne forms only weak dative bond with the Au site. It is amazing that the single point vacancy in the atomically precise metal cluster can exhibit distinctly predominant roles in catalytic reactions. Exploring the mechanisms for point vacancy involvement appears meaningful for the catalysis of atomically precise metal clusters, and further investigations are recommended.

10.3.3 Metal Cluster Surface

Atomically precise metal clusters with comparable sizes can possess different structural motifs. For example, the core of the thiolate-protected gold cluster can be bulk-like [fcc (face-centered cubic) or hcp (hexagonal closed packed)] or non-bulk-like (icosahedral or Marks decahedral). Correspondingly, the surface structures of metal clusters can be different as are the adsorption configurations of protective ligands.

Xu et al. [59] studied the selective hydrogenation of CO_2 over two thiolated Au_{28} catalysts with kernel homology. The two catalysts, $Au_{28}(SPh\text{-}Bu)_{20}$ (Ph-Bu = 4-*tert*-butylphenyl, denoted as $Au_{28}T$) and $Au_{28}(SC_6H_{11})_{20}$ (C_6H_{11} = cyclohexyl, denoted as $Au_{28}C$), have the same fcc Au_{20} kernel but different types and arrangements of surface $Au_x(SR)_{x+1}$ staples. Despite the fact that the two $Au_{28}(SR)_{20}$ clusters differ only by the outer staples, significant differences in their catalytic selectivity were observed. The $Au_{28}T$ reactions led to an approximately 100% yield of the CH_3OH C_1 product (Fig. 10.7A), while $Au_{28}C$ reactions led to 75% yield of the CH_3OH C_1 product and 25% yield of the $HCOOCH_3$ C_2 product (Fig. 10.7B). The different selectivities of $Au_{28}T$ and $Au_{28}C$ suggest that the configurations of the staples determine the reaction pathways. Excellent durabilities were observed for both catalysts in reaction cycling, and thus the thiolates are mostly preserved during the reactions. DFT calculations were performed to understand the reaction mechanisms. In terms of staple composition and connecting position, $Au_{28}C$ contains three types of staples, two $Au_3(SR)_4$ and a $Au(SR)_2$, and $Au_{28}T$ contains two types of staples, a $Au_2(SR)_3$ and a $Au_3(SR)_4$ (Fig. 10.7C). These sites exhibit different chemical and electronic properties, as indicated by the predicted hydrogenation reaction energy (Fig. 10.7C) and atomic charge (Fig. 10.7D) per surface Au site. DFT calculations suggested that H_2 can activate the surface Au by forming H–Au–(H)–S structure at the $Au_2(SR)_3$ or $Au_3(SR)_4$. The formed H–Au–(H)–S structure was found to be an initial intermediate for the reaction toward the formation of the CH_3OH C_1 product via a reaction pathway akin to the reverse water–gas shift reaction (Fig. 10.7E). At the $Au(SR)_2$ site, the reaction of the first H_2 forms the Au–(H)–Au bridge and a leaving HSR, while the reaction of the second H_2 forms terminal Au–H and a leaving HSR. The as-formed asymmetric AuH_2 (with one terminal H and one bridge H) at the vertex of a Au_3 triangle is an initial intermediate for the reaction toward the formation of the $HCOOCH_3$ C_2 product (Fig. 10.7F). The asymmetric adsorption of two H atoms leads to the subsequent asymmetric adsorption of two CHO ligands. The terminal CHO is more prone to reduction than the bridge CHO, which accounts for the formation of CH_3OCHO via adsorbate recombination.

It was proposed that formation of the CH_3OH C_1 product occurs at the Au atoms of $Au_2(SR)_3$ and $Au_3(SR)_4$ present in both $Au_{28}T$ and $Au_{28}C$. The 2:3 ratio of the methanol-favoring sites in $Au_{28}C$ and $Au_{28}T$ was in line with the CH_3OH conversion ratio between $Au_{28}C$ and $Au_{28}T$ (Fig. 10.7A,B). The formation of the C_2 product over $Au_{28}C$ was likely due to formation of the Au_3 active site and asymmetric dual-ligand adsorption on it. DFT calculations suggested that the Au_3 acitve site forms only between the kernel and the $Au(SR)_2$ staple of the $Au_{28}C$ catalyst, possibly requiring a metallic kernel Au and the ionic staple Au (Fig. 10.7D). It was also demonstrated that the C_2 product formed at the Au_3 acive site was exclusively methyl formate.

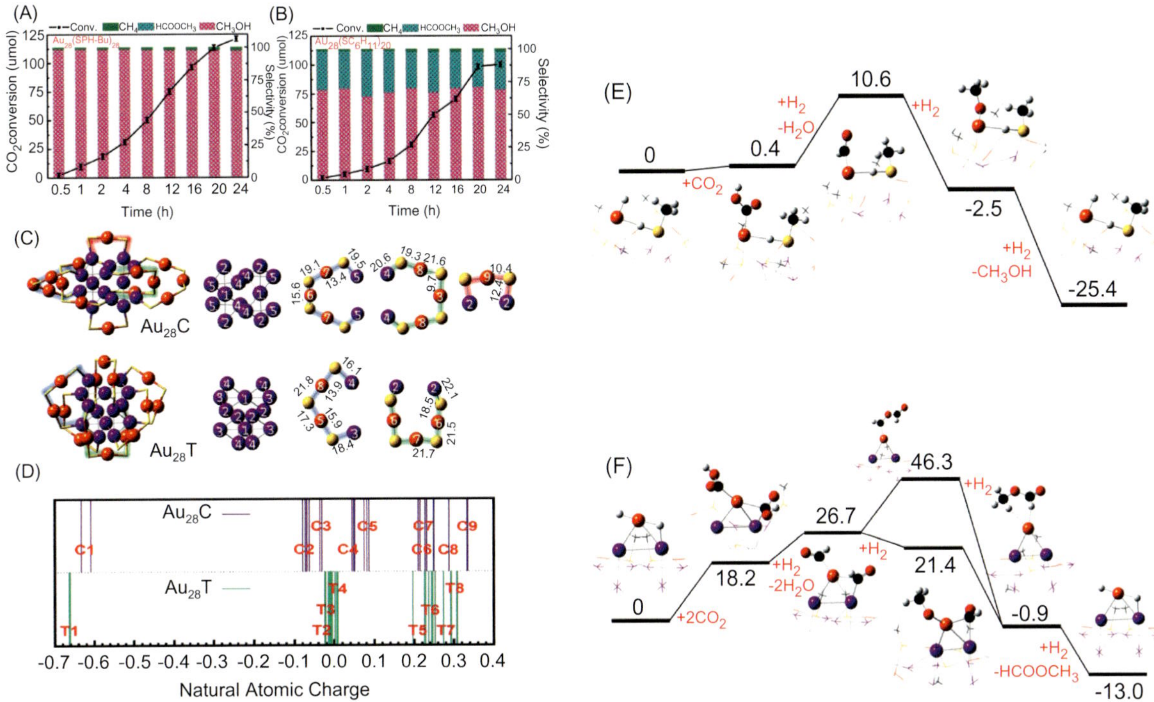

Figure 10.7 Conversion and selectivity for (A) $Au_{28}T$ and (B) $Au_{28}C$. (C) Au_{14} core (Au in purple) and $Au_x(SR)_{x+1}$ staples (Au in orange and S in yellow) of $Au_{28}C$ and $Au_{28}T$. Hydrogenation energy per surface Au site at the DFT level is in kcal/mol. (D) Natural atomic charge for the Au atoms of $Au_{28}C$ and $Au_{28}T$ at the DFT level. (E) Reaction energy profile (in kcal/mol) for CO_2 hydrogenation toward CH_3OH. (F) Reaction energy profile for CO_2 hydrogenation toward $HCOOCH_3$ [59].

Some atomically precise metal clusters naturally contain unsaturated surface metal sites and, therefore, exhibit extraordinary catalytic activities. The catalytic behavior of $Au_{22}(L^8)_6$ nanocluster [L^8 = 1,8-bis(diphenylphosphino)octane] with in situ uncoordinated Au sites on several oxide supports was investigated by Wu et al. [60]. HAADF-STEM (high-angle annular dark field scanning transmission electron microscopy) and EXAFS measurements showed that the precise structure of the $Au_{22}(L^8)_6$ was preserved when loaded on the oxide surfaces (Fig. 10.8A,B). The as-synthesized loaded $Au_{22}(L^8)_6$ showed an ultra-low light-off termperature for the CO oxidation reaction (Fig. 10.8C). DFT calculations showed that adsorptions of both CO and O_2 are exothermic at the unsaturated surface gold site (Fig. 10.8D). In contrast, the intact $Au_{25}(SR)_{18}$ catalyst was inactive for CO oxidation [61] with all the surface Au sites being coordinately saturated. Removal of the thiolates from $Au_{25}(SR)_{18}$ activated the surface sites of $Au_{25}(SR)_{18}$.

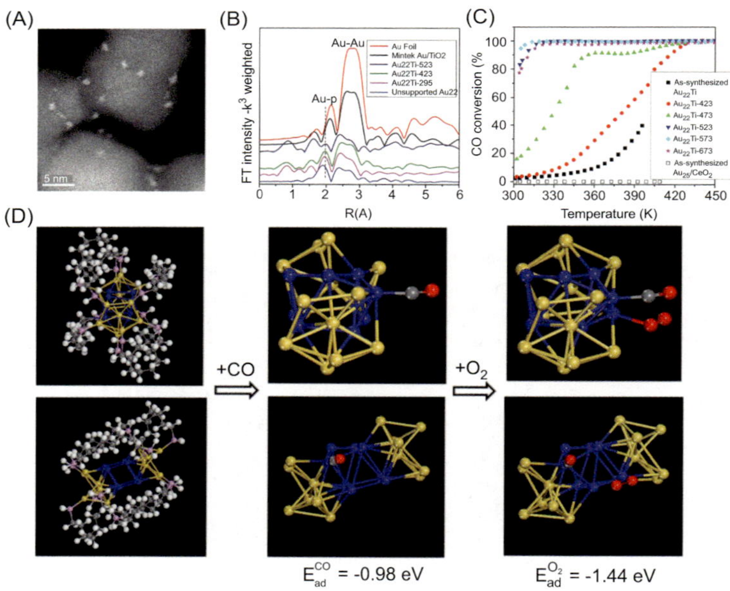

Figure 10.8 (A) HAADF-STEM images of as-synthesized $Au_{22}(L^8)_6/TiO_2$. (B) EXAFS spectra of $Au_{22}(L^8)_6/TiO_2$ samples pretreated at different temperatures in O_2. (C) CO oxidation light-off curves for $Au_{22}(L^8)_6/TiO_2$ samples treated at different temperature. (D) CO and O_2 adsorption on $Au_{22}(L^8)_6$ (ligands are omitted) [60]. Reprinted with permission from Ref. [60], Copyright 2016, American Chemical Society.

The versatility in the surface structures of atomically precise metal clusters allows for functional design of such catalysts for selectivity and reactivity. DFT predictions may provide guidance for such processes.

10.3.4 Roles of Protective Ligands

Protective ligands passivate the metal cluster core and is key to stabilize the chemical structure and morphology of atomically precise metal clusters during synthesis. In catalysis, protective ligands play diverse roles. Activation of the ligand-protected catalysts often requires removal of the protected ligands or cleavage of the metal–ligand bonds. However, under mild conditions, some of the protective ligands can remain to protect the catalyst by inhibiting aggregation of the metal clusters and by preventing activity loss due to severe reconstruction of the cluster [62]. DFT calculations suggest that the dangling thiolate formed from Au–S bond cleavage can bind H^+ through a dative bond and form reaction intermediates that facilitate H^+ transfer reactions, for example, the hydroamination of aminoalkyne [57]. Large protective ligands such as SC_2H_4Ph and PPh_3 also induce steric hindrance for two-site recombination of adsorbates for the icosahedral Au_{24} and Au_{25} reactions [48].

Alfonso et al. [63] performed DFT calculations using continuum solvation models to study the electroreduction of CO_2 to CO catalyzed by $[Ag_{25}(SR)_{18}]^-$. They compared the calculated free energy profiles for the reactions over the intact $[Ag_{25}(SR)_{18}]^-$ and dethiolated $[Ag_{25}(SR)_{17}]^-$ and suggested that the dethiolated catalyst was likely responsible for the reduction reaction. The active site was identified to be the Au atom of the dethiolated staple, which forms Au–Au bond with the icosahedral Au_{13} kernel upon dethiolation.

An "on-and-off" activation mechanism involving the exchange of protective ligands by the precursor was proposed by Yun et al. [64] for the aerobic oxidation of benzyl alcohol reaction over the $[Pd_3Cl(PPh_2)_2(PPh_3)_3]^{+/0}$. DFT calculations were carried out for the α-C-H abstraction reaction. The substitution of $PhCHO^-$ for PPh_3 for an activation energy of ~20 kcal/mol, as well as the subsequent substitution of PPh_3 for $PhC{=}O^-$, occurs spontaneously with an exothermicity of ~13 kcal/mol. The ligand exchange between the reaction intermediates and the protective PPH_3 has dual effects for the reaction, increasing the activity while maintaining the stability of the catalyst.

It was recently found that partially removing some of the protective ligands was all that was required for the catalytic reactivity of $Ni_6(SR)_{12}$ (R = C_2H_4Ph) clusters in the hydrodesulfurization (HDS) of thiophene by a combined experimental and computational study of Cheng et al. [65] Thermogravimetric analysis of $Ni_6(SR)_{12}$ clusters suggests that they gradually lose their thiolate ligands at temperatures above 150 °C (Fig. 10.9A). $Ni_6(SR)_{12}$ clusters supported by CeO_2 were found to have much higher HDS activity than the corresponding Ni and NiS catalysts, indicating the superior performance of activated $Ni_6(SR)_{12}$ clusters to the Ni and NiS nanoparticles (Fig. 10.9B). It is thus clear that partial removal of the thiolate ligands in $Ni_6(SR)_{12}$ clusters leads to the formation of the active site for the HDS reaction.

Nevertheless, it is not obvious from the experiment how the $Ni_6(SR)_{12}$ cluster was activated at these modestly elevated temperatures. DFT calculations were first performed for the full $Ni_6(SR)_{12}$ cluster (Fig. 10.9C) to explore the possible mechanism of cluster activation, and breaking the Ni–S bond was predicted to be very endothermic by 26.4 kcal/mol compared to the exothermic cleavage of the C–S bond, suggesting that the latter might be more likely to occur (Fig. 10.9D). Due to the high computational cost of using the full cluster as the catalyst model, further calculations of the PES for the HDS reaction were carried out using a much simplified model of $Ni_6(SEt)_{12}$ (Fig. 10.9E). The HDS reaction of thiophene was predicted to be proceeded by activation of the cluster by its reaction with H_2 (Fig. 10.9F), which was calculated to involve a quite high energy barrier of 64.7 kcal/mol, although the relatively high reaction temperature and the high H_2 partial pressure favor the cluster activation. The active structure of the catalyst (CAT in Fig. 10.9F) is formed upon the formation of a S vacancy (S_v), as the Ni–S_v–Ni bridge site is the active site for the adsorption of the thiophene. This is followed by two sequential hydrogenation steps to break the ring structure with much lower energy barriers of 33.8 and 2.1 kcal/mol. Further, hydrogenation of the resulting linear structure leads to the cleavage of the remaining C–S bond and the regeneration of the active structure.

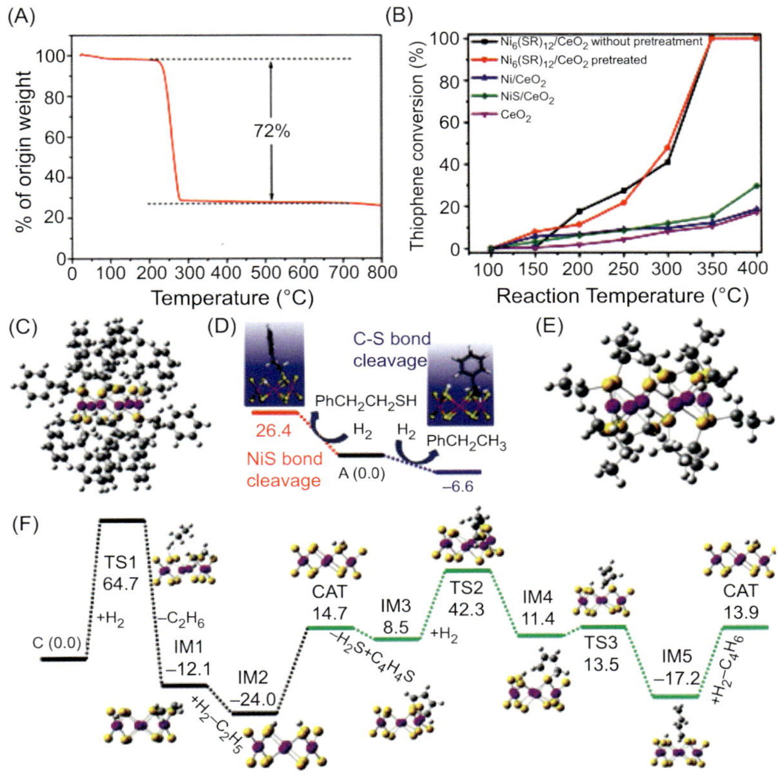

Figure 10.9 (A) Thermogravimetric analysis of $Ni_6(SR)_{12}$ (R = C_2H_4Ph) clusters. (B) Catalytic performance of HDS of thiophene over different catalysts. (C) Fully optimized structure of the $Ni_6(SR)_{12}$ cluster. (D) Free energy changes for Ni–S and C–S bond scissors using the full cluster model. (E) Optimized structure of the simplified $Ni_6(SEt)_{12}$ (Et = CH_2CH_3) cluster model. (F) Free energy surface for the activation of the simplified cluster model (black lines) and the HDS reaction catalyzed by the activated cluster (green lines). Pink for Ni, yellow for S, gray for C, and white for H. Only R or Et groups located at the active site are explicitly shown in (D) and (F) with energies given in kcal/mol at 0 K [65].

Despite the great efforts made in studying the on-and-off dynamics of the protective ligands in catalysis by atomically precise metal clusters using a variety of experimental and computational tools, the nature of the active site, its formation, and catalytic mechanisms cannot be said to be completely understood.

Nevertheless, it is clear that the roles of the protective ligands are dependent not only on the chemical structures of the ligands and cluster surfaces, but also on the activation conditions of the catalyst. The deep understanding of the various roles is that the protective ligands are crucial for better utilization and future design of the ligand-protected metal cluster catalysts, especially for controllable activation of catalysts. To thoroughly elucidate such relationships, more extensive computations, particularly using DFT methods, are required.

10.3.5 Structural Evolution of Catalyst

One important merit of the atomically precise metal cluster catalyst is the versatile reactivity under different activation conditions. Activation of the atomically precise metal cluster catalysts often involves changes in the protective ligands, but for some catalysts, the metal cluster core can also evolve toward specific catalytic activity.

Chai et al. [66] reported a combined experimental and computational study on the hydrogenation reaction of nitriles toward primary amines over the double-crown-like $Ni_6(SR)_{12}$ catalyst (R = SC_2H_4Ph) (Fig. 10.10A). The reaction was carried out at 45 °C and initiated with a H_2 pressure of 6 atm. Among various alkali additives, NH_3 (dissolved in the solution phase) increased the conversion of 4-cyanotetrahydropyran from 9% to 100% with a >99% selectivity toward primary amines, whereas other alkalis (e.g., Cs_2CO_3, K_2CO_3, and Na_2CO_3) provided no improvement in the conversion (Fig. 10.10B). The study also reported that the type of thiolate ligands did not affect the catalytic activity and selectivity of $Ni_6(SR)_{12}$ clusters for the studied catalytic reactions. In comparison, nickel boron (NiB) and Ni nanoparticles exhibited significantly lower conversion and worse selectivity of the amine product with or without NH_3 under the experimental conditions.

NH_3 is of particular importance for the enhanced activity of the $Ni_6(SR)_{12}$ catalyst. To elucidate the specific enhancement of the hydrogenation of nitriles on $Ni_6(SR)_{12}$ by the addition of NH_3, DFT calculations were performed with a simplified double-crown $Ni_6(SCH_3)_{12}$ model, and the acetonitrile (CH_3CN) hydrogenation was chosen as the model reaction. The PESs for the hydrogenation

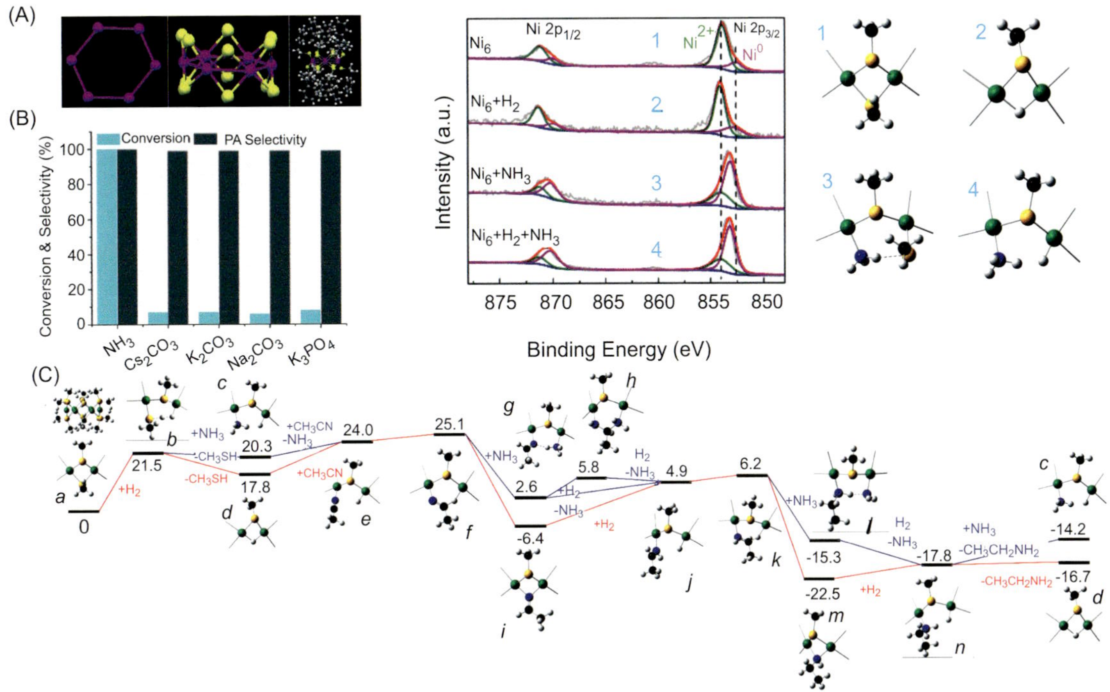

Figure 10.10 (A) Anatomy of the $Ni_6(SC_2H_4Ph)_{12}$ catalyst. Magenta for Ni, yellow for S, grey for C, and white for H. (B) The catalytic performance of 4-cyanotetrahydropyran over the $Ni_6(SC_2H_4Ph)_{12}$ catalyst with different alkalis. (C) Predicted reaction potential energy surfaces for the hydrogenation of nitrile toward amine over the Ni_6 catalyst. The reaction pathway in red occurs with or without NH_3, and the pathway in blue only occurs with NH_3. Green for Ni, yellow for S, black for C, blue for N, and grey for H. (D) Ni 2p XPS profiles of the initial $Ni_6(SC_2H_4Ph)_{12}$ catalyst and the treated catalyst under different reaction conditions and the corresponding molecular structures from DFT calculations [66].

reactions of the CH_3CN to form $CH_3CH_2NH_2$ amine were predicted (Fig. 10.10C). The reaction can be divided into three stages, that is, the activation of the $Ni_6(SR)_{12}$ catalyst, the formation of imine (CH_3CHNH), and the formation of amine ($CH_3CH_2NH_2$). The fresh $Ni_6(SR)_{12}$ cluster was found to be inert due to the superior stability of the thiolate-protected double-crown structure (complex ***a***, Fig. 10.10C). None of the H_2, NH_3, or CH_3CN can form strong physisorption at either Ni or S site. Activation of the catalyst is required in order to initiate the hydrogenation reactions. It was found that the activation of the $Ni_6(SR)_{12}$ was achieved by the cleavage of the Ni–S–Ni bridge bonds of the Ni_6S_{12} network with a H_2 inserted between Ni and S, leading to the open-bridge complex ***b***. The Ni–H–H–S chain of the open-bridge complex ***b*** is metastable, and the open bridge tends to close to form the closed-bridge complex ***d*** with a Ni–H–Ni bridge following desorption of CH_3SH. The closed-bridge complex ***d*** regains the double-crown structure and is thus very stable. Interestingly, in the presence of NH_3, the open-bridge configuration of the Ni_6 complex ***b*** can be stablized after CH_3SH desorption, leading to the open-bridge complex ***c***. The role of the NH_3 is to prevent relaxation of the Ni_6 catalyst into the stable planar square framework. It was found that all of the hydrogen transfer steps involving hydrogen transfer from Ni–H to C or N of Ni–C–N could take advantage of this mechanism (***f*** $\rightarrow$ ***g***/***i***; **k** $\rightarrow$ ***l***/***m***). When NH_3 is present, the rate-limiting steps for the catalyst activation, imine formation, and amine formation stages are ***c*** $\rightarrow$ ***e*** (ΔH = 3.7 kcal/mol), ***g*** $\rightarrow$ ***j*** (ΔH = 2.3 kcal/mol), and ***n*** $\rightarrow$ ***c*** (ΔH = 3.6 kcal/mol), respectively, which are less endothermic than the corresponding rate-limiting steps in the absence of NH_3, that is, ***d*** $\rightarrow$ ***e*** (ΔH = 6.2 kcal/mol), ***i*** $\rightarrow$ ***j*** (ΔH = 11.3 kcal/mol), and ***m*** $\rightarrow$ ***n*** (ΔH = 4.7 kcal/mol).

Ni 2p XPS profiles were measured for the pure Ni_6 catalyst, $Ni_6 + H_2$, $Ni_6 + NH_3$, and $Ni_6 + H_2 + NH_3$ (Fig. 10.10D) under the same conditions as the studied catalytic reactions. The XPS profile for $Ni_6 + H_2$ was found to be comparable to that for the pure Ni_6 catalyst, while the XPS profiles for $Ni_6 + NH_3$ and $Ni_6 + H_2 + NH_3$ exhibited apparent peak shifts from the XPS profile for the pure Ni_6. At first glance, this was surprising, as the H_2 was likely more reactive than NH_3 for reacting with Ni_6. A reasonable explanation was provided based on DFT calculations. The predicted favorable products for the $Ni_6 + H_2$, $Ni_6 + NH_3$, and $Ni_6 + H_2 + NH_3$ reactions are closed-bridge Ni–H–Ni, open-bridge Ni–NH_3|Ni–SR, and Ni–NH_3|Ni–H complexes, respectively. Therefore, the chemical shift in Ni $2p_{3/2}$ spectra for the $Ni_6 + NH_3$ and $Ni_6 + H_2 + NH_3$ reactions might result from the open-bridge coordination of Ni. This could serve as the indirect evidence for the proposed catalytic mechanism. The structural evolution from the closed-bridge double-crown to open-bridge complex is the key for not only catalyst activation but also maintaining the catalytic reactivity. Therefore, the catalyst that participates in the catalytic reaction can be structurally different from the fresh catalyst. For the mechanistic study of atomically precise metal cluster catalysts, it is important to identify the actual structure and the relevant state of the catalyst.

10.4 Conclusions and Future Perspectives

Even though the current usage of DFT methods in computational catalysis for atomically precise metal clusters is limited, they have proven to be effective in a number of case studies and thus are promising for future applications, in particular, in elucidating the effects and catalytic mechanisms for various types of structural design factors. Computational complexity is still the major factor that limits a wider utilization of DFT methods in making quantitative predictions for catalytic reactions of atomically precise metal clusters. Solution to this may come from further advances in computer hardware and the development of linear-scaling DFT algorithms. Currently, the growth rate of computing power of the computer hardware is still far behind the $O(N^3)$ scaling of DFT; the linear-scaling DFT methods also have many drawbacks that limit their applications (e.g., linear-scaling DFT works only for a limited range of chemical systems; linear-scaling DFT may not be suitable for evolving chemical systems). Nevertheless, the limit of the computational capacities is continuously being pushed forward. We are hoping to see more extensive and systematic computational studies for the further understanding of atomically precise metal cluster catalysts. A few daunting tasks for future DFT studies are as follows:

(1) Full-scale atomistic modelling of the catalytic reactions with the inclusion of the all-atom catalyst, precursor, medium, and support and with considerations of reaction conditions and durations
(2) Predicting the activation conditions for directed catalytic reactions or in situ activation
(3) Systematically predicting new catalytic reactions for existing atomically precise metal clusters
(4) Predicting new synthesizable atomically precise metal clusters and their synthesis conditions
(5) Deeper understanding of catalytic mechanisms from model catalysts and model reactions, and designing functional materials (that may or may not be atomically precise metal clusters) that benefit from the elucidated mechanisms

Acknowledgments

We acknowledge the support of National Natural Science Foundation of China (91845104 and U1930402).

References

1. Daniel, M.-C. and Astruc, D. (2004). Gold nanoparticles: Assembly, supramolecular chemistry, quantum-size-related properties, and applications toward biology, catalysis, and nanotechnology, *Chem. Rev.*, **104**, pp. 293–346.
2. Jin, R., Zeng, C., Zhou, M. and Chen, Y. (2016). Atomically precise colloidal metal nanoclusters and nanoparticles: Fundamentals and opportunities, *Chem. Rev.*, **116**, pp. 10346–10413.
3. Chakraborty, I. and Pradeep, T. (2017). Atomically precise clusters of noble metals: Emerging link between atoms and nanoparticles, *Chem. Rev.*, **117**, pp. 8208–8271.
4. Joshi, C. P., Bootharaju, M. S., Alhilaly, M. J. and Bakr, O. M. (2015). $[Ag_{25}(SR)_{18}]^-$: The "golden" silver nanoparticle, *J. Am. Chem. Soc.*, **137**, pp. 11578–11581.
5. Du, Y., Sheng, H., Astruc, D. and Zhu, M. (2020). Atomically precise noble metal nanoclusters as efficient catalysts: A bridge between structure and properties, *Chem. Rev.*, 120, pp. 526–622.

6. Parr, R. G. and Yang, W. (1989). *Density-Functional Theory of Atoms and Molecules*, Oxford University Press, New York, USA.
7. Jones, R. O. (2015). Density functional theory: Its origins, rise to prominence, and future, *Rev. Mod. Phys.*, **87**, pp. 897–923.
8. Hohenberg, P. and Kohn, W. (1964). Inhomogeneous electron gas, *Phys. Rev.*, **136**, pp. B864–B871.
9. Kohn, W. and Sham, L. J. (1965). Self-consistent equations including exchange and correlation effects, *Phys. Rev.*, **140**, pp. A1133–A1138.
10. Baroni, S., de Gironcoli, S., Dal Corso, A. and Giannozzi, P. (2001). Phonons and related crystal properties from density-functional perturbation theory, *Rev. Mod. Phys.*, **73**, pp. 515–562.
11. Runge, E. and Gross, E. K. U. (1984). Density-functional theory for time-dependent systems, *Phys. Rev. Lett.*, **52**, pp. 997–1000.
12. Lopata, K. and Govind, N. (2011). Modeling fast electron dynamics with real-time time-dependent density functional theory: Application to small molecules and chromophores, *J. Chem. Theory Comput.*, **7**, pp. 1344–1355.
13. Garzón, L., Rovira, C., Michaelian, K., Beltran, M. R., Ordejorn, P., Junquera, J., Sanchez-Portal, D., Artacho, E. and Soler, J. M. (2000). Do thiols merely passivate gold nanoclusters? *Phys. Rev. Lett.*, **85**, pp. 5250–5251.
14. Akola, J., Walter, M., Whetten, R. L., Häkkinen, H. and Grönbeck, H. (2008). On the structure of thiolate-protected Au_{25}, *J. Am. Chem. Soc.*, **130**, pp. 3756–3757.
15. Wang, P., Xiong, L., Sun, X., Ma, Z. and Pei, Y. (2018). Exploring the structure evolution and core/ligand structure patterns of a series of large sized thiolate-protected gold clusters $Au_{145-3N}(SR)_{60-2N}$ (N = 1–8): A first principles study, *Nanoscale*, **10**, pp. 3918–3929.
16. Gurman, S. J. and McGreevy, R. L. (1990). Reverse Monte Carlo simulation for the analysis of EXAFS data, *J. Phys. Condens. Matter*, **2**, pp. 9463–9473.
17. Winterer, M. (2000). Reverse Monte Carlo analysis of extended X-ray absorption fine structure spectra of monoclinic and amorphous zirconia, *J. Appl. Phys.*, **88**, pp. 5635–5644.
18. Iwasa, T. and Nobusada, K. (2007). Theoretical investigation of optimized structures of thiolated gold cluster $[Au_{25}(SCH_3)_{18}]^+$, *J. Phys. Chem. C*, **111**, pp. 45–49.
19. Lopez-Acevedo, O., Akola, J., Whetten, R. L., Grönbeck, H. and Häkkinen, H. (2009). Structure and bonding in the ubiquitous icosahedral metallic gold cluster $Au_{144}(SR)_{60}$, *J. Phys. Chem. C*, 113, pp. 5035–5038.

20. Zhu, M., Echenhoff, W. T., Pintauer, T. and Jin, R. (2008). Conversion of anionic $[Au_{25}(SCH_2CH_2Ph)_{18}]^-$ cluster to charge neutral cluster via air oxidation, *J. Phys. Chem. C*, **112**, pp. 14221–14224.

21. Zhu, M., Aikens, C. M., Hendrich, M. P., Gupta, R., Qian, H., Schatz, G. C. and Jin, R. (2009). Reversible switching of magnetism in thiolate-protected Au_{25} superatoms, *J. Am. Chem. Soc.*, **131**, pp. 2490–2492.

22. Zhu, Y., Qian, H. and Jin, R. (2010). An atomic-level strategy for unraveling gold nanocatalysis from the perspective of $Au_n(SR)_m$ nanoclusters, *Chem. Eur. J.*, **16**, pp. 11455–11462.

23. Jin, R. (2010). Quantum sized, thiolate-protected gold nanoclusters, *Nanoscale*, **2**, pp. 343–362.

24. Foster, J. P. and Weinhold, F. (1980). Natural hybrid orbitals, *J. Am. Chem. Soc.*, **102**, pp. 7211–7218.

25. Mulliken, R. S. (1955). Electronic population analysis on LCAO-MO molecular wave functions. I, *J. Chem. Phys.*, **23**, pp. 1833–1840.

26. Bader, R. F. W. (1990). *Atoms in Molecules — A Quantum Theory*, Oxford University Press, Oxford, UK.

27. Hirshfeld., F. L. (1977). Bonded-atom fragments for describing molecular charge densities, *Theor. Chim. Acta*, **44**, pp. 129–138.

28. Weinhold, F., Landis, C. R. and Glendening, E. D. (2016). What is NBO analysis and how is it useful? *Int. Rev. Phys. Chem.*, **35**, pp. 399–440.

29. Yamamoto, H., Yano, H., Kouchi, H., Obora, Y., Arakawa, R. and Kawasaki, H. (2012). *N,N*-Dimethylformamide-stabilized gold nanoclusters as a catalyst for the reduction of 4-nitrophenol, *Nanoscale*, **4**, pp. 4148–4154.

30. Nieto-Ortega, B. and Bürgi, T. (2018). Vibrational properties of thiolate-protected gold nanoclusters, *Acc. Chem. Res.*, **51**, pp. 2811–2819.

31. Zhu, M., Aikens, C. M., Hollander, F. J., Schatz, G. C. and Jin, R. (2008). Correlating the crystal structure of a thiol-protected Au_{25} cluster and optical properties, *J. Am. Chem. Soc.*, **130**, pp. 5883–5885.

32. Shibu, E. S., Muhammed, M. A. H., Tsukuda, T. and Pradeep, T. (2008). Ligand exchange of $Au_{25}SG_{18}$ leading to functionalized gold clusters: Spectroscopy, kinetics, and luminescence, *J. Phys. Chem. C*, **112**, pp. 12168–12176.

33. Bauernschmitt, R. and Ahlrichs, R. (1996). Treatment of electronic excitations within the adiabatic approximation of time dependent density functional theory, *Chem. Phys. Lett.*, **256**, pp. 454–464.

34. Lopata, K. and Govind, N. (2011). Modeling fast electron dynamics with real-time time-dependent density functional theory: Application to small molecules and chromophores, *J. Chem. Theory Comput.*, **7**, pp. 1344–1355.

35. Matsushima, N. and Yamauchi, J. (2019). First-principles X-ray photoelectron spectroscopy binding energy shift calculation for boron and aluminum defects in 3C-silicon carbide, *Jpn. J. Appl. Phys.*, **58**, 031001.

36. Cavigliasso, G. and Chong, D. P. (1999). Accurate density-functional calculation of core-electron binding energies by a total-energy difference approach, *J. Chem. Phys.*, **111**, pp. 9485–9492.

37. Artyushkova, K., Matanovic, I., Halevi, B. and Atanassov, P. (2017). Oxygen binding to active sites of Fe–N–C ORR electrocatalysts observed by ambient-pressure XPS, *J. Phys. Chem. C*, **121**, pp. 2836–2843.

38. Zhao, F., Gao, S., Pujari, P., Zuilhof, H. and Teplyakov, A. V. (2017). Universal calibration of computationally predicted N 1s binding energies for interpretation of XPS experimental measurements, *Langmuir*, **33**, pp. 10792–10799.

39. Wu, Z., Gayathri, C., Gil, R. R. and Jin, R. (2009). Probing the structure and charge state of glutathione-capped $Au_{25}(SG)_{18}$ clusters by NMR and mass spectrometry, *J. Am. Chem. Soc.*, **131**, pp. 6535–6542.

40. Barone, V. (1996). Electronic, vibrational and environmental effects on the hyperfine coupling constants of nitroside radicals. H_2NO as a case study, *Chem. Phys. Lett.*, **262**, pp. 201–206.

41. Häkkinen, H., Walter, M. and Grönbeck, H. (2006). Divide and protect: Capping gold nanoclusters with molecular gold-thiolate rings, *J. Phys. Chem. B*, **110**, pp. 9927–9931.

42. Craig, C. F., Duncan, W. R. and Prezhdo, O. V. (2005). Trajectory surface hopping in the time-dependent Kohn-Sham approach for electron-nuclear dynamics, *Phys. Rev. Lett.*, **95**, pp. 163001–163004.

43. Parandekar, P. V. and Tully, J. C. (2005). Mixed quantum-classical equilibrium. *J. Chem. Phys.*, **122**, 094102.

44. Yu, C., Li, G., Kumar, S., Kawasaki, H. and Jin, R. (2013). Stable $Au_{25}(SR)_{18}/TiO_2$ composite nanostructure with enhanced visible light photocatalytic activity, *J. Phys. Chem. Lett.*, **4**, pp. 2847–2852.

45. Senanayake, R. D., Akimov, A. V. and Aikens, C. M. (2017). Theoretical investigation of electron and nuclear dynamics in the $[Au_{25}(SH)_{18}]^{-1}$ thiolate-protected gold nanocluster, *J. Phys. Chem. C*, **121**, pp. 10653–10662.

46. van Santen, R. A. and Niemantsverdriet, J. W. (1995). *Chemical Kinetics and Catalysis*, Springer, New York.

47. Fernando, A. and Aikens, C. M. (2015). Ligand exchange mechanism on thiolate monolayer protected $Au_{25}(SR)_{18}$ nanoclusters, *J. Phys. Chem. C*, **19**, pp. 20179–20187.

48. Cai, X., Saranya, G., Shen, K., Chen, M., Si, R., Ding, W. and Zhu, Y. (2019). Reversible switching of catalytic activity by shuttling an atom into and out of gold nanoclusters, *Angew. Chem. Int. Ed.*, **58**, pp. 9964–9968.

49. Xu, W. W., Gao, Y. and Zeng, X. C. (2015). Unraveling structures of protection ligands on gold nanoparticle $Au_{68}(SH)_{32}$, *Sci. Adv.*, **1**, e1400211.

50. Chorkendorff, I. and Niemantsverdriet, J. W. (2003). *Concepts of Modern Catalysis and Kinetics*, Wiley-VCH, Weinheim, Germany.

51. Walter, M., Akola, J., Lopez-Acevedo, O., Jadzinsky, P. D., Calero, G., Ackerson, C. J., Whetten, R. L., Grönbeck, H. and Häkkinen, H. (2008). A unified view of ligand-protected gold clusters as superatom complexes, *Proc. Natl. Acad. Sci.*, **105**, pp. 9157–9162.

52. Nobusada, K., and Iwasa, T. (2007). Oligomeric gold clusters with vertex-sharing bi- and triicosahedral structures, *J. Phys. Chem. C*, **111**, pp. 14279–14282.

53. Liu, Y., Chai, X., Cai, X., Chen, M., Jin, R., Ding, W. and Yan, Z. (2018). Central doping of a foreign atom into the silver cluster for catalytic conversion of CO_2 toward C–C bond formation, *Angew. Chem. Int. Ed.*, **57**, pp. 9775–9779.

54. Hu, G., Tand, Q., Lee, D., Wu, Z. and Jiang, D.-E. (2017). Metallic hydrogen in atomically precise gold nanoclusters, *Chem. Mater.*, **29**, pp. 4940–4947.

55. Park, S. and Lee, D. (2012). Synthesis and electrochemical and spectroscopic characterization of biicosahedral Au_{25} clusters. *Langmuir*, **28**, pp. 7049–7054.

56. Das, A., Li, T., Nobusada, K., Zeng, Q., Rosi, N. L. and Jin, R. (2012). Total structure and optical properties of a phosphine/thiolate-protected Au_{24} nanocluster, *J. Am. Chem. Soc.*, **134**, pp. 20286–20289.

57. Wang, L., Shen, K., Chen, M. and Zhu, Y. (2019). One-core-atom loss in a gold nanocluster promotes hydroamination reaction of alkynes, *Nanoscale*, **11**, pp. 13767–13772.

58. Goh, J.-Q., Malola, S., Häkkinen, H. and Akola, J. (2013). Role of the central gold atom in ligand-protected biicosahedral Au_{24} and Au_{25} clusters, *J. Phys. Chem. C*, **172**, pp. 22079–22086.

59. Xu, J., Chen, M. and Zhu, Y. (2019). Selective hydrogenation of CO_2 dictated by isomers in $Au_{28}(SR)_{20}$ nanoclusters: Which one is better? *Chem. Eur. J.*, **25**, pp. 9185–9190.

60. Wu, Z., Hu, G., Jiang, D., Mullins, D. R., Zhang, Q.-F., Allard, L. F., Wang, L.-S. and Overbury, S. H. (2016). Diphosphine-protected Au_{22} nanoclusters on oxide supports are active for gas-phase catalysis without ligand removal. *Nano Lett.*, **16**, pp. 6560–6567.

61. Wu, Z., Jiang, D., Mann, A. K. P., Mullins, D. R., Qiao, Z.-A., Allard, L. F., Zeng, C., Jin, R. and Overbury, S. H. (2014). Thiolate ligands as a double-edged sword for CO oxidation on CeO_2 supported $Au_{25}(SCH_2CH_2Ph)_{18}$ nanoclusters. *J. Am. Chem. Soc.*, **136**, pp. 6111–6122.

62. Zhang, B., Fang, J., Li, J., Lau, J. J., Mattia, D., Zhong, Z., Xie, J. and Yan, N. (2016). Soft, oxidative stripping of alkyl thiolate ligands from hydroxyapatite-supported gold nanoclusters for oxidation reactions, *Chem. Asian J.*, **11**, pp. 532–539.

63. Alfonso, D. R., Kauffman, D. and Matranga, C. (2016). Active sites of ligand-protected Au_{25} nanoparticle catalysts for CO_2 electroreduction to CO, *J. Chem. Phys.*, **144**, 184705.

64. Yun, Y., Sheng, H., Yu, J., Bao, L., Du, Y., Xu, F., Yu, H., Li, P. and Zhu, M. (2018). Boosting the activity of ligand-on atomically precise Pd_3Cl cluster catalyst by metal-support interaction from kinetic and thermodynamic aspects, *Adv. Synth. Catal.*, **360**, pp. 4731–4743.

65. Cheng, X., Chai, X., Hu, W., Li, S. and Zhu, Y. (2019). The on-and-off dynamics of thiophene on a nickel cluster enables efficient hydrodesulfurization and excellent stability at high temperatures, *Nanoscale*, **11**, pp. 4369–4375.

66. Chai, X., Li, T., Chen, M., Jin, R., Ding, W. and Zhu, Y. (2018). Suppressing the active site-blocking impact of ligands of $Ni_6(SR)_{12}$ clusters with the assistance of NH_3 on catalytic hydrogenation of nitriles, *Nanoscale*, **10**, pp. 19375–19382.

Index